NORTH CAROLINA
STATE BOARD OF COMMUNITY COLLEGES
LIBRARIES
ASHEVILLE-BUNCOMBE TECHNICAL COLLEGE

WIND ENERGY

DISCARDED

JUN 2 4 2025

WIND ENERGY
How to Use It

Paul Gipe

Stackpole Books

Copyright © 1983 by Stackpole Books

Published by
STACKPOLE BOOKS
Cameron and Kelker Streets
P.O. Box 1831
Harrisburg, PA 17105

All rights reserved, including the right to reproduce this book or portions thereof in any form or by any means, electronic or mechanical, including photocopying, recording, or by any information storage and retrieval system, without permission in writing from the publisher.
All inquiries should be addressed to Stackpole Books, Cameron and Kelker Streets, P.O. Box 1831, Harrisburg, Pennsylvania 17105.

Printed in the U.S.A.

Library of Congress Cataloging in Publication Data

Gipe, Paul.
  Wind energy.

  Bibliography: p.
  Includes index.
  1. Wind power. I. Title.
TJ825.G56   1983   621.4'5   83-13667
ISBN 0-8117-2273-2

To my family and friends—whose encouragement has given me the freedom and the strength to choose my own path.

Contents

	Preface	9
	Acknowledgments	11
1	Introduction	13
2	Applications	23
3	Siting	40
4	Measuring the Wind	69
5	Estimating Output and Payback	119
6	Evaluating the Technology	150
7	Towers	210
8	Cutting Costs	228
9	Buying a Wind System	248
10	Installation and Maintenance	258
11	Safety	315

Appendixes		332
A	Air Density Corrections for Temperature	332
B	Capital Recovery and Compound Amount Factors	333
C	Wind Data	336
D	Sources of Materials and Information	384
Notes		391
Bibliography		393
Index		396

Preface

After having written numerous but brief articles on the subject of wind energy, the task of writing a book appeared awesome. It seemed particularly difficult in light of the well-written and thorough books by Jack Park and Don Marier. However, my contact with many homeowners, farmers, and small businessmen through my lectures and from the inquiries I receive indicated that their needs were not being met.

There was a gap among the books written for those wanting to build their own wind machine, those devoted to battery charging wind systems, and those surveying the whole field. There wasn't a book for the people I dealt with that realistically answered the questions they raised. They wanted to know what they could do themselves to obtain a working wind machine and not an experimenter's toy. They also wanted to know about wind systems interconnected with the utility and what requirements the utility would place on them. They wanted to know how to get the most for their money and how to avoid being taken. They wanted practical information to help them get started in the right direction.

As a consultant, my work has taken me from an Eskimo village on the Yukon, to the suburbs of Los Angeles, to the windblown rangeland of the Texas Panhandle, and to the rocky coast of Brittany. I have had ample opportunities to learn how not to do it, from my own experience and from the experience of others whose successes and failures I have chronicled over the years.

I find myself in the unique position of being at the same time an observer of the wind industry and a participant. As an observer, I have traveled throughout the country and through northern Europe reporting on the technology and on the industry's key figures. But I've also hunted windchargers in Montana and know firsthand the dangers and excitement of prowling the range for a 600-pound beast balanced precariously atop sixty feet of rusted steel. I have installed small wind machines and know the feeling of frustration when, as the turbine dangles overhead, a part doesn't fit quite like it is supposed to and the truck unexpectedly runs out of gas. It is at such a time that the situation demands all the wits—and strength—you can muster. And I have felt the cold and exhaustion after working dawn to dusk in ten-degree weather raising a tower.

From these experiences, I prepared a daylong lecture on the pitfalls and the promise of wind energy that I have given at colleges and universities across the country. People who attended seemed to appreciate the frank approach. This book grew out of that seminar.

I have tried to keep the same tone in this book. I've collected my earlier works, brought them up-to-date, and added sections based on my experience as a consultant and installer. I have also gathered tips, advice, and recommendations from leaders in the field. It is my hope that this book answers the practical questions of those wishing to use wind energy, that it will prevent some of the mistakes I as well as others have made, that it will lead to the installation of more wind machines, and that it will result in better performance and safer operation of those to be installed.

Acknowledgments

I am indebted to Karl and Mike Bergey and to Ward Slager of Bergey Windpower Company for the opportunity to attend their dealer training seminars and for their support during the planning and installation of several small wind machines in Pennsylvania. I am also grateful for the use of illustrations and techniques contained in their fine installation manual.

Bob Sherwin, of Enertech Corporation, and Jim Sencenbaugh, of Sencenbaugh Wind Electric, also deserve a note of thanks for their ready cooperation in my search for further installation details and illustrations. As a couple of old hands in the wind business, they have fielded thousands of requests from writers, consultants, academics, and the public, yet still find time to handle one more.

I am especially grateful to Farrell Seiler, editor of *Wind Energy Report,* for his patience and understanding in his attempts to teach me the art of writing, for his listening ear, for the sharing of insights while covering the third International Symposium on Wind Energy Systems in Copenhagen, and for the unforgettable experience of barreling down the back roads of Denmark in search of wind machines. Farrell, a superb journalist, is a true storehouse of information on the wind industry both here and abroad. I am thankful for the opportunities I have had to work with him.

In my travels, I've been fortunate to encounter a number of professionals who both know what they are doing and who are a pleasure to work with: Tom Gray, American Wind Energy Association; Paul Vosburgh, Forecast Industries, Inc.; Vaughn Nelson, Alternative Energy Institute, West Texas State University; Mark Newell, Wind Systems Engineering; Jon Traudt, Windcatcher Company; Bill Hopwood, Springhouse Energy Systems; Tom Werking, Windworks; and Gil Morrissey, Advance Electrical Service.

My thanks to Capitola Reece, Gene Heisey, and Art and Maxine Cook for sharing their experiences; to Carl Judy, Susan Schillmoeller, and Diane Warriner for their research; to Alma Hemperly for her preparation of the manuscript; and to Donna and Andrew Pitz, and Susan Nelson for their faith in the future.

And my thanks to the following for the use of information and illustrations:

A. B. Chance Company
Aeolian Kinetics, Inc.
Aermotor Div., Valley Industries
Alcoa
Atlas Safety Equipment Company
Battelle Pacific Northwest Laboratory
Danish Ministry of Energy
Panhandle-Plains Historical Museum
Helion, Inc.
Holec (USA), Inc.
JBF Scientific
Low Energy Systems, Ltd.
Mitre Corp.
Natural Power, Inc.
New Mexico State University
New York State Energy Office
Northwind Power Company
P.I. Specialists, Ltd.
SINCO Products, Inc.
University of Massachusetts
Vestas a/s
WindMaster Corp.

1

Introduction

This book is not a treatise on why you should use wind machines. You already know or you would not have picked it up. It is about how to get the most for your money. It is about doing it right.

This book is not a discourse on large government-sponsored wind machines and how they are the future hope of mankind, though some are mentioned. Nor is it about the worldwide development of wind energy, though the technologies being used outside the United States are discussed. Nor is it about experimental wind machines and how — one day — they will bring wind energy within the reach of all. It is about small wind machines, primarily those suited for the home and small business, that are available today.

This is not another how-to-build-your-own windmill book, though much of what it contains provides essential information for building and installing one safely. It is about what you need to know to make intelligent choices (for example, whether or not to build one yourself). It is about how to proceed in a logical and methodical manner to determine if you can use a wind machine and (where you can) the kind that is right for you.

Wind machines are not tomorrow's technology. They are here today.

All across the country people are installing wind machines in a continuing effort to stabilize rising utility bills, to reduce them, to eliminate them entirely, or to reverse the tables and sell energy to the utility for a profit. They are doing so because it makes economic sense—because wind works.

As you will see in the chapters ahead, though, wind machines are not for everyone. You must have a good site, have enough wind, and you must have the right wind machine. You will also need a little something extra. You need, as my coach used to say, intestinal fortitude—guts.

Installing a wind machine is never completely risk free. There will always be some uncertainty. The purpose of this book is to help you minimize that risk. It will maximize the probability that you will succeed in erecting a wind machine that will work reliably, safely, and provide you with a good return on your investment.

This book differs from others on the subject because it not only describes the technology, but also tells how to evaluate it, what to look for, what is important, and what is not. But at some point a decision must be made that only you can make. You must weigh the options, then act. It takes courage. Wind machines are not cheap, and whether you install it yourself or contract a dealer to do it, the installation of a wind machine is a major undertaking.

The people who use wind energy are prudent, but they are doers. This book was written with them in mind. It was written for people who ask questions, who like to know what is going on around them, and who like to do what they can themselves, but who also know when to stop and call in help. It is organized in the same way that such a person would approach the question, "Is wind energy for me?"

How Can I Use Wind Energy? In chapter 2 we look at the many ways in which the wind can be used. The emphasis of this book is on wind machines interconnected with the utility, but much of the information within applies to any kind of wind system, whether it is used to pump water or charge batteries.

Where Can I Put a Wind Machine? In chapter 3 we discuss where you can put them, where you should not, and why. You will also learn about potential zoning conflicts, how to avoid them, and (where you can't) how to deal with them. You will also learn about the utility's concerns and what you can and cannot do with an interconnected wind system.

Do I Have Enough Wind? The wind—what it is, the importance of wind speed, and how to find out what you have—is the subject of chapter 4. You will learn how to use the most up-to-date information on wind speeds and power available in this country. You will learn when it is necessary to measure wind speed at your site and when it is not. Where you want to measure wind speed, you'll discover what equipment to use, how to use it, and how to interpret the resulting information.

Will It Pay for Itself? Chapter 5 describes two methods for estimating the energy production from any wind machine and how to use published information on the annual energy output. It illustrates a simple method for calculating how long it will take the wind machine to pay for itself—its payback. It discloses the subtleties behind wind machine economics and why there is so much interest in wind energy today. You will also learn the best way to compare the cost-effectiveness of various wind machines and what ways not to use.

Which Wind Machine is Right for Me? Chapters 6 and 7 explore the technology, where we have been, where we are today, and where we are headed. The advantages and disadvantages of the various technologies in use today are explained. You will get a feel for what is on the market and how to evaluate it.

Can I Do It for Less? How to cut the costs of installing a wind system is the subject of chapter 8. If you think that building the wind machine yourself is the answer, this chapter asks you some hard questions about your purpose in doing so. It suggests using an old windcharger instead, and it offers suggestions on how you can locate and rebuild one yourself. But if you are looking for a trouble-free wind machine, your options are narrowed to installing a small commercial wind machine yourself from a kit or to preparing the site for a dealer-installed wind machine.

How Can I Evaluate Them? Chapter 9 tells you how to put it all together, how to buy a wind machine by evaluating the manufacturer and their dealer, what to include in a contract, and what to expect from the dealer.

How Are They Installed? Chapter 10 was designed for those who would like to do some of the installation work themselves. But it is not only for them; it is useful to anyone for tips on different methods of installation, anchoring, and wiring. There is also a section on how to operate and maintain a modern wind machine and how to monitor its performance.

How Can I work with the Wind Safely? The topic of Chapter 11 is safely working with the wind, a discussion of which is long overdue.

Before we plunge into the text, there are a few preliminaries we need to deal with first. You may wonder, for example, why I have an affinity for the term wind machine instead of something fancier like WECS, or better yet SWECS, for small wind energy conversion system.

NOMENCLATURE

Wind Machines. I use the term *wind machine* to describe these kinematic devices because it is simple and unpretentious. It does the job with the least fuss. *Wind turbine* is equally descriptive, but I try to apply it more specifically to the spinning blades that comprise the rotor. When referring to the wind machine, tower, and ancillary equipment as a whole, *wind system* works well. *Windmill* is reserved for the multiblade, water-pumping wind machine (American farm windmill) and for the European wind machine used before the age of steam (Dutch windmill), because that is what nearly everyone calls them. There are plenty of terms that can and have been used, as the following list attests:

aero-electric generator	wind engine
aerogenerator	wind generator
aeromotor	wind machine
aeroturbine	windmill
windcharger	wind plant
wind driven engine	windpower generator
wind dynamo	wind turbine
wind-electric generator	wind turbine generator
wind energy conversion system	

Some of these, particularly *wind energy conversion system* (WECS) and *wind turbine generator* (WTG), will be found in technical literature and promotional brochures describing wind machines.

Power and Energy. In casual conversation we use the terms interchangeably; but when we describe what a wind machine can do for us — what it can produce — we must be more specific. Power and energy, though closely related, have separate and distinct meanings. They are both related to work.

In the technical sense, *work* is performed when a *force* acts through some *distance*. It is expressed as:

$$\text{Work} = \text{Force} \times \text{Distance}$$

When you push a stalled car, for example, you are applying a force. If the car moves ten feet, we say the force (in pounds) you applied acted through a distance of ten feet. For work to be done, something must be accomplished: an object moved, lifted, or turned. If the object does not move, no work is accomplished. In our example, if the car did not move, no matter how hard you grunted and groaned to move it, no work was performed.

Energy is defined as the ability to do work or the amount of work actually performed. Both use the same units and are given in the same terms. When the wind strikes the blades of a wind turbine, it imparts a thrust or force that turns the rotor. A finite amount of energy in the flowing wind has been converted to rotational energy in the spinning rotor. From this we can say that when a force does work on an object, energy is transferred from one form to another.

Now couple that spinning rotor to a generator. Work is accomplished when electrons flow from the generator to a load. This flow can be used to heat a wire as in a toaster (thermal energy), or to cause a wire to glow as in a light bulb (photo energy), or to turn a motor (mechanical energy) as in a fan. Now, take a closer look at the electric motor. The flow of electrons, what we call electricity, transfers the rotational energy imparted by the wind to the wind turbine rotor through wires or power lines to spin the shaft of the motor giving us, once again, rotational energy. Because the motor spins a fan, work is accomplished. We have gotten something out of it.

The conversion from one form of energy to another is never 100 percent efficient; that is, you cannot get out as much as you put into it. There is always some energy lost in the process. Friction is the chief culprit. Though no one has ever built a machine that can convert 100 percent of energy in one form to another, people keep trying. Their perpetual motion machines appear every so often as a "startling discovery" in the popular press. Wind machines are a favorite target of this breed. Such machines never operate as claimed, nor do they ever produce useful work.

As a person who uses energy, you are concerned about the actual work accomplished and the amount of energy transferred. It is the bottom line of the technical balance sheet. You measure the performance of a wind machine by what it does for you, the work it actually performs, the energy it actually transfers and puts to use.

Where does power come in? *Power* is the rate at which work is performed, the rate at which energy is transferred (changed or released), the rate at which the energy in the wind passes through a unit of area. Power is energy per a unit of time. Power is given in *watts* (W) or in *kilowatts* (kW) (1,000 watts).

Energy and work are related to power by:

$$\text{Energy} = \text{Power} \times \text{Time}$$

Energy is given in units of *kilowatt-hours* (kWh). When you buy a gallon of oil, a cubic foot of natural gas, or a kWh of electricity, you are paying for the energy that can be released as heat in the case of oil and gas, or as heat and mechanical energy in the case of electricity, not the power that could be instantaneously delivered.

In this book I try to stress that wind machines produce energy or that a home consumes energy. But I will occasionally refer to the power that can be drawn from a generator or the power demanded by a circuit in the home.

In an electrical circuit, the flow of electrons in a wire is known as *current*. It is given in units of *amperage* or *amps*. We perceive *voltage* as the pressure trying to force the electrons to flow (the electrical force) in a wire. But no flow takes place and no work is done unless a load like a toaster is attached to the circuit. (No electricity flows out of a receptacle until an appliance is plugged into it.) Power (P) in watts is given by the product of voltage and current.

$$\text{Power} = \text{Voltage} \times \text{Current}$$

This equation represents the instantaneous rate work is being done when a force (voltage) moves electrons some distance through the circuit (current). A toaster, for example, will operate at 110 volts (household voltage) and will draw 10 amps or

$$P = 110 \text{ volts} \times 10 \text{ amps} = 1,100 \text{ watts} = 1.1 \text{ kW}$$

This is equivalent to eleven 100-watt light bulbs. Yet in the average household, a toaster uses much less energy (power × time) than lighting because it is used so infrequently (only a few minutes every morning). Lights, on the other hand, are used for hours on end. One 100-watt light bulb operating eleven hours will burn as much electricity, use as much energy, as the toaster operated for one hour.

$$100 \text{ W} \times 11 \text{ hours} = 1,100 \text{ Wh} = 1.1 \text{ kWh}$$
$$1,100 \text{ W} \times 1 \text{ hour} = 1,100 \text{ Wh} = 1.1 \text{ kWh}$$

Levers. Whenever you struggle with the lug nuts as you are changing a tire, you are faced with two choices: increase the force applied or increase the length of the wrench. We are only so strong and our backs will only take so much strain before we begin to look at our options. We could stand on the wrench thus applying our full weight (increasing the force), or we could go for a longer wrench. That is what most of us do — find a longer wrench. This gives us a mechanical advantage (the use of less force to do the same job) by increasing the leverage we exert on the nut.

The span from the nut to the end of the lever is the *moment arm*. A longer moment arm must act through a greater distance to turn the nut the same amount as a short moment arm (wrench). We must push the end of the wrench farther than before. This illustrates an important principle; where we gain a mechanical advantage we must make up for it by increasing the distance through which it acts. (It also demonstrates a concept summed up by Barry Commoner; there is no such thing as a free lunch. We gain by giving up something. Our hope is to come out ahead — on balance.) The work we wish to perform remains the same. We have only altered the relationship between force and distance — decreasing the one, increasing the other. This principle remains the same whether we are talking about levers, transmissions, or pulleys.

The force applied to the lever or moment arm times the span from the nut to where the force is applied is called *torque*. An understanding of torque and moment arms is helpful when considering what wind machine designers face when trying to prevent a 100-foot tower from buckling in a 120 mph wind. The wind's thrust on the wind turbine at the top of the tower applies a force, and the 100-foot span from the ground to the top of the tower is the moment arm through which it acts. Another example is the torque applied to the shaft attaching the blade to the rotor. More torque can be delivered by a long blade than a short one, where all other conditions are the same. If the blade moves through the same arc, more energy can be delivered by a long blade than a short one with the same force applied. This affects design by requiring that a heavier (stronger) shaft be used, but this disadvantage is offset by the greater amount of energy delivered to the generator.

Application of such basic principles of physics is valuable when trying to understand why wind machines differ, why one designer approaches the problem one way while another takes a different tack. If need be, pull out your old high school physics book and give yourself a quick review as you tackle some of the material ahead.

EQUATIONS AND TABLES

The relationship between quantities can be expressed in three ways: by equations, by tables, and by graphs. Equations express the mathematical relationships between quantities, and tables and graphs display the results of variations between them. Equations tell us what is significant and how much so. If we are calculating the power in the wind, the equation for power tells us why wind speed is so important, as you will see in chapter 4. Wherever an important equation appears in the text, I have also included an accompanying table to summarize the results. You need not use the equations if you don't want to, just go to the tables.

I have chosen to leave the equations in the text rather than mislead you into relying on the tables alone. An equation reveals its weakness — its Achilles heel — for all to see. By studying the equation given, you will learn when you can use it and when not. You will become aware of the underlying assumptions, and when they are no longer appropriate you can substitute what you believe is correct.

Tables are useful because they are easy to read, but trends are hard to spot in a table. Where necessary, graphs are used to illustrate trends and general relationships, such as the distribution of wind speeds over a year's time.

APPROXIMATE ARITHMETIC

One consequence of my work measuring wind speed (it could be any other phenomenon, say air quality or the energy output from a wind machine) in the real world is that the results I have obtained seldom follow the orderly relationships shown in the textbooks. If you were plotting a graph of this data, the points representing your measurements would not all fall in a neat line. Rather, they would be scattered. They would lie above and below where the line should be. When your job is to interpret this information, you draw a line that best fits the data. The line is an approximation of what happened; it does not say exactly what happened.

Knowing this, I have always found it annoying when the results of an imprecise calculation are presented as precise, especially the potential output from a wind machine. It is absurd to say that a wind machine will produce 495 kWh per month, when the data used in the estimate indicates there is a likely range from 450 to 550 kWh per month. Why not say simply 500 kWh per month and be done with it, or better yet present the estimates as a range from 450 to 550 kWh/month. Then you know that the estimates are only a gross approximation of what can actually occur.

This false accuracy, I believe, is a carry-over from marketing hype. It is the same thinking that labels a $4,000 wind machine $3,995. Maybe the users of such numbers are afraid that you will think less of their product if they acknowledge the difficulty of estimating output accurately.

Since the advent of pocket calculators, the results of calculations are often presented in meaningless detail. Back in the days of the slide rule, every student learned that the calculation was only as accurate as the divisions on the rule, as accurate as the least accurate number in the calculation. You could not carry a number out to ten decimal places if you wanted to. The concept is still valid today. The results of a calculation are only as accurate as the least accurate number used. The precision of a number is indicated by its significant figures.

Consider average wind speed as an example. It is normally presented to one decimal place (e.g., 10.5 mph). It has three significant figures: two to the left, and one to the right of the decimal. If we were to use this average speed in a calculation to estimate the energy production from a wind machine, the results should likewise be presented to no more than three significant figures. Say the calculation resulted in 22,525.49 kWh on your calculator. We need to round off the number to three significant figures instead of the seven indicated. Therefore, the result, 22,500 kWh, is more realistic considering the accuracy of the numbers used to derive it. Ignoring the concept of significant figures leads to false precision.

Most scientists and engineers are accustomed to *approximate arithmetic*. They round off numbers every chance they get. It makes their work easier. More importantly, it allows them to get quickly to the heart of a calculation without wasting time on needless detail. The use of approximate arithmetic is how an engineer can take a seemingly complex problem, such as estimating the potential energy production from a wind machine, and solve it mentally or scratch it out on the back of an envelope.

I have liberally rounded the numbers in the calculations and the tables presented in the following chapters where I thought it appropriate to indicate approximation.

INTANGIBLE BENEFITS

People use wind machines for economic reasons. The knowledge that you are saving money—in some cases earning it—is sufficient reward for taking the plunge and installing one. But there is more to it than that. There is an intangible side that is not quantifiable.

There is a feeling from working with the wind that is hard to describe. You need not install a wind machine yourself (most don't) to gain a sense of satisfaction from it. You can't escape the excitement that an operating wind machine creates. (I have yet to meet someone who was not thrilled at the site and sound of a sleek turbine whirring in the wind.) Wind machines have fascinated us for centuries and will continue to do so. In a sense, they beckon.

And, if you are as fortunate as I and install a wind machine with your own hands, you will understand what I mean when I speak of the feeling you get hanging from an 80-foot tower on a warm sunny day gazing at the countryside below, a view that few can share, the feeling of seeing your wind machine spinning overhead for the first time, the feeling of accomplishment one gets from a job well done, and the feeling of camaraderie that develops between you and your friends after several days of arduous work together (the hearty backslapping variety reminiscent of barn raising in another era). There is nothing quite like it except—possibly—climbing a mountain and basking in the achievement.

2

Applications

Look around you. How do you currently use energy? To run your appliances, to heat and light your home? In almost every example, a wind machine can be used to supplement or replace your consumption of conventional energy. We will glance at some of the methods for doing so in this chapter. I will also explain how you can determine the amount of energy you presently consume.

WATER PUMPING

Throughout history, wind machines have been used to pump water. They still are. The American farm windmill, a common site along rural roadsides, was and still is used primarily for watering stock. The mechanical action of the farm windmill dependably pumps low volumes of water from shallow wells for this purpose. The farm windmill is capable of pumping up to ten gallons per minute from depths of 100 feet.

Many applications for pumping water, notably irrigation on the Great Plains, require considerably more capacity than the farm windmill can supply. High discharge irrigation wells, for example, in the Texas Panhandle, may pump upwards of 1,000 gallons per minute from depths approaching 400 feet. The farm windmill is not big enough even to dent such a load.

Modern wind machines can be used for irrigation by either pumping water mechanically or by producing electricity to run a well motor. Because fluctuations in the wind would cause the well's output to be sporadic and unpredictable, the wind machines must drive the well pumps in conjunction with a conventional energy source. This is not difficult to accomplish with an electric well pump but it is with the engine-driven pumps more commonly used.

West Texas State University in cooperation with the Department of Agriculture employs an ingenious device for mechanically coupling the varying output of a wind machine with an irrigation pump: the overrunning clutch. When the wind is strong and the wind machine is producing full output it mechanically drives the well pump entirely on its own, via the overrunning clutch (see fig. 2-1). When the winds are weaker, the wind machine assists in driving the pump. The conventional power source makes up the difference. During a calm spell the conventional power source operates the pump alone. The overrunning clutch assures that a constant volume of water is pumped regardless of the wind, while also taking advantage of the wind when it is present to reduce consumption. This concept has been successfully tested at the USDA's experiment station near Amarillo, Texas.

Using the wind to assist irrigation pumping is even simpler when the pump is driven by an electric motor. Wind machines producing utility-grade electricity can be connected directly to the well motor. When the wind blows, output from the wind machine offsets consumption of electricity from the utility.

SUPPLYING REMOTE POWER

For homesteads without utility service, such as in remote areas of Alaska and Canada, wind machines are used to provide electrical power for lighting and appliances. Because the wind machine's output varies with wind speed, it cannot be relied upon as a power supply unless used with battery storage (see fig. 2-2). The batteries store energy during windy days for use during calms. The direct current (DC) output from the wind machine or the batteries is used either directly in DC appliances or inverted to alternating current (AC).

Appliances that operate on DC are obtainable through specialty houses serving the recreational vehicle market. Most appliances in everyday use, however, operate only on AC. To use these appliances on a remote

Applications • 25

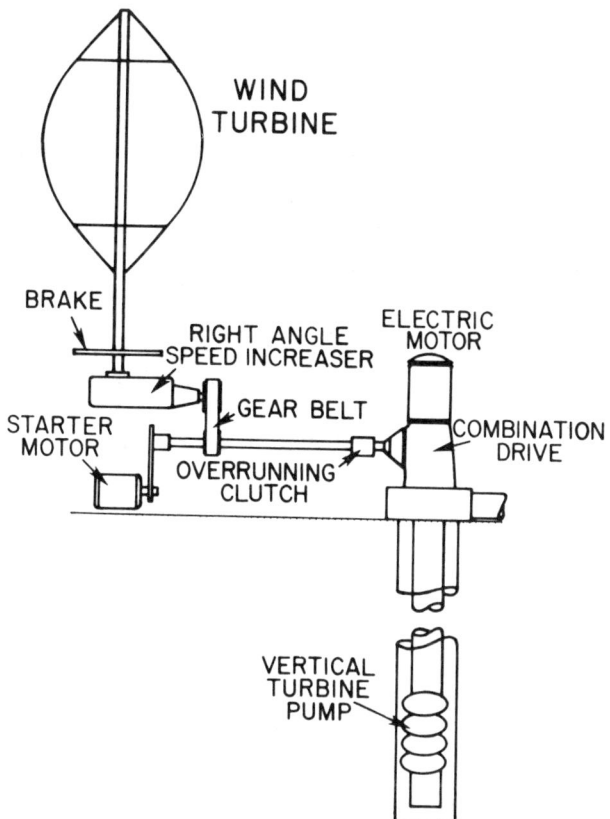

Fig. 2-1. Wind-assisted irrigation. The Darrieus turbine mechanically drives the well pump via the overrunning clutch. (Alternative Energy Institute, WTSU)

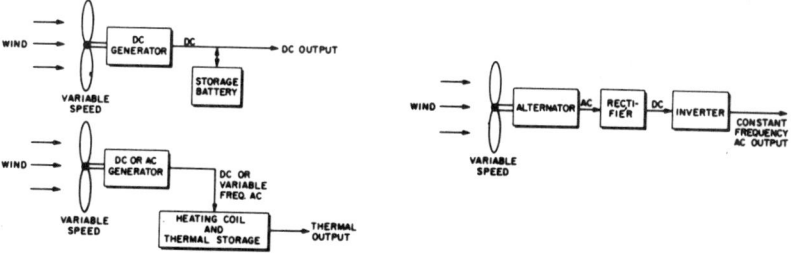

Fig. 2-2. Schemes for harnessing wind energy without the aid of a utility interconnection. (Mitre, DOE)

power supply requires that DC be converted to AC. Employing a rotary inverter is one way in which this is done (see fig. 2-3). Power from the batteries drives a motor on one side of the inverter. The motor in turn drives an alternator to produce 60-cycle AC power like that from the utility. Static inverters can also be used to produce AC electronically.

Wind machines for powering remote homesteads were more widely applicable during the thirties than they are today. In 1936, only 10 percent of the farms in this country were served by central station power. Electric utilities now reach 98 percent of rural families. Wherever there is existing utility service, a battery charging wind system is impractical. Batteries are expensive, demand regular maintenance, and can be dangerous. (Anyone who has jump started stalled cars in midwinter knows that a spark can ignite a battery explosion.) Battery charging comes into its own when there is little likelihood that your site will ever be served by a utility or if the cost of stringing a power line is prohibitively expensive.

Battery charging wind systems are usually competitive with gasoline or diesel generators. If these are your only other source of power, battery charging may be practical. Don Marier's *Wind Power for the Homeowner* and Terrance Paul's *How to Design an Independent Power System* are excellent references for putting together your own home light plant.

One note of warning. If you build your homestead around a battery charging wind system, decide early whether you will use AC or DC for your lighting and appliances. Which you use will determine how you wire your house. You should never use DC with wiring and switches rated for AC only. Direct current has a tendency to arc when you throw a switch or pull a plug out of a receptacle. The arc jumps across the contacts, and if, for example, the switch was not designed to handle it, the arc could melt the switch and start a fire. I have seen articles in popular magazines that extolled the wonders of using a DC wind system in a conventionally wired house. Not once did they warn their readers about the hazards involved; it would have detracted from the blissful tone. Designing an independent power system requires that you do your homework before you begin work on your home.

Wind machines producing DC can also be used as independent power systems for mountaintop telecommunications stations (microwave relay, radio transmitters, and so on) and for cathodic protection on pipelines.

HEATING

Heating domestic water and the living space comprise most of a home's energy demand. It is not surprising then that there have been several attempts to use wind machines strictly for heating. Another mark in their

Applications • 27

Fig. 2-3. Batteries are used to store wind-generated energy in remote applications. For use in modern appliances, the direct current stored in the batteries must be converted to 60-cycle alternating current by a rotary inverter as shown in the foreground, or by a solid-state static inverter.

favor for such an application is the coincidence of wind with the demand for heat. Demand is greatest during the winter months when, for much of the country, winds are also highest.

This coincidence of supply with demand differs from that of solar collectors where the greatest heat gain is during the summer, not the winter. Yet wind machines can be used in much the same way. Small wind machines heat domestic hot water like two panels of a solar system, or a larger machine can be used to heat the whole house like a larger array of solar collectors.

The wind furnace concept (see fig. 2-4) takes advantage of these similarities and integrates a wind machine with solar collectors for space heating. It is believed that with the two systems operating together, they can eliminate the need for backup power. On cloudy days, for example, the wind may be blowing, or when the winds are weak the sun may be shining. If each were operating independently, conventional power would be needed to make up the difference. Together, they are thought to reinforce each other. In practice, however, homeowners can usually afford one or the other, but not both.

Like a solar collector, the wind machine can heat air or water depending upon the type of space heating system in use. If you have a hydronic (water) heating system, then the wind is used to heat water to augment your existing furnace. If you use forced hot air, then the wind machine will heat air. And, as with solar, the economics of a wind-powered heating system can be improved with thermal storage: either as heated water or as heated rocks.

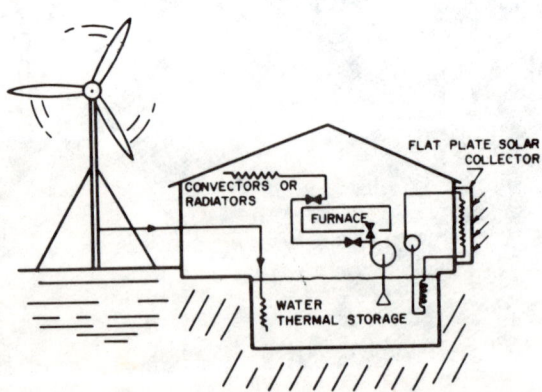

Fig. 2-4. Wind furnace. The wind machine produces low-grade electricity to supplement conventional heating sources such as oil, gas, or electricity, or to supplement heat from solar collectors as illustrated. (University of Massachusetts)

The wind machine can be designed to generate heat electrically with resistance elements (like those in your toaster, only much larger) or mechanically with a churn. The University of Massachusetts successfully demonstrated the use of electric resistance heating with their 10-meter wind turbine (see fig. 2-5). This wind machine drives an alternator at variable speeds, depending upon the wind's strength, and produces variable voltage, variable frequency electricity. The output is ideal for applications, such as heating, that can use low-grade electricity. (This form of electricity is not compatible with that from the utility.) An attempt at commercializing the concept, however, failed.

Heat can also be generated mechanically by spinning paddles in a tank of water (see fig. 2-6). Heated water from the churn is then piped to your home or business. There have been several experiments demonstrating this method for space heating. Though the water churn is commercially available, no vendors currently offer a package including the wind machine and the water churn.

Some manufacturers of small battery charging wind systems have adapted their products for heating domestic hot water with electric resistance elements. As with a solar system heating domestic water, the wind machine is used to preheat the cold water supply to the existing hot-water tank. This is accomplished by using the wind machine's DC output to heat water in a second tank. Warm water from this tank is then fed to the conventional water heater where it is brought up to the desired temperature.

One argument in favor of generating low-grade energy with a wind machine for heating either with an alternator or churn, is that it can be done at less cost than producing the same amount of utility-compatible electricity. It would also be simpler, proponents add, than trying to generate the high-grade electricity demanded by the utility. It hasn't worked out that way. Experience has shown that in most cases it is cheaper and easier to interconnect the wind system directly with the utility and produce high-grade electricity.

INTERCONNECTING WITH THE UTILITY

If you are going through all the trouble and expense of putting a wind machine in the air, it is my view that you want to get the most out of it. To me, that means high-grade electricity like that supplied by the utility: constant voltage, constant frequency AC. You can do anything you want with it from that point on. It is adaptable to many uses. It will run your clock, as well as your stereo. It will run your refrigerator or it can be stepped down a notch in quality and be burned in your lights or in your

Fig. 2-5. This 10-meter (33-foot) downwind turbine was used by the University of Massachusetts to demonstrate the wind furnace concept. The rotor drives an alternator, the voltage and frequency of which vary with wind speed.

Applications • 31

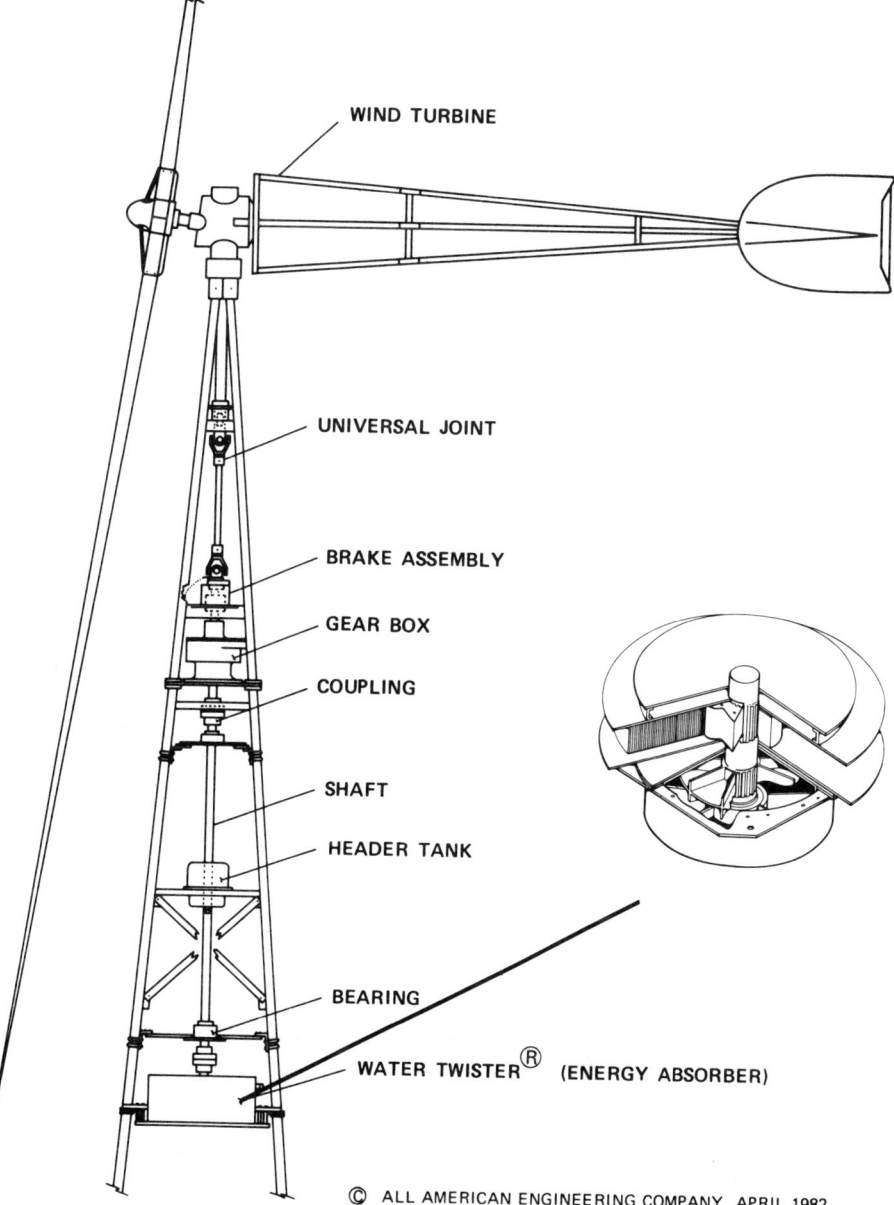

Fig. 2-6. A wind turbine modified to transmit mechanical power (torque) down the tower to spin the shaft of a water churn. Water heated by the agitation of the spinning paddles within the churn flows to a conventional hot water heating system. This concept is undergoing tests on the west coast of Ireland. (Courtesy of All American Engineering)

32 • *Wind Energy*

hot-water tank. If you are a farmer, it will run your milkers or your feeders. In short, you use it wherever you presently use electricity.

The only disadvantage is that you must deal with the utility company, and if there is a power outage your wind machine is idled as well.

Small wind machines use either induction generators or synchronous inverters for producing utility-compatible power (see fig. 2-7). (These will be discussed more fully in chapter 6.) No matter which is used, their interconnection with the utility is the same. Wires from the control box are connected to terminals within your service panel. The wind system becomes a part of your home's electrical circuit like that for an electric range.

When the wind is blowing, the wind machine produces energy. This energy will flow to your service panel and seek out those circuits where electricity is being consumed. It is first come, first served. It will deliver energy to the first circuit it comes across where it is needed. If more energy is being generated than can be used by the first circuit, it seeks out the next and so on. When the wind machine can not deliver as much energy as needed for a circuit, the utility makes up the difference. There is no fancy electronics controlling which circuit gets energy. It is all accomplished silently and effortlessly. That is the beauty of electricity. You never know from whence it came, but it is there nonetheless.

As with individual circuits, so too with the service to your whole house. If you are not using electricity when the wind machine is producing it, or if you are not using as much as it is producing, the excess flows from your service panel through your meter out to the utility's lines and on down the road. It seems mysterious, but it works. It does so neatly, cleanly, and without fuss.

You can use the wind machine to reduce your consumption of electricity for lighting and appliances. Or you can use it to supplement your

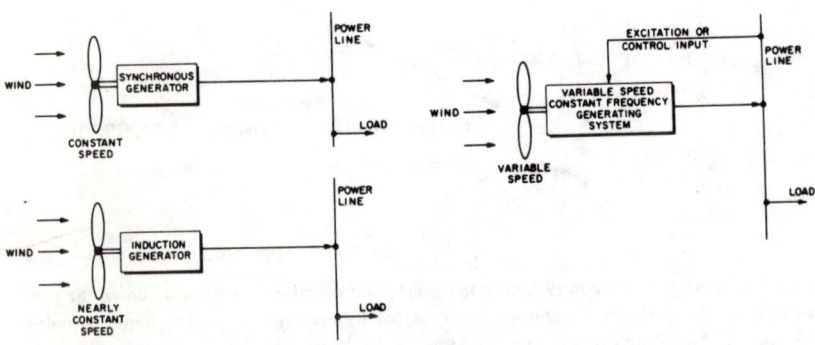

Fig. 2-7. Methods for interconnecting wind machines with the utility. (Mitre, DOE)

Applications • 33

lighting and appliance loads as well as electric heat. The only difference is that a bigger wind machine would be needed to make a proportionate reduction in your bill.

ESTIMATING DEMAND

How much electricity do you consume? Do you know? If you are like most people, you know how much you pay each month but you don't have the foggiest idea of how much electricity you use. To find out, you don't have to do anything more complicated than pull out your old electric bills.

There are at least two items of importance on them. One is your consumption in kWh. The other is the total cost for this energy. From this, you can calculate what you are paying for each kWh of electricity.

If you are lucky, the utility will clearly tell you how much electrical energy you consumed last year. The bill in figure 2-8 does just that. On the right side just below the total amount is the total kWh used for the past twelve months: 3,363 kWh. If your bill does not have this feature (an electric use profile), and many do not, you will have to tally up monthly consumption yourself for the entire year.

Look for the meter readings and the resulting monthly consumption. In this example, it is 494 kWh in February. Frequently, the utility estimates your consumption from past usage and balances your account whenever they get around to reading your meter.

If you have one of those so-called budget accounts, where you pay so much a month regardless of immediate consumption, the utility's estimate of monthly or annual consumption should be shown on the bill somewhere.

The average household without electric heat uses approximately 20 kWh per day or 600 kWh per month. Such a household will consume 7,200 kWh per year. (This is twice as much as I use each year, as seen in the sample bill.) Homeowners with electric heat use considerably more. Depending upon the climate, the size of the building, and how well it is insulated, an all-electric home will use from 20,000 to 80,000 kWh per year.

The same technique can also be used with any other form of energy besides electricity. Say your primary concern is the cost of heating with oil. Pull out the oil bills. Find out how many gallons of oil you use each year and what it costs. You can then convert the oil consumed to kWh of equivalent electricity. (An example is given in chapter 5.) You can do this for natural gas or even wood. The method is the same.

You will note that I have presented electrical consumption on an annual basis rather than monthly. There is a reason for this. Wind data

34 • Wind Energy

is usually given for the entire year as is the output of wind machines. Wind data is presented as the *average annual wind speed* or power. The performance of wind machines is reported as the *annual energy output* at various average annual wind speeds. By presenting them in the same annual format, you can make direct comparisons.

The energy use profile in figure 2-8 has another good feature. It tells me how much electricity I used each month and compares my current consumption with that of one year earlier. This is helpful in my conservation efforts. The profile tells me two things — First, that my consumption peaks in the winter. Knowing this, I can look for the reason why (an electric space heater in my workplace). Second, it tells me that I used more this February than I did before. Again, I can seek an explanation. It could be that my refrigerator is running too often, for example, and that I need to repair it. In this case my typewriter has been running continuously.

If your electric bill is not this detailed, you can still use it to help

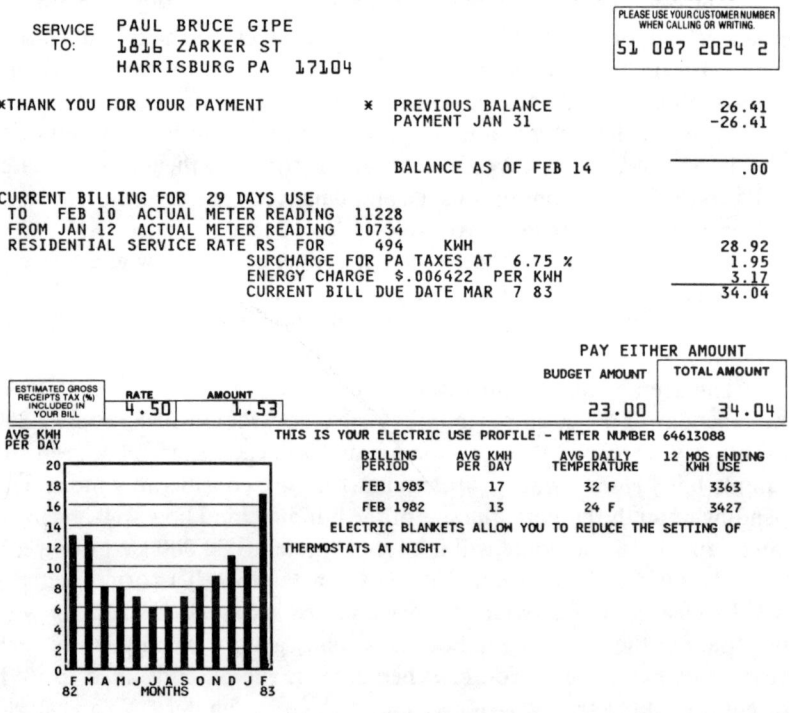

Fig. 2-8. Sample electric bill.

limit consumption. By keeping good records, you can detect abnormal changes in consumption. In the process, you will become more attuned to how you use electricity and you will begin to find ways to use less.

If you find an appliance where you may be able to save electricity, you can easily estimate how much the appliance is costing you now and how much you can save. There is a label on every appliance giving its power rating in watts (W) or its current draw in amps (A). You are after watts because they enable you to calculate the amount of electricity the appliance consumes. If the label gives amperage instead, it will also list the voltage. The product of voltage and amperage will give you watts. Got all that? Consider my electric space heater. It is rated at 1,000 watts or one kilowatt (kW). This is the same as saying the 110-volt heater draws 9 amps. For residential customers, the utility does not charge for power, only for energy. If I operated this one-kilowatt space heater for one hour it would consume one kilowatt-hour (kWh) of electrical energy.

The cost per kWh can be found by dividing the total amount ($34.04) in the electric bill in figure 2-8 by the consumption for the month (494 kWh). The cost, in this example, is 6.9¢/kWh. So, if my space heater ran one hour per day, in a month's time it would cost me:

$$\$0.07/\text{kWh} \times 1 \text{ kWh/day} \times 30 \text{ days} = \$2.10.$$

You can apply this technique to any appliance or load to gauge its effect on your total electric bill.

Consumption not only varies from month to month, but also by time of day (see fig. 2-9). For a typical home without electric heat, the use peaks with peak activity around the house: early morning as everyone readies themselves for work or school, again at lunch, and then in the evening around dinner.

Because wind speed fluctuates throughout the day, output from the wind machine also varies (see fig. 2-10). There is no assurance that the output from the wind machine will match consumption. The wind may howl at night when consumption is lowest and be calm during the day when use peaks. In an independent power system, this mismatch is tempered by storing excess energy in batteries. In an interconnected application, excess energy flows to the utility. During light winds, the utility makes up any shortfall. This is of interest because the utility may not pay as much for the energy flowing into their lines as they charge for what they sell. Consequently, it may be necessary to size the wind machine so that it only occasionally produces excess energy or to use as much of the excess on-site as possible where it is most valuable.

If you determine that the sale of excess energy is not in your interest,

36 • Wind Energy

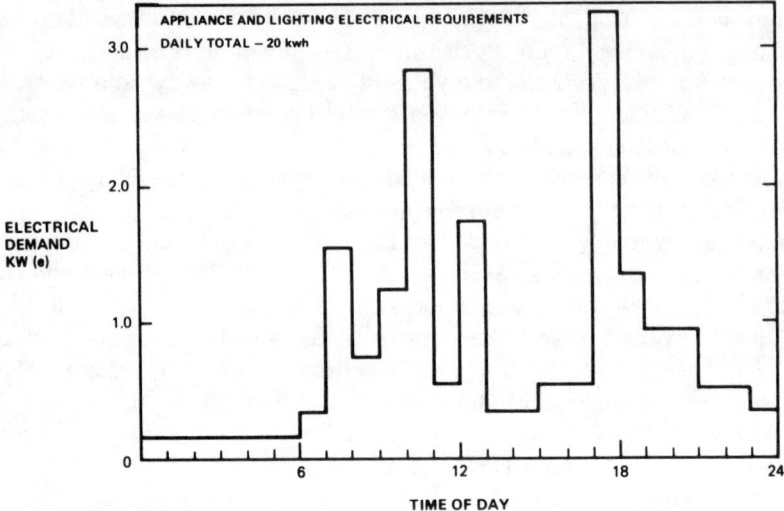

Fig. 2-9. Typical daily electrical consumption without electric heat. Consumption varies with time of day. (JBF Scientific, DOE)

then a wind machine meeting only one-half your demand may be more appropriate than a larger one displacing all your current consumption. The other option is to use the larger (and more cost-effective) wind machine, but use a *dump circuit* (see fig. 2-11) to direct the excess energy to a standby load. For example, you could install another tank for your domestic hot water. As in the wind furnace concept, wind generated elec-

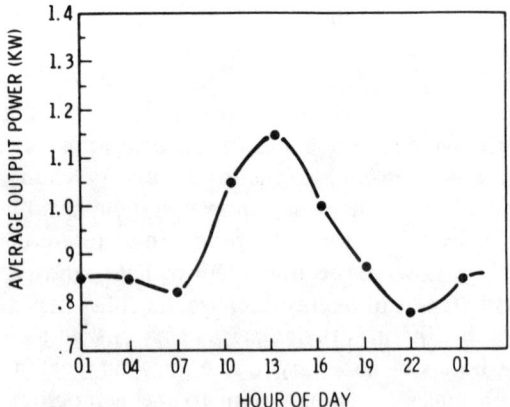

Fig. 2-10. Typical daily wind machine output. Generation also varies with time of day. (Battelle PNL, DOE)

Applications • 37

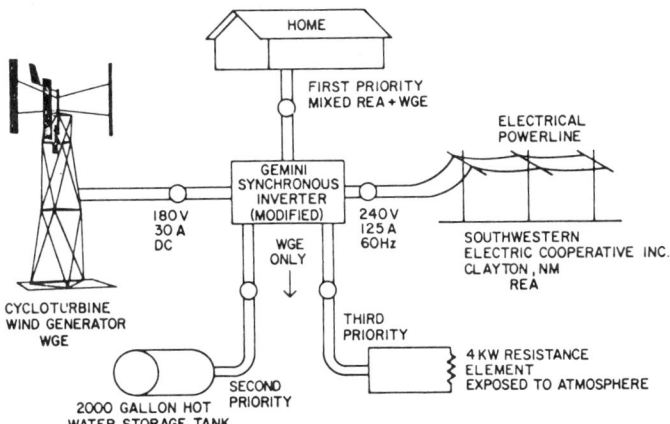

Fig. 2-11. Dump circuit. Excess electricity is dumped or diverted to alternate loads rather than being sold back to the utility at less than the retail rate. (New Mexico State University)

tricity is used to heat water in the tank which is then fed to your hot-water heater. The preheated water reduces your consumption of conventional energy whether you heat water with electricity or with natural gas. The extra tank allows you to dump excess electricity whenever you have it into thermal storage. Your water heater can then draw from this tank whenever it needs makeup water. In this way, you have stored electricity as low-grade heat. It is much simpler and less costly than storing energy in batteries.

You may wonder why you need the dump circuit and an extra tank if you already have an electric hot-water heater. When the tank's thermostat calls for electricity to heat the water to a certain temperature, it draws automatically from the house's circuits, which the wind machine is supplying. That is the problem. It only draws power when needed. It is the boss. We want to use the wind-generated power when it is available, and that is not always at the same time when there is a demand for it. (That is why we have an excess, right?)

Once you have mastered the concept of using excess electricity to preheat domestic water, you can turn your attention to other loads. If you heat your house with hot water then you can preheat the water in your furnace. Just put in a bigger storage tank. Voilà, we are right back where we started from with the wind furnace concept. We have the same advantages as before, plus the wind machine supplies all our regular electrical loads. It is more useful overall than when we produce only low-grade heat.

If you are fortunate, the utility will pay enough for the energy they

38 • Wind Energy

buy that you will not have to go through the trouble of adding the dump circuit. You can then shop for the wind machine that is most cost-effective. The economics usually improve with increasing size. So you will probably be seeking a wind machine that meets as much of your consumption as your budget will allow.

FARMING THE WIND

One wind machine may not be enough. You may need more than one. For homeowners this is not always practical, but for many businesses, it is. As explained in the next chapter, it may be profitable to sell power to the utility—as much of it as you can. Hundreds of wind machines are being installed in California and elsewhere for this purpose (see fig. 2-12).

Fig. 2-12. Wind farm. Belgian WindMasters on Altamount Pass wind farm east of San Francisco. The 22-meter (72-foot) diameter rotor drives a 150-kW generator interconnected with the utility. The wind machines produce energy solely for sale to the utility. (Courtesy WindMaster Corp.)

The utilities have conceived their own schemes for integrating wind machines with their systems. One popular idea is pumped storage. When the wind machine's output is not needed on the grid, the excess energy is used to pump water uphill into a storage reservoir. When demand from the grid is higher than can be met by the wind machine, water is released from the reservoir to drive hydroelectric turbines. The reservoir acts as a huge battery, storing energy when there is an abundance and releasing it when there is not.

3

Siting

Now that you know how a wind machine can be used, you need to determine if you have the site for it. Your site may not be suitable for a wind machine because of physical limitations restricting safe erection or operation, or because of legal limitations restricting how you can use your land and your wind machine.

PHYSICAL RESTRICTIONS

Foremost is, do you have the space for a wind machine? Not only must there be sufficient space for the tower but you must have the room to install it. I have seen guyed towers used on small city lots where an anchor was placed in each corner of the backyard. And I have seen free-standing towers installed in equally small spaces. But there is a limit. In one case, a homeowner wanted to install a free-standing tower in his cluttered backyard. His plans included, among other gems, running the hoisting cable through his garage. A crane would have been a better choice. But it would have had a difficult time maneuvering the narrow alleys leading to the site. In another installation, the crane lifted the tower over the house to set it on the foundation. All of which is to say you can install a wind machine almost anywhere if you are willing to go through the trouble and you are willing to take the risk.

Clear Zone

As an installer, my gut feeling is to keep a clear zone around the tower equal to its height. In other words, site the tower far enough from your house and property lines such that if it fell it would not damage the building or extend beyond your property lines. This is more important for guyed towers than it is for free-standing towers. Yet, I acknowledge, we think nothing of other man-made and many natural hazards that pose a similar threat. We all have seen homes sheltered beneath the branches of an old oak or maple where, occasionally, a storm-weakened limb crashes down onto the roof. But we accept this hazard as the price we pay for the benefits the tree provides (shade and appearance). We act the same with citizen band (CB) radio towers. In many ways they are similar to towers for wind machines. They are made of metal, extend visually above the roof line, and can fail. We have grown to accept them, and because their failure rate is so low we often mount them right next to occupied buildings. Of course there is a difference. The effect from a several hundred pound wind machine crashing onto the roof would be more severe than that from a lightweight CB tower. Tower failures are rare, but it is a possibility that should be considered. Wind machines are not for everyone. This space requirement excludes many of us.

Service Zone

Do not be concerned about a loss of power by siting the tower a safe distance from your house. This is much less a problem with modern wind machines operating at 110 and 220 volts than it was for the low-voltage windchargers of days past. You can place the tower up to 1,000 feet from the house before incurring a significant penalty. (Wiring is discussed at length in chapter 10.)

Exposure and Turbulence

Positioning the tower away from buildings allows you more freedom to take advantage of sites with better exposure to the wind. These sites will also experience less *turbulence*—rapid changes in wind speed and direction caused by buildings and trees obstructing wind flow. Turbulence is like the eddies swirling around a rock in a stream. It is more critical to modern wind turbines than to water-pumping windmills, which were sited haphazardly, because they use long slender blades operating at high speeds. Turbulence can wreak havoc on a wind machine, shortening its life or, under severe conditions, destroying it.

Buildings and trees hinder the wind. The lower wind speeds in the wake of obstructions will drastically reduce the energy that can be generated by a wind machine. To maximize your investment, to get as much energy from the wind machine as you can, it must be sited where the effects from obstructions are minimal.

The tower must be located either far enough upwind or far enough downwind to avoid the turbulent zone around a building (see fig. 3-1). When that is impractical, a taller tower must be used to elevate the wind machine above the turbulence. If neither approach alone is sufficient, some combination of siting and a taller tower must be used. For example, if the peak of your roof is 30 feet above ground (H), then the tower should be a minimum of 60 feet tall ($2 \times H$) and preferably taller for the wind machine to be above the zone of maximum turbulence. If not, the tower must be placed 60 feet upwind ($2 \times H$) or 600 feet downwind ($20 \times H$).

Turbulence also occurs around trees in a shelter belt. Though they appear more open to the wind than a building, their effect on turbulence is similar. The same siting rules apply.

From years of experience, a general rule of thumb has evolved: the wind machine should be 30 feet above any obstruction within 300 feet (the length of a football field). Preferably, this is applied to the tip of the rotor which, on the larger diameter turbines, may be 10 to 15 feet below the top of the tower. If you have determined, for example, that a group of trees along a fence-row is 60 feet tall (see chapter 4 for techniques to use) you will need at least a 90-foot tower.

Fig. 3-1. Zone of disturbed flow. Wind speeds decrease and turbulence increases in the vicinity of obstructions. The effects are most pronounced downwind but also occur upwind as the air piles up in front of the obstruction. The flow over a hedgerow or group of trees in a shelter belt is disturbed in a similar manner. (Courtesy New York State Energy Office)

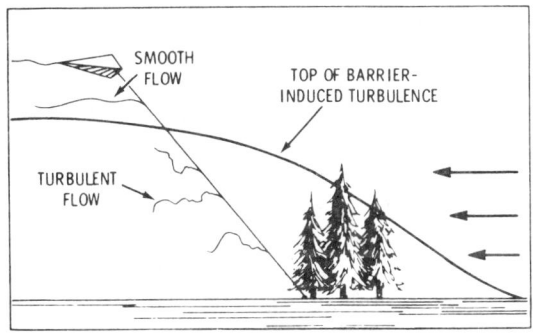

Fig. 3-2. Detecting turbulence. Trailing streamers from a kite is a simple yet effective way of detecting turbulence. (Battelle PNL, DOE)

When you are uncertain about the amount of turbulence over your site—go fly a kite. As in figure 3-2, tie streamers to the kite string and note how they flutter in the wind. Check turbulence this way for winds from several different directions until you are satisfied you have covered all possibilities. Note the site and height where turbulence is minimal. That is where you will want the tower.

Avoid sites at the bottom of creeks, draws, or ravines and sites at the foot of hills. If there is a hill on your property with a well-exposed summit, consider siting the wind machine there instead of lower on the slope. I had a site where a hill rose 200 feet above a farmhouse at its base. Even though it was 1,000 feet from the house, the hilltop was a far better site than one at the farmhouse. (In fact, it was the only site, as far as I was concerned, where the wind machine would work reliably.)

For more information on siting, consult Battelle Pacific Northwest Laboratory's *A Siting Handbook for Small Wind Energy Conversion Systems* (see the Bibliography for details).

Final Siting

Once you have selected the area where the tower will be erected, note how the power will be delivered to your house. Will you bury the cables, for example, or run them through the air? At this stage, you need to anticipate any problems that may develop later. They are easier to avoid than to solve. For example, a buried telephone line crossing your path may preclude digging a trench for an underground cable or at least make it more complicated. (The phone company now becomes involved.) Similarly, you may have trouble stringing an aerial cable if the electric utility's lines cross your path.

The electric service from the wind machine, ideally, should enter the building near where the utility's lines enter. (The reason for this is explained in chapter 10.) If the service entrance and meter are on the other side of the house from where you are planning to erect the tower, how will you route the power cables? It is easier to route them around an obstacle in a trench than it is when stringing them from poles.

Are there any sidewalks, driveways, or roads in your path? How will you cross them? To bury a line, it is not necessary to dig a trench through a driveway, for example. You can "point" a hole underneath it by hand or with power tools made for this task.

These are important questions because the answers affect the cost of installing the wind system. They also determine how difficult it will be to meet certain institutional restrictions such as the National Electrical Code. If you are planning, for example, to string the power cables across the roof of a building, they must have a specified clearance on all sides to meet the code's requirements. If they do not, you may lose your insurance coverage. (Wiring and code requirements are treated more thoroughly in chapter 10.)

INSTITUTIONAL RESTRICTIONS

Imagine buying a wind system only to find that you cannot—are not permitted—to install or operate it. This disaster has confronted more than one hapless person. But it can be avoided by thoroughly doing your homework. In this section we will examine the possible legal restrictions on where and how you use your wind system. Our principal concern is with zoning ordinances because they can prevent you from erecting a wind machine. In a subsequent section, we will take a detailed look at the issues surrounding the interconnection of the wind machine with the utility, because they can affect how you operate it.

Zoning

Most people who have installed wind machines have had little, if any, problems with zoning restrictions. Either their property was not covered by zoning ordinances (laws) or permission was quickly and easily obtained where it was. Many rural areas in the country are not zoned at all. And where they are, there are practically no restrictions on land that is zoned agricultural.

Problems arise in more densely populated areas where the right to swing your fist ends where your neighbor's nose begins. Zoning (or more broadly, restrictions on land use) is not part of a Commie plot to hamstring red-blooded Americans. It is a responsibility entrusted to local governments by the public to protect the general health and welfare from someone swinging his fist in disregard of the consequences to others.

My experience with zoning issues has spanned the full range of possibilities. In a rural area, for example, a client made one telephone call. That was it. No written permission was required. In a more settled, but still rural area, another client had to obtain a variance. Once again there was no problem. The variance was granted as a matter of course. The client had ample room for the wind machine, was on good terms with his neighbors, and was respected in the community. At the other extreme, I have found myself in a highly charged legal battle where the client made every mistake you could think of before calling in help.

At times, I admit, I have railed against the ignorance and the narrow-minded view of some zoning officials. But for the most part, they are honest, hardworking people, called upon to perform a civic chore. Treat them cordially. One thing is certain: if you need a building permit or a zoning variance, you want them on your side. Much of zoning enforcement is left to subjective interpretation.

Building Permit

Where zoning ordinances apply, you must conform to the law. No excuses are accepted. Find out what the requirements are in your area by calling the local building inspector, board of supervisors, or planning office. You want to know what is necessary to obtain a building permit (where required) and who is responsible for issuing it (usually the building inspector). Get details. Whoever is responsible should provide a list of what you must do: the forms to fill out, the fees to pay, where and when to file, and any other information you must supply. Then, methodically go through it.

The intent of this process is to alert the public to your project. Take the initiative and contact anyone who might be affected, especially your neighbors. You have a right to your wind machine. They also have a right to live in peace and quiet and to live without the fear that your towering wind machine could some day come crashing through their roof. Speak to them early in the project so that they feel consulted rather than pressured into backing you. It is much better to talk with them informally than in court or in a shouting match at a public hearing. If you get along well, there should be no problems. But, if you have consistently driven over their prize rose bush, better make amends. Objecting to your building permit is a great opportunity to even the score. You can head off conflict by considering the needs of your neighbors and respecting them. It is your responsibility as a member of the community.

The building inspector will probably require a plot plan (map) showing the dimensions of your lot and where the tower will be located. You can prepare this yourself. Drawings of the wind machine, tower, and foundation with their specifications may also be required. The dealer or manufacturer can supply these to you.

Zoning ordinances follow either of two approaches. One allows you to do whatever you want, unless specifically prohibited. This laissez-faire attitude is common in the Midwest where I was raised. The other approach prohibits you from erecting any structure unless it is specifically permitted. This is the practice in the Northeast. Where the latter approach is used, your building permit could be denied because no one has ever installed a wind machine before. No decision has yet been made whether or not to allow them.

In communities where this is the situation, you can sometimes get permission for a wind machine by bringing it under a permitted category such as CB towers, TV antennas, or chimneys. The building inspector may be empowered to make such a determination. If not, formal action of the zoning appeal board is necessary. They must determine if your use conforms to the intent of the ordinance. Where it does not or where the ordinance specifically excludes wind machines or similar structures, you must obtain a variance.

The zoning appeal board or board of adjustments is the arbiter of zoning disputes. They are a political body, and if there is a public outcry they will respond accordingly within the limits of the law. *Variances*—variations from the law—give the zoning appeals board flexibility in meeting the goal of zoning: the protection of the common good without undue restrictions.

They will want to know whether or not your wind machine detracts from your neighbors' use of their land or whether it poses any kind of threat to your neighbors or the community. You must convince them that it will not. The burden of proof is on you, the petitioner.

Often the granting of a variance is little more than a formality. You may not even need to be present. But if the board has questions that you have not answered previously, or if the variance is contested, you will need to be present and you will need to be well prepared. When contested, the hearing can take on the appearance of an expensive court room battle with opponents bringing in their own "expert" witnesses to counter your assertions. It can be rough if you are unprepared.

Height Restrictions

The most frequent problems involve height restrictions, setback requirements, and neighborhood concern about safety and noise. In most residential zones, there is usually a 30- to 40-foot height limit to preserve the appearance of the neighborhood. Variances to such ordinances can be obtained by pointing out other sites where structures taller than the limit have been allowed or exist despite the zoning ordinance: CB towers, chimneys, or utility poles. (The utility has the right of eminent domain; they can do what they please. The zoning board has no control over utilities.)

When you are within a mile from an airport or heliport, the Federal Aviation Administration (FAA) may impose lighting or special painting requirements. Normally, your tower's potential interference with aircraft is never a question—they operate well above the height of most towers—and no lights or painting is needed. You are required to report the existence of your tower to the FAA only when it exceeds 200 feet in height or when it is within the flight path to a nearby airport. Most wind machines operate well below this height. If there is any doubt, the building inspector may forward a notification to the FAA or advise you to do so.

Safety

In some communities, towers must be set back from the property line a distance equal to their height. For guyed towers, this is a reasonable precaution as I suggested earlier (less so for free-standing towers). This assures your neighbors that should the tower fail, it could not fall onto their property. If your lot is too small to permit this, I suggest you reconsider wind power. But I believe you have the right, if you choose, to install a tower on such a lot just as you now have the right to grow a tree of equal height without setback restriction. If this is your desire, you must convince the zoning board that the wind machine does not pose any unusual hazard to your neighbors.

The zoning board will be concerned that your tower could collapse. You must show them that the tower meets certain design standards (as indicated by the manufacturer) and applicable building codes, and that similar towers operate throughout the country in a host of severe environments without incident. Though towers have failed, the occurrence is rare and no more frequent than that of falling trees, utility poles, or similar objects.

They will also be concerned that the wind machine could throw a blade. Once again, you must convince them that the wind machine has been designed and built to accepted standards and that there is little likelihood that the wind machine will throw a blade into the midst of a neighbor's lawn party. Wind machines have been destroyed in this manner. It has happened more frequently than the failure of a tower, but it is still uncommon. Your only assurance that it will not happen is to buy a wind that has proven reliable.

This reasoning assumes that your tower and wind machine can indeed do all you state that it will and that you have evidence to support your position. If you don't, because you built it yourself or it is a prototype—one of only a few built—you better find a good attorney. You will need one.

The fear that the tower could become an "attractive nuisance" is a related concern. Generally, a property owner is not liable for accidents to trespassers, but a different test is applied to the acts of children. Swimming pools are thought (in some communities) to entice or attract children to trespass. Because children cannot discern the hazard presented by the pool, the community views it as a public nuisance, and if an accident occurs, a court can hold the owner liable. Zoning ordinances permit attractive nuisances when they have met requirements designed to prevent accidents. Swimming pools can be fenced; towers can also be fenced.

Fencing is not the only way to prevent someone from climbing the tower. Look around you. How does the utility do it? On their transmission towers they remove the climbing rungs to a level 10 to 12 feet above the ground. You can do the same. Or you can wrap the base of the tower in sheet metal or wire mesh. These alternatives to fencing should be acceptable to a zoning board. They accomplish the same goal less obtrusively and at less cost. You do not see the utility company erecting chain link fences around their utility poles or transmission towers, do you?

Aesthetics

For some, the appearance of the wind machine on the skyline becomes the cause célèbre—the issues of safety and noise being used to buttress their opposition. The concern about the visual effect the wind machine will have on a neighborhood should not be dismissed lightly. To disregard the issue is a mistake. Try to incorporate the community's wishes when you are considering the type of tower to use. For example, there is much less opposition to the clean lines of a tapered tube tower than to a wooden utility pole.

At the same time, point out similar structures on the horizon that we have learned to tolerate: billboards, chimneys, and utility poles. You

Noise

have as much reason to erect a tower for a wind machine as the utility does to string a transmission line. It is true that the utility serves the whole community, but each individual uses the power for his personal benefit just as you plan to do with your wind machine.

Noise

Concern about the potential noise from a wind machine is also common. Because most people are unfamiliar with the terms used to describe noise—unwanted sound—we will examine noise in more detail than the other questions that might be raised by a building inspector or zoning board.

Sound is given in units of *decibels* (dB). The decibel scale is relative, not absolute. It spans the range from the threshold of hearing (0 dB) to the threshold of pain (140 dB). See table 3-1 for the sound level of common noises on this scale.

Note that for most discrete sources, such as a jet engine, the distance from the source is given. This is important when considering the level of sound produced by a wind machine or the limits tolerated by a local noise ordinance. The distance must always be specified because sound level decreases with distance. Noise measurements on wind machines are usually given at the base of the tower; for ordinances, at the property line. For example, say you installed a wind machine on a 100-foot tower 200 feet from the nearest property line. How would this affect the noise level? The distance from the source (the wind machine) to the property line is roughly

TABLE 3-1
Typical Sound Levels, dBA Scale

dBA		dBA	
140	Threshold of pain	60	*(continued)*
	Siren (100 ft.)		Large transformer (200 ft.)
130	Jet aircraft (200 ft.)		Wind in trees (40 ft.)
120		50	Light traffic (100 ft.)
110			Average residence
100	Jack hammer	40	
90		30	Soft whisper (5 ft.)
80	Inside sports car		Sound stage
70	Feight train (100 ft.)	20	Still forest
	Vacuum cleaner (10 ft.)	10	
60	Office	0	Threshold of hearing

NOTE: Typical wind machine at base of tower = from 45 dBA to 60 dBA.

double that to the base of the tower. For every doubling of distance the sound level decreases by 3 dB. (The scale is logarithmic not linear.) If the manufacturer provided the sound level for its turbine at the base of the tower you would need to lower the value by 3 dB to find the level at the property line.

It works the other way, too. If you were to install two machines at the same site, doubling the strength of the noise source, you would increase the sound level by only 3 dB.

The apparent loudness varies not only with the sound level but also with the *frequency* (or pitch) of the sound and its nature. Our ears pick up high pitched sounds much easier than they do low pitched sounds. Ambulance sirens emit high pitched wails so they are more noticeable. The sound of a complex machine like that of a wind turbine is composed of sounds from many sources: the wind over the blades, the whirring of the generator, and so on. Each sound has a characteristic pitch making it distinctive from the others. Those with a high pitch are the most noticeable. When measuring noise we try to take into account the way the human ear perceives sound by using a scale weighted for those frequencies we hear best. The A scale is most commonly used.

The apparent loudness of impulsive sounds is also greater than those of a constant level. Kick a garbage can or bang a pair of cymbals and you will get someone's attention a lot quicker than you will by humming even if the sound levels are equivalent.

It is a fairly simple task to measure the sound level produced by a wind machine and give it a dB rating on the A scale. However, interpreting this information is another matter. Many subjective factors affect our perception of what is and is not noise. If your neighbors are opposed to your wind machine because it looks ugly to them, or because you never invite them to dinner, they are more likely to find the sound produced by it to be objectionable than you are. On the other hand, if your community has had a few spats over rate increases with the local utility, then the sound of your wind machine may warm their hearts. As an example, a client of mine could hear a low hum on her radio when the family's wind machine was running. This electronic "noise" did not bother her. But, one morning she awoke and the hum was not there. That did disturb her—the absence of the noise. She looked for the hum as a sign they were lowering their utility bill. It was not noise to her. (It wasn't quite music to her ears either, but it was acceptable.)

Background noise is another consideration. If you live near an airport, your wind machine will hardly create a noise problem no matter how loud it is. The noise generated by the wind turbine must be placed within the context of other noises around it. The most obvious is the wind itself.

We site wind machines where it is windy. And there is always going to be noise from the wind whenever the machine is running. Most standards for measuring noise require that measurements be made only when the winds are less than 15 mph because the wind noise interferes with the measurement. The wind begins to mask the noise from the source; in our case, a wind turbine. Studies of wind in trees, for instance, have found the noise level to be from 51 dBA to 53 dBA at 40 feet when the winds are approximately 15 mph.

Some wind machines, though, do produce frequencies and impulsive sounds that make them noticeable. The generator or transmission may whine, for example. While the transmission may whine at a nearly constant level, the passage of the blades may generate more discrete sounds. The swish-swish-swish of small three-bladed rotors is a common sound. On larger, two-bladed machines a distinctive whop-whop can be heard as the blades pass behind the tower. Being noticeable, however, is not the same as being objectionable. I notice motors whirring in refrigerators wherever I go. (I instinctively look to see if the lights flicker as a sign of undersize circuits.) I do not find the sound objectionable, but I do notice it.

So what does it all mean? When evaluating the noise from a wind machine, you must consider the background noise level, who will hear the noise, and how far away they are. The noise from wind machines range from 45 dBA to 60 dBA at the base of the tower. True, this is quite a range. As seen in table 3-1, the upper end of the range corresponds to light traffic in a residential neighborhood and wind in the trees when the wind speed is from 25 mph to 30 mph. In the latter case, the wind turbine produces the most noise when the wind is the noisiest, and at this wind speed the wind can mask the noise from the wind machine.

Noise ordinances vary significantly from one place to the next. If you are in an area zoned as commercial or industrial, you will have no problem from noise. In a residential zone, the noise from a wind machine should be well within most standards. In the winter when the winds are usually the highest and the wind machine is operating more frequently, the noise will be the greatest. But with the windows closed the noise will not be audible within the house; that is, you will not be able to hear it. During the summer you may be able to hear the machine indoors with the windows open, but this should not prove to be any more objectionable than the refrigerator humming in your kitchen.

Protective Covenants

In suburban housing developments or planned communities, there may be restrictions in the deed on how the land is used. These restrictions are intended to preserve the identity of the neighborhood. Take a look at your deed. Or call your attorney, realtor, or mortgage company. (They can check the deed for you.) If there are any restrictions, they will know how to best deal with them. (For example, the restrictions may be unenforceable.)

Also note the location of any easements on your property for utility rights-of-way. You may not be able to encroach on these easements, though you own the land.

Summary

Much of the opposition to installing a wind machine is due to the fear of the change this new technology may bring to the community. Wind machines are as uncommon today as utility lines were 100 years ago. They are an unknown. Because we as a community do not know how they will fit in with our lives, we are naturally suspicious. But just as we grew to accept (and now demand) the utility's intrusion on the landscape to serve our needs, it is likely that in twenty years we will have grown to accept wind machines in the same way for the same needs.

The community should not apply more stringent standards to wind machines than they would to any other similar structure or device now standing. You are not asking for special treatment, but equal treatment.

Whatever you do, don't bypass the building inspector or the zoning ordinances. In one case, a homeowner bought a wind machine to install in his backyard. After he paid for it, he learned that he needed a building permit before he could start construction. His application was rejected because his use did not conform with those permitted in the residential neighborhood and because his lot was too small. He hired an attorney and then engaged in a lengthy and expensive effort to get a variance. His neighbors objected vociferously. Then amidst the glare of television lights and a packed hearing room, his permit was denied. His troubles did not end there, however. The dealer refused to buy back the wind machine and the homeowner had to sell it at a loss. He took it on the chin as they say. He did not do his homework.

His site was ill suited for a wind machine in the first place, and the one he bought could not have produced nearly as much power as he thought it would. He should never have attempted it, but he can be excused because he did not know any better.

I know so-called wind farm developers who have committed similar blunders. The only difference is the sums of money involved: not thousands as in the homeowner's case, but hundreds of thousands. These self-styled professionals were planning to erect several wind machines that had never worked where they were used previously, and they were planning to install them in a residential neighborhood—without a building permit. They were about to begin construction when the local news media uncovered the story. (There was also some backroom politics involved.) The scheme was killed in a lively public hearing before the local zoning board.

These cases are the exception. They illustrate how not to do it. There are many examples of where the appropriate approvals have been obtained in an orderly and businesslike manner and the wind machine successfully installed. Consider the example of an upper income suburb of Pittsburgh. Fox Chapel Township has a reputation for strict interpretation of its zoning ordinances. "They'll never let you put one up there," some said. Yet the dealer and his client were thoroughly prepared. (They had to get a variance just to erect the anemometer. So they were familiar with the process.) They answered all questions forthrightly, allayed the zoning board's fears and, to the surprise of cynics, were granted the variance. The wind machine was installed without incident and is operating today.

UTILITY INTERCONNECTION

You may jump all the zoning hurdles only to find the utility is slower than molasses when it comes to granting your request to interconnect with their lines. This was the case in the suburb of Fox Chapel. The dealer had contacted the utility about the same time he began the zoning variance application. Yet even though the zoning board took months to make their decision and it took another couple of months to install the wind machine, the utility still had not made up its mind how to proceed. Once the utility does become familiar with interconnected wind machines, the possibilities are limitless. Consider the following scenario.

Night has fallen and the sky is clear. December's chill winds whip the shoreline. Drifting snow swirls about the fence posts and outbuildings of George McClain's small farm. The whistling wind rises in crescendo and then dies away in an unpredictable ebb and flow. A faint whirring, rhythmic and ever present, can be heard. Dark, saberlike shapes sweep the starry sky.

"Looks like it's going to be a cold one tonight," George wonders aloud. His two kids, scampering around in their flannel pajamas, flee their mother as she readies them for bed.

Darlene, both mother and partner in the McClains' dairy, responds, "George, don't you think we ought to turn the heat up? I feel like I'm coming down with something."

"Yeah, Daddy," the kids chime in, "just turn it up to seventy like we used to."

"Now you kids know better than that," he says, "Christmas will be here soon and we want to get Mommy that dishwasher she's been waiting for so long, don't we," he winks at Darlene. "We only get one more check from Pennelec before the new year and I want to sell them just as much power as we can. Darlene, you know that on a night like this when everybody's going to be switching on their electric heaters, demand's going to be high. We need to save every kilowatt we can. The more we save, the more we can feed to Pennelec. They'll pay a fortune for it. I'll bet we can make fifty dollars by morning, more, if this weather holds. Those turbines will really be turning out the juice in winds this high. Just listen to 'em hum."

Far-fetched? A family that awaits winter's winds, and looks to the local electric utility as a source of income? Certainly not. The day may soon be near when farmers such as the fictional McClains sell a new cash crop: energy. But the McClains will not be alone. Anyone who has access to the wind, has the land, and can afford an investment in the future could find himself selling power to the utility (see fig. 3-3). Times have indeed changed.

Yes, it is true. You can interconnect your wind system with your existing electric service; that is the emphasis of this book. And yes, you can sell power back to the utility. As you can imagine, though, it is not as simple as it first sounds.

This strange state of affairs is largely due to congressional passage of the National Energy Act in 1978. One bill of the many that became law is the Public Utilities Regulatory Policy Act — PURPA, for short.

Many electric utilities had already seen the handwriting on the wall. It was inevitable, they felt, what with the public clamor over high rates, that they would have to permit interconnections with home wind systems — at least as a goodwill gesture. Utilities in serious financial or capacity binds even welcomed these small power producers onto their lines. But a few fought every inch of the way and PURPA was enacted to bring them — willingly or unwillingly — into the postembargo age of scarcity. PURPA is downright revolutionary. (When pondering its potential, one cannot help but wonder how a law such as this passed the congressional gauntlet of special interests. It is enough to revive one's flagging faith in democracy.) Though multifaceted, PURPA is most widely known for Section 210 which, in essence, states that most utilities must buy power from, and sell power to, cogenerators and small power producers at reasonable rates.

Fig. 3-3. Small wind farm. Three Enertech E-44's in use at a small wind farm on Nantucket Island off the coast of Massachusetts. The high cost of electricity and a good wind resource (14 mph to 16 mph) offer property owners an attractive investment opportunity. The output from these wind machines is sold to the local utility. (Courtesy Enertech Corp.)

Congress entrusted the Federal Energy Regulatory Commission (FERC) with responsibility for drafting the regulations resulting from the act. The regulations are implemented by state regulatory authorities (public utility or public service commissions) who rule on each individual utility's compliance with PURPA.

PURPA affects nearly all utilities, both regulated (those that come under the purview of state utility commissions) and unregulated (those utilities exempted from state jurisdiction). Investor-owned utilities (IOUs as they are known), public power corporations such as the Tennessee Valley Authority, and small rural cooperatives all have to comply with PURPA's provisions.

In one step, PURPA removed two major barriers to more widespread use of small wind machines and other alternative sources of electricity. First, it exempts small power producers from restrictions of the Federal Power Act. Previously, a home wind system could have been considered a utility and regulated as such by the state public utility commission. The paperwork burden alone would have buried many small power producers. Second, PURPA assured wind system users of backup power, and it stipulated how utilities were to charge for this standby supply.

Prior to PURPA some utilities charged discriminatory rates for backup service. In effect, they said, "Sure we'll sell you power when you need it, but, boy, are you going to pay for it." PURPA puts a stop to such antics. Utilities now cannot penalize you by charging unreasonable rates for standby service.

Through PURPA, Congress sought to encourage small power production by removing barriers to its use. PURPA also creates a powerful financial incentive which was absent before. Utilities must not only allow small power producers to generate electricity in parallel with their own system, but they must also pay more for that power than they have in the past.

Most utilities that were purchasing excess power from home wind systems paid only a fraction of the retail rate. If a homeowner was buying power at \$0.05/kWh, the utility would offer roughly one-half that amount, sometimes even less, for the power that they bought back. Utilities argued that the wind generator only offset their need for fuel. It did not reduce their need for the equipment it takes to generate and distribute the power. The lines, transformers, and "spinning reserve" were all still needed in case the wind died down and the customer demanded power.

The best that a homeowner could hope for would be permission to run the utility's watt-meter backwards. When running the meter in reverse, the wind system owner would be selling power at the retail rate (the same price the utility would charge him, in turn, for using the power). But under

PURPA the buy-back rate must reflect the costs the utility "avoids" by not having to generate the power itself. This is a major departure from common utility pricing practice. Utilities have to pay small power producers the incremental cost of energy as well as capacity the utility avoids using to provide the same amount of power. No longer will they be able to use their average cost of fuel as the criterion. Avoided costs in some cases will exceed the retail rate.

The avoided cost standard opens the door for commercial sales of power from wind systems to utilities. This is particularly true of those utilities, such as those in New England and in southern California, heavily dependent on older oil-fired power plants. The incentive will be less in areas served by coal-fired utilities. The potential is also good in the public power systems that have exhausted the capabilities of large scale hydro (e.g., in the Pacific Northwest).

The regulations go even further. They allow state utility commissions to take into account escalating avoided costs over the life of the contract when ruling on a utility's purchase rate. If the state utility commission chooses to encourage alternative energy, it may establish a "levelized" buy-back rate. In the case of a wind system designed to run for twenty years, the levelized rate would be much higher than today's avoided cost. For example, assume a buy-back rate today of $0.05/kWh and in the year 2000 of $0.25/kWh. The levelized rate could be set at $0.15/kWh over twenty years. Levelized rates offer much greater revenue in the early years than is available from rates based on escalating avoided costs. This accelerates payback and increases the return on investment.

PURPA fundamentally changes the way we look at power generation, conservation, and supplemental power sources. It encourages decentralization and offers decentralized energy investment opportunities as well. Anyone who can afford a wind machine and has a good site can get into the utility business. As in the opening scenario it alters our view of energy conservation from one of conserving to save money to conserving to make money.

PURPA also affects the size of wind systems home owners or farmers may choose. As a supplemental power source, wind systems were looked upon originally as a means of reducing utility bills and they were sized accordingly. Sales to the utility at greater than the retail rate encourage the potential user to seek the most economical wind system on the market without regard to its size. Thus, a homeowner may choose to install a 10-meter wind turbine even though it will produce two to four times more power than needed.

It is only a short step from buying one turbine larger than that required to buying two, three, or even more units. Space, the level of risk one is willing to take, and the availability of capital are the only limits. Like the McClains, farmers who began looking at the wind as a way of relieving their utility bills may now recognize another resource that can be tilled and a new crop raised—a cash crop on contract at that.

Multiple machine installations—*wind farms*—offer several advantages that should not be overlooked. Discounts may be obtained on quantity purchases, and economies can be gained by performing all the site work for multiple units at one time. In all, thousands of dollars could be saved.

Direct sales to the utility also give you more flexibility in siting. Under the law you may purchase power simultaneously from the utility while also selling power to them. Let us say you own some rolling farmland. Your house and barn rest snugly at the base of a tall hill with trees all around. It is a beautiful setting but a lousy place for a wind machine. Don't abandon the idea yet. It just so happens that the utility line to your house (or to anyone else's for that matter) runs across the top of the hill (which has been cleared for a pasturage). The top of the hill is ideal but a good ways from the house. What to do? Install the wind machine on the hill and sell all your power directly to the utility. At the same time you will continue to buy power from them for your house and barn. The revenues from the wind machine will offset your bill much like it would have if you had installed it near the house.

Practical Application of PURPA

Instead of getting carried away with the idea of selling power back to the utility let us look at how PURPA affects the majority of wind system users who simply want to reduce their electric bills.

To summarize, PURPA gives you—the small power producer—a little bargaining power where there was none before. If you want to install a wind system and it meets certain safety standards, to be discussed in a moment, then the utility must permit you to interconnect with their lines and they must offer you standby or supplemental power at reasonable rates. These rates cannot discriminate against you because you are using a wind generator. They must also buy any excess power you produce.

The specifics vary from state to state and from utility to utility. But there are provisions common to all. One of the more important is the regulation permitting the utility to charge for the interconnection costs they incur. This may be in the form of a one-time bill for the installation of additional equipment and a service charge levied against your account periodically for administrative costs caused by greater billing complexity, or both. These charges vary and can be sizable in some cases.

Output from the wind machine fluctuates wildly over time as the wind gusts and subsides. Superimpose this output on your varying consumption of electricity, as explained in the preceding chapter. Depending on the size of the wind system relative to your consumption there will be times when excess power is produced. The excess power will run your existing kilowatt-hour meter backwards.

Most utilities, even today after the implementation of PURPA, pay less for power they buy back than what they will sell it for—it's the nature of the business. So, they do not like their meters to run backwards. They will send someone out to ratchet it, after which the meter will run only one way measuring the power you consume. If the output from your machine is substantially less than your consumption (say you are installing a 3- to 4-meter turbine at your all-electric home), this does not present a problem (you may never feed excess power back into their lines). But, if you want the utility to pay you for the power you sell back to them, then at least one more meter will be needed. One measures the energy you consume; the other measures the excess energy flowing back into the grid. Some utilities may wish to install more meters (they like meters).

Meters are not expensive. But for every additional meter, the utility will usually tack on a service charge to your monthly bill. It will not be much, but the charge adds up and erodes the economic benefit from small machines.

Ideally, you would like to run your meter backwards, selling any excess energy at the retail rate and avoiding any additional charges for the second meter. In several states, regulatory commissions encourage just such an approach with *net energy billing*. In principle, it allows you to run your kilowatt-hour meter backwards. In practice, two ratcheted meters will probably be used so the utility can keep track of what is happening.

The utility pays the retail rate for any excess energy you produce until the net account is zero. That is, you can produce excess energy at the retail rate until the amount of energy sold back equals that consumed. You do not owe them any money and they don't owe you. Above this amount the utility pays a rate based on their avoided cost. In some places, this could be more than the retail rate, but for much of the country it will be less.

Power Quality and Safety

Wind systems using synchronous inverters or induction generators produce line-quality power (explained in chapter 6), and the systems on the market today do so without endangering the utility's equipment or personnel. However, technical questions remain and in some cases special precautions may need to be taken.

The utility has a franchise from the state to supply electric power within a restricted territory. The company is required to provide reasonably reliable service to its customers by this franchise. They are responsible only to the point of delivery known as the *service drop*. This is where the utility's lines reach the building or premises. From this point on, the customer is responsible for the installation, operation, and maintenance of all other wiring—including that of a wind system.

The tariff under which the utility operates allows it to refuse service when the safety of their equipment or linemen is threatened or when they believe service to other customers may be interrupted or impaired. The utility's only means of insuring a safe interconnection with a wind machine is to refuse service when it deems that a hazard has been created or their level of service to other customers degraded.

The safety of their linemen is the utility's principle concern. They fear that an interconnected wind system could *energize* (deliver power) to a downed line during a storm-related power outage or during normal repairs and electrocute a lineman. This fear is not unfounded. Linemen have been injured by improperly wired emergency generators.

Synchronous inverters and induction generators are line synchronized. Without the utility's lines present they cannot generate power. Nevertheless, it is possible for them, in rare circumstances, to self-excite. To preclude this from occurring, *relays* (a form of electrical switch) are placed on the utility side of the wind machine's inverter or control panel. When utility power is present, the AC relay, called a *contactor,* is energized completing the electrical circuit. If utility power is lost for any reason, the spring-loaded relay is *de-energized* (turned off) and breaks open the circuit disconnecting the wind system from the utility line. Proposed standards of the American Wind Energy Association require that all wind machines designed for interconnection with the utility must disconnect themselves from a utility during a line outage and must not be able to self-excite.

Self-excitation is only a problem where capacitors have been installed on the wind machine side of the contactors for power factor correction. Utilities frequently use capacitors for this purpose on their lines. (*Capacitors* are rectangular metal cans mounted in racks on utility poles. Transformers, in contrast, are cylindrical.) Capacitors on the utility side are disconnected from the wind machine whenever the contactors open. Any capacitors used with a wind system for power factor correction should be placed on the utility side of the interconnection or should be designed to bleed down (discharge) after power is removed. This prevents self-excitation as well as any shock hazard to those servicing the wind machine.

Why does a utility or wind generator need capacitors to correct power factor? To understand the answer you need to know the difference between true and apparent power. (No, I didn't make this up; they both exist and there is a difference.) Power in watts, as you recall, is the product of voltage and amperage. This is generally true, yes, but when we are dealing with alternating current (AC) as produced by the utility we need to add another parameter to the equation: *phase angle*.

Do not let phase angle scare you. It merely describes the degree with which rising and falling voltage is in phase with the rising and falling current given in amperes (see fig. 3-4). If current rises from zero at the same time voltage rises from zero, the two wave forms are said to be in phase.

When this occurs, the cosine of the phase angle (0 degrees) equals unity (1), and *true power* is the product of voltage and amperage. In this case true power and apparent power are equal and the *power factor* — the ratio of true power to apparent power — is one. This is the ideal. Unfortunately, the real world is never this tidy.

Loads on the utility's lines cause the current waveform to shift slightly. Current either leads (starts rising earlier) or lags (starts rising later) voltage. In rural areas, where power must be transmitted a long distance, the length of the line itself is sufficient to cause current to lag.

At this point you may cry out, "Who cares." The utility cares. And they care a lot. When current and voltage are out of phase, true power decreases (the cosine of the phase angle becomes less than one), yet *apparent power* remains the same. The utility is called upon to deliver apparent power. Their entire system from generators to transmission lines must be sized to meet the demand for apparent power. Generators (including wind generators), by the way, are rated by their ability to deliver apparent power in units of kilovolt-amperes (kVA) and not, usually, true power in kilowatts.

The utility's trusty watt-hour meter measures only true power. Therein lies the dilemma. They must generate apparent power while only getting paid for true power. And true power is always less than apparent power (the power factor is either one or less than one). They get shortchanged. The utility tries to correct this — to keep the power factor as close to unity as possible — by adding banks of capacitors to the line (see fig. 3-5).

The foregoing applies to wind systems as well as it applies to electric utilities. It is of importance to you because the power factor of your wind generator may cause the utility to take corrective action by installing capacitors — and then charging you for them. Power factor is also a favorite whipping boy for opponents of wind power within utility companies.

If problems arise, you need to remember that the utility pays only for true power when you sell to them. (You are now in the utility's shoes; that is, trying to maximize the production of true power by getting your power factor as close to one as you can. Values above 90 percent are

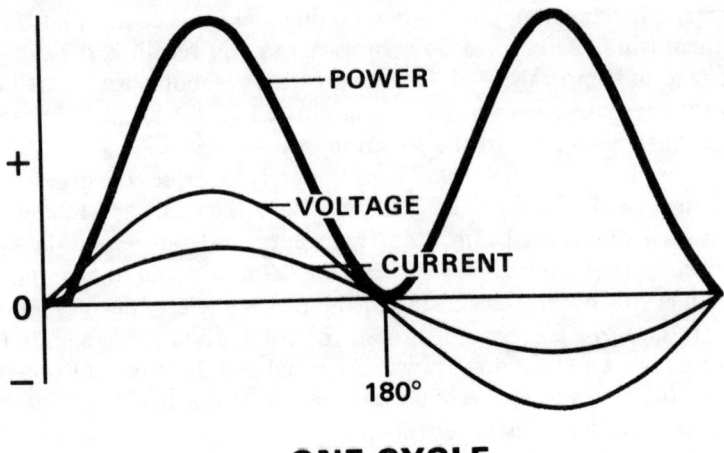
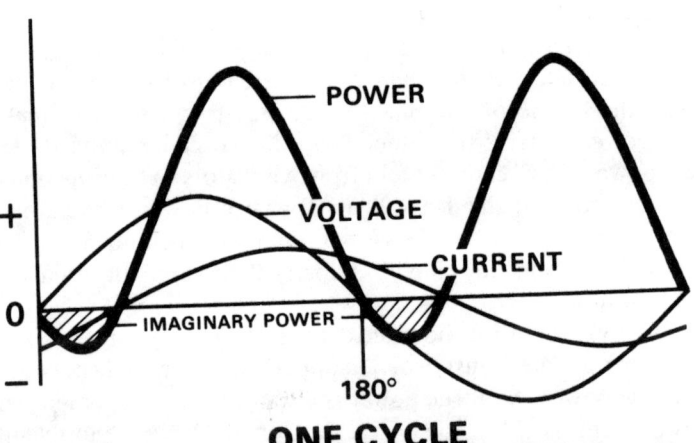

Fig. 3-4. True and imaginary power. *Top,* when voltage and current are in phase, power is the product of the two; *bottom,* when voltage and current are out of phase, true power is less than apparent or imaginary power. The utility must deliver apparent power, but it gets paid only for true power. Thus, the utility is concerned about power factor—the ratio of true to apparent power.

desirable.) Do not think you are selling them horse meat and getting paid for prime beef as has been implied by some utility spokesmen. Moreover, your wind system appears to the utility as just another load (albeit a negative load) calling for power factor correction. It is no different than if you added a couple of freezers or new power tools to your home as far

Fig. 3-5. The utility uses capacitors to correct the power factor in its distribution system.

as your impact on the utility is concerned. They do not charge you for power factor correction when you install a new freezer, do they? No, of course not. And they should not charge you for power factor correction when you install a wind generator either. It is a normal cost of doing business.

Another problem arises with certain wind systems using induction generators. On some models the rotor is locked in place until the cut-in wind speed is reached. The rotor is then motored up to synchronous speed by drawing power from the utility. The effect on the utility is similar to the start-up of a compressor motor in a refrigerator or freezer. There is a momentary surge of current until the rotor gets up to speed and the wind begins driving it. Because the power rating of wind generators is usually larger than most household appliances, the magnitude of the in-rush current can be large enough to cause a slight voltage drop in the line. The result may be voltage flicker. Your lights, for example, may dim briefly whenever the wind generator starts up. This is not a serious problem but can be annoying. The utility, though, may charge that an induction wind machine will detract from the level of service they offer to other customers and require what the utility calls a *dedicated transformer* to mitigate the problem. The transformer isolates the voltage drop to the customer using the wind machine.

In sparsely populated rural areas most homes or farms already have a dedicated transformer; that is, only one customer is served by the transformer. One transformer, on the other hand, may be used for several customers in the suburbs. You can tell if you have a dedicated transformer by taking a look out your window. The utility *primaries,* also called high voltage lines or sometimes just high lines, are carried at the top of the utility pole. Leads to the primaries are attached to the top of the transformer. The low voltage lines or *secondaries* are attached to terminals on the side of the transformer. The service drop to your home is always from the secondary (low voltage) side of the transformer. If the secondaries are strung directly to your house and to no other, the transformer is dedicated to your service drop. But, if the transformer is located several poles away and the low voltage lines serve several other customers, the transformer is communal (see fig. 3-6).

Not all induction wind machines have this problem. Many do not motor the rotor up to speed. Instead the rotor is free wheeling. When there is sufficient wind to turn the rotor at synchronous speed, the AC line contactors are energized and the wind system is brought on the line. There is no in-rush current, and there is no voltage flicker. Consequently, the utility should not impose a blanket requirement for a dedicated transformer without first demonstrating the need for such.

Fig. 3-6. Transformers step down voltage from the "high line" or high-voltage line at the top of the pole to voltages that can be used safely in the home. This is a communal transformer. The secondaries or low-voltage lines below the transformer serve several customers, an arrangement common in cities, towns, and suburbs.

Synchronous inverters are not without their power quality faults. They can produce current harmonics in various degrees. This does not affect most electric appliances but, theoretically, can cause electromagnetic interference with television, radio, and telephones. The degree of interference depends upon the inverter, its size, location, and the level of noise on the utility's lines with which it is connected. (Power from the utility is not always free from harmonics itself.) One wind machine owner I know thought his machine was performing normally, when his wife walked into the kitchen and said it was not producing any power. Amazed, he asked how she knew. She said she could not hear the faint humming it made on their radio. He proceeded to check the control panel and sure enough the inverter was down with a blown fuse.

With well over one thousand inverters operating successfully in wind systems around the country, current harmonics have not presented a significant problem—not even annoyance. Again, a transformer can be used to isolate any disturbance to the customer's immediate vicinity. But as with the case of voltage flicker, the utility should not make a blanket requirement for a dedicated transformer without proving there is a need.

If a transformer is necessary, so be it. They are expensive and can add several hundred dollars to the cost of an interconnection and as a result should only be used when absolutely required.

Dealing with the Utility

Utilities are not inherently evil as often depicted. They have no moral sense. They are a business and their employees are charged with helping it make money. In the case of most IOUs it is a large corporation. And, like any big bureaucracy, the right hand frequently does not know what the left is doing.

Corporate policy as expounded by its managers may be opposed adamantly to the interconnection of small wind systems. Yet, the word may not have drifted down to the lower levels where the work gets done. You may call up and be surprised to find the staff friendly and curious about your project. They may go out of their way to be helpful (I've run into this situation with one notorious Pennsylvania utility). Of course, there is the other extreme. The utility makes supportive pronouncements about wind energy. They brag about all they are doing to promote it. But, when you contact them, it is a different story. During contract negotiations, for example, they will try to take the shirt right off your back. (I've found this type, too.)

The fact that the utility is a bureaucracy also means that nothing happens quickly. You should notify them of your plans weeks, even months, in advance. They may have a clearly defined policy and may be able to give you the answers you seek promptly. More often the particular person handling your account has not dealt with a small power producer before. They will not want to take any chances and will cover themselves every step of the way with time consuming approvals. (It's the nature of the beast.)

You need to appreciate the utility's point of view because you are still dependent on their cooperation. The difference today is now you have legal recourse if they do not cooperate. Yet you want to avoid exercising your rights. To do so entails seemingly endless meetings between you, your attorney, regulatory officials, and the utility. Neither legal counsel nor your time comes cheaply.

Utility executives dread nothing more than a self-righteous customer strutting into the office thumping his chest demanding that the utility kneel down and graciously give him a trouble-free interconnection. If you want a fight, there is no better way to find one.

Strike a balance. You are making a reasonable request and you should not be intimidated by the utility. But, do not be pompous. The utility is not there to serve your interests alone. Interconnection is a business transaction and, like any other, it can be handled in a cordial manner.

After you have passed the technical gauntlet there are a few more obstacles to overcome: liability insurance, other contract provisions, and redundant disconnect switches. Some utilities, principally the rural electric cooperatives (rural co-ops), insist that the small power producer purchase one million dollars in liability insurance. Seems outrageous, doesn't it. But here is their rationale. Say a lineman is injured by a wind machine. The utility's Workmen's Compensation Insurance pays the lineman's claim. The insurance company who covered the utility's compensation sues the small power producer for recovery of their money. (The lineman may also sue for damages.) The small power producer loses in court, cannot pay, goes bankrupt, and the insurer has to absorb the loss. The utility's insurance rates rise dramatically and you know who pays the bill in the end: all the other customers. This scenario has not taken place with a wind system but electric utilities have been sued for a lot less when they were not at fault—and have lost. The rural coops, because they are small and cannot afford mistakes, are gun shy when it comes to liability.

The cost for this much insurance is not out of reach, but it can cut deeply into the total revenues of a small wind system. Before you acquiesce, determine how the utility treats other customers with emergency or standby generators. Are they required to have the same insurance? If not, why not? If they are not required to have insurance the utility may be discriminating against you. The line contactors on your wind system respond in much the same manner as automatic transfer switches on standby generators. (Technically, anyone with a standby generator would fall into this category whether using a manual or an automatic transfer switch.)

Another contract provision that might raise your ire is the "hold harmless" clause. Read the contract's fine print. If you do not understand what it says, get an attorney to look it over. One contract I reviewed for a client held the company harmless against all claims arising from damage or injury due to the interconnection whether the fault of the small power producer or the fault of the utility. We inserted the following to make the clause more equitable: ". . . so long as the acts giving rise to the claims were not due to the negligent conduct of (the utility), its employees and agents." In this way, the utility becomes responsible for any damage they cause.

Some contracts may make you liable for any damage to anyone by anybody on the utility's entire system. They cover all bases when they do that. Do not let them get away with it. You should be responsible only for damages you cause.

As with transformers and capacitors, the utility may load the interconnection down with unneeded equipment and raise the cost of the interconnection prohibitively. Another item they may add is a redundant disconnect switch. For example, they may require an accessible and lockable disconnect switch between their billing meter and your service panel. This means they want a standard disconnect switch below the meter on the outside of the building.

For a wind system with any kind of electrical storage, this is a reasonable measure. The utility's linemen need only throw the switch to assure that no power can feed back into the line. It is unnecessary and a burden, though, on an interconnected wind system without storage—especially, if a lockable disconnect switch on the wind generator's service to the building is installed as suggested in chapter 10. (The switch should be placed adjacent to the utility's service drop so it is visible and easily accessible.) It can also be discriminatory.

When the wind system is disconnected from the customer's service panel, no power can flow from the wind generator to the utility's lines. With the wind generator disconnected, the customer's service is like that of any other utility customer. If the small power producer is forced to install another disconnect switch, this one between the billing meter and the service panel, then it is only reasonable to assume that all other customers must do the same. You can bet the utility will not do that.

Should a dispute with the utility develop, consult the state regulatory commission. They should have a consumer affairs office that provides mediation. Your state energy office may also be helpful. Be patient. But, if all else fails—and the issue is worth it to you—"Sue the bastards."

4

Measuring the Wind

In this chapter we will discuss the wind, what it is, how local climate and terrain effect it, and how it changes over time. We will explore the meaning of wind power and how to estimate it, and how wind speed and power increase with height. You will learn where to find wind information for your area and how to determine the winds at your site. For those with an aversion to math, easy to use tables are presented wherever a formula appears.

WIND – WHAT IS IT?

The atmosphere is a huge, solar-fired engine that transfers heat from one part of the globe to another. Large-scale convection currents set in motion by the sun's rays carry heat from lower latitudes to northern climes. The rivers of air that pour across the surface of the earth in response to this global circulation are what we call the wind, the working fluid in the atmospheric engine.

Local and regional effects caused by differential heating of the earth's surface can modify large-scale atmospheric motion. And, these effects may be strong enough at times to dominate the winds of local weather.

When the sun strikes the earth it heats the soil near the surface. In turn, the soil warms the air lying above it. Warm air is less dense than cool air, and, like a helium-filled balloon, rises. Cool air flows in to take its place and is itself heated. The rising warm air eventually cools and falls back to earth completing the cycle.

This cycle is repeated over and over again rotating like the crankshaft in a car, as long as the solar engine driving it is in the sky. *Thermals,* those rising currents of warm air that boil up over land during bright daylight hours are as much sought after for soaring by man as by hawks. The cumulus clouds of summer are a sign of this *convective circulation* that causes winds to strengthen in late afternoon. If you are a pilot you probably prefer to fly in the early morning hours before the sun has had a chance to warm the surface and when winds are light. On the other hand, if you are making a trip to inspect a wind machine, late in the day is better. You are more likely then to find it running than at any other time because of convective circulation.

Winds are higher and more frequent along the shores of large lakes and along the coasts because of differential heating of the land and the water. During the day, the sun warms the soil on land much quicker than it does the surface of the water because of water's higher specific heat. The air above the land, once again, is warmed and rises. Cool air over the water flows landward replacing the warm air, and a large convection cell is generated. At night the flow reverses as the land cools more quickly than the water. The air above the land cools, sinks, and flows out to sea where it is warmed by the sun's heat stored in the water near the surface. In late afternoon, when the sea breezes are strongest, winds can reach 10 mph to 15 mph on an otherwise calm day. Land-sea breezes are most pronounced when general winds are light. Millions along the eastern seaboard enjoy the refreshing benefit of land-sea breezes during the dog days of late summer. The influence of land-sea breezes diminishes rapidly inland and is insignificant more than two miles from the beach.

The winds along the shore are also higher because of the long unobstructed path over the water. Hills, trees, and buildings stand in the way of the wind on land, slowing it down. The shores of the Great Lakes and much of the nation's seacoasts have average wind speeds of approximately 12 mph, partly due to these effects.

The mountain-valley breeze is another example of local winds caused by differential heating (see fig. 4-1). Mountain-valley breezes occur when the large-scale wind over the area is weak and there is strong heating and cooling. These breezes are found principally in the summer months when solar radiation is the strongest. During the day, the sun heats the floor and sides of the valley. The warm air rises up the slopes and towards the

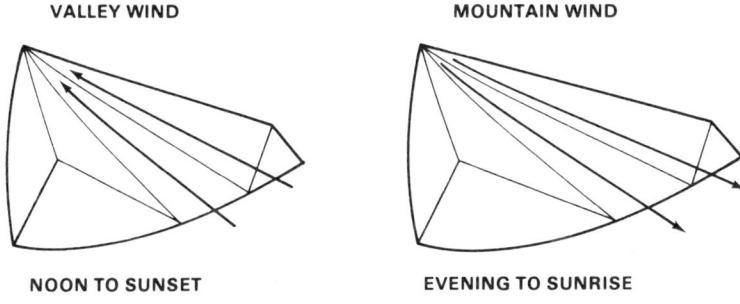

Fig. 4-1. Mountain-valley winds. (Battelle PNL, DOE)

head of the valley. Cooler air is drawn up from the plains below causing a valley breeze. At night the situation reverses and the mountains cool quicker than the lowlands. Cool air falls down the slopes and is channeled through the valley out onto the plains forming a mountain breeze. Mountain breezes are stronger than valley breezes with winds reaching speeds of 25 mph in valleys with steeply sloping floors located between high ridges or mountain passes.

Mountain-valley breezes become insignificant when strong prevailing winds cross the valley. The prevailing winds overpower this local wind, but mountain breezes can be reinforced by the prevailing winds when they are flowing in the same direction.

The effects of channeling are most pronounced on the Pacific coast where coastal winds are funneled into narrow passes between mountains. Where interior deserts cause large differences in temperature from those along the coast, convective flow reinforces the channeling. Average wind speeds above 20 mph are typical, with some seasons averaging well above that. The tremendous wind potential in the passes near Portland, Oregon, and near San Francisco, Los Angeles, and San Diego, California, has sparked the interest of wind farm developers.

Many of these passes lie in a generally east-west direction. San Gorgonio, for example, funnels winds from along the coast around Los Angeles, through the San Bernadino Mountains, into the deserts surrounding Palm Springs. Similar flows occur from the San Joaquin Valley, across the Tehachapi Mountains, onto the Mojave Desert.

Long ridges across the path of the wind generally enhance air flow. Wind speeds may double as the flow accelerates up the gradual slopes of a long ridge. This enhancement occurs only in the last third of the slope near the crest (see fig. 4-2).

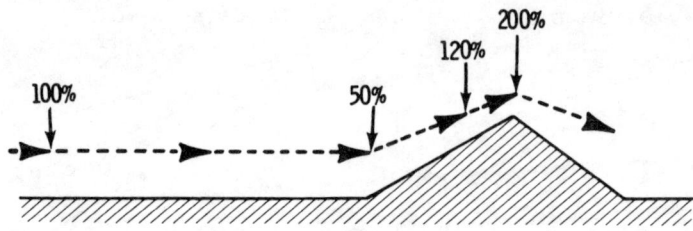

Fig. 4-2. Percentage variation in wind speed over a ridge. Wind speed increases near the summit of a long ridge lying across the wind's path. (Battelle PNL, after Park and Schwind)

A similar terrain enhancement of the wind has been found on the island of Oahu in Hawaii. There the westerly winds sweep around the ends of a long ridge that runs the length of the island from north to south. The winds are accelerated as they pass over the ridge, but more so where they pass around the ends. Kahuku Point, at the northern end of the ridge, is a prime prospect for wind farming because of this local concentration of the wind.

Mount Washington, in the White Mountains of New Hampshire, is one of the few areas on the East Coast where terrain enhancement is well known. In fact, this phenomenon was first observed on Mount Washington and the Green Mountains of Vermont. Mount Washington has an average speed of approximately 27 mph and has recorded the highest winds ever measured; 231 mph, on April 12, 1934.

The Green Mountains are representative of a long ridge astride the prevailing winds. Because of potential speed enhancement, they were chosen as the site for the experimental Smith-Putnam wind turbine during the forties. The long unobstructed fetch across Lake Champlain to the west of the Green Mountains made them even more attractive.

Mountains and ridges offer higher winds for reasons other than channeling. Prominent peaks often pierce temperature inversions that can blanket valleys and low-lying plains. Temperature inversions cause a stratification of the atmosphere near the surface. Above the inversion layer normal air flow prevails, but below, the air is stagnant. Winds are light or calm. The air below may be completely cut off from the circulation normal for the weather system moving through the area. Concentrations of air pollutants can grow to dangerous levels. Temperature inversions are a problem in hilly or mountainous terrain, such as southern California and western Pennsylvania, both areas notorious for their severe air pollution episodes.

The inversion itself may act to accelerate winds above what they would be normally. Temperature inversions have been characterized as lakes of stagnant air. Winds in the upper zone blow across this surface unimpeded by the hills, trees, buildings, and other features that would generally slow the wind. Ridge tops may not only experience more frequent winds because they are above inversions, but they will also experience higher winds because they are exposed to winds less hindered by the surface.

WIND SPEED AND TIME

As everyone knows, the wind is inconsistent — calm one day, howling the next. Wind speed and direction vary widely over almost all measuring periods whether it is a year, month, day, or hour. Because wind speed fluctuates, it becomes necessary to average wind speed over a period of time. That most commonly used is the average speed over an entire year.

The average annual wind speed itself is not constant. It varies from year to year. The average speed can change as much as 25 percent from one year to the next. This can amount to more than 2 mph in a moderate wind regime where an average of 10 mph is the norm. Meteorologists have found that it takes ten years or more of data before they feel they have captured all the yearly cycles and can arrive at a reliable average annual wind speed.

Average wind speeds vary by season and by month. March roars in like a lion and goes out like a lamb is a popular adage signifying that early spring is windy, while summer is not. Generally, winds are light during summer and fall and increase during the winter, reaching their maximum in the spring. In figure 4-3, the curves for the interior cities of Amarillo, Texas (average annual speed, 13.7 mph) and Harrisburg, Pennsylvania (average annual speed, 7.7 mph) reflect this relationship. For San Francisco, California (average annual speed, 10.5 mph), however, the situation is reversed. Wind speeds are highest during the summer and lowest during the winter. The average annual speeds for these cities are derived from more than ten years of records. At these three sites (see fig. 4-4) the average annual power in the wind follows the same relationship, though the difference in average power changes more dramatically from one month to the next than does average speed.

When we looked at the differential heating of the earth's surface and its effects on local winds, we saw that wind speeds generally increased during late afternoon after convective circulation had been set in motion. This tells us that wind speeds vary by time of day, not only due to changing weather, but also due to convective heating. Figure 4-5 illustrates the change in average wind speed over twenty-four hours for each season.

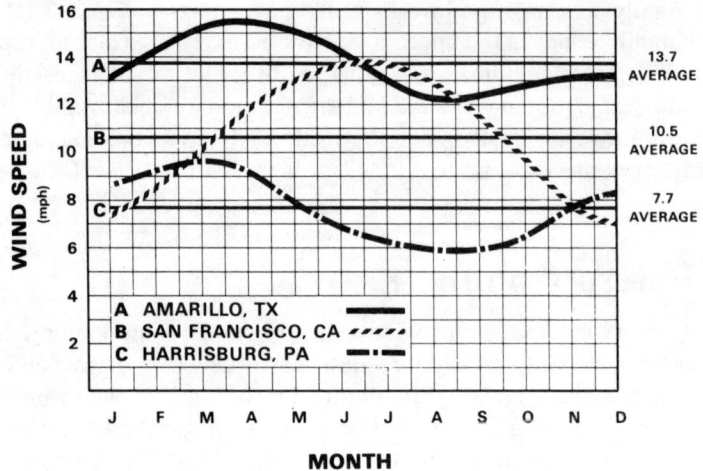

Fig. 4-3. Average annual wind speed.

As you would expect, wind speeds reach a peak at about 3:00 P.M. (1500 hours).

The effects of local convective winds are greatest during the summer when the winds aloft are light and solar heating is strongest. The range of speeds, from the minimum during the night to the maximum during the day, is almost 6 mph. In contrast, the difference between maximum and minimum average speeds during the winter is 4 mph. Winter speeds

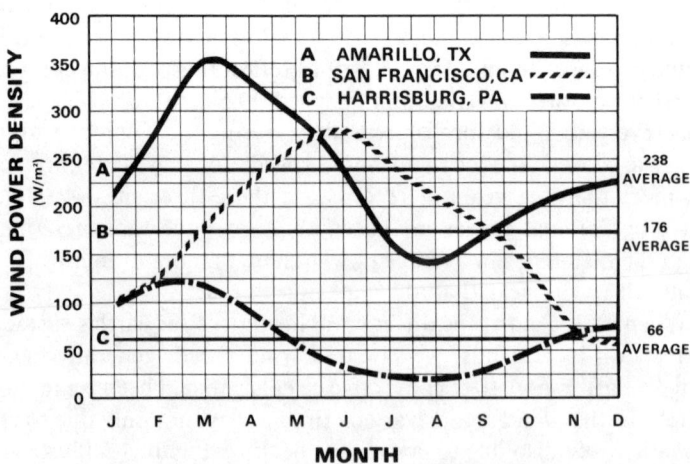

Fig. 4-4. Average annual wind power.

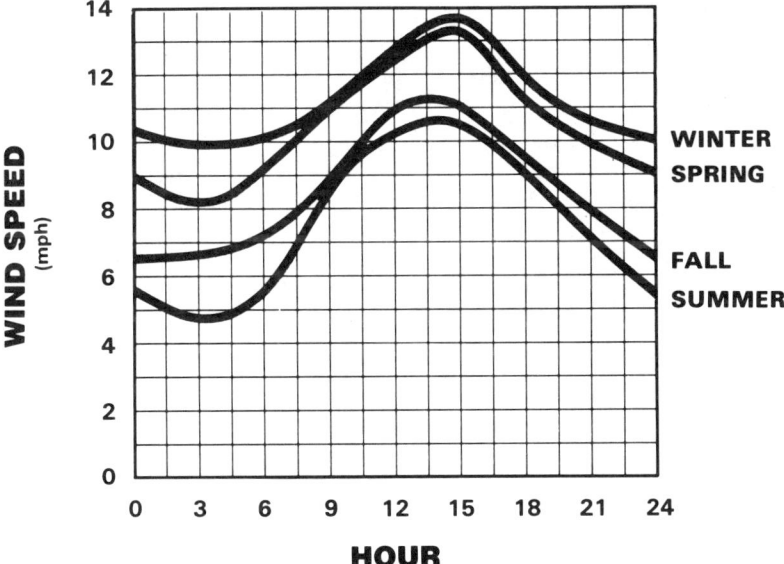

Fig. 4-5. Average daily wind speed by season.

are higher than those for summer because of the generally higher winds during the winter, but the change in speed from morning to night is less marked because of less convective circulation.

POWER IN THE WIND

For the sake of thoroughness, we will start right at the beginning. Wind is air in motion. (Good so far.) The air about us has *mass* (think of it as weight if you are unfamiliar with the term). Though extremely light, it has substance. A gallon of air is similar to a gallon of water, but the gallon of air is lighter. It has less mass than that of water because air is less dense—it is more diffuse. Like any other moving substance, whether it is a gallon of water plummeting over Niagara Falls or a car speeding down the highway, this moving air—the wind—contains *kinetic energy.* This energy of motion gives the wind its ability to perform work.

When the wind strikes an object, it exerts a force or pressure attempting to move it out of the way. Some of the wind's kinetic energy is given up or transferred, causing the object to move, and is said to perform work. We can see this when leaves skitter across the ground, trees sway, or the blades of a wind turbine spin.

The amount of energy in the wind is a function of its speed and mass. At higher speeds more energy is available. Much like cars on the highway, if you increase speed or mass, you increase the energy of motion. A car driven at a high speed contains more energy than a car of equal size it passes. It takes more effort — energy — to stop a car driven at 70 mph than it does one at 50 mph. Likewise, heavy cars contain more energy than light cars traveling at the same speed. This relationship between mass, speed, and energy is given in the following equation:

$$\text{Kinetic Energy} = 1/2 \, m \, S^2$$

where m represents the air's mass, and S is speed.[1]

We can calculate the air's mass from the product of its density (d) and its volume. Because the air is constantly in motion, the volume must be found by multiplying the wind's speed (S) by the area (A) through with which it passes. Thus,

$$m = dAS.$$

When we substitute this value for mass into the earlier equation, we can find the kinetic energy in the wind.

$$\text{Wind Energy} = 1/2 \, (dAS) \, S^2$$
$$= 1/2 \, dAS^3$$

We have gone through this derivation for a reason. Equations are the language of science and in that terse, compact script the fundamentals of wind energy are precisely stated. But before we go over them, let's complete one more step. Power, as you may remember, is the rate at which energy is available, or the rate at which energy passes through an area per unit of time. Therefore, power (P):

$$P = 1/2 \, dAS^3$$

This equation states that wind power is directly proportional to air density, the area intercepting the wind, and wind speed (*velocity*). Increase any one and you increase the power in the wind.

Air Density. The wind is a diffuse source of power because air is less dense than most common substances. (Water, for example, is 800 times denser than air. This is why so much more effort, comparatively, has been expended to harness hydropower than wind power.) Air density decreases even further with increasing temperature and increasing altitude. Air is less dense in summer than in winter. Density can vary 10 to 15 percent from one season to the next and have a subsequent effect on the power in the wind. On a yearly average, though, changes in temperature have such a slight influence that we can safely disregard them. We will assume for all calculations in this book that we are in an area where conditions approximate those at a standard temperature of 60°F for our values of air density. Of course, if you plan to install a wind machine on the North Pole or on the sands of the Sahara, average temperature can make such a difference in air density that it should be taken into account. If this is the case, see appendix A for temperature corrections to air density.

Increasing altitude has a more substantial effect than temperature. For example, at the same wind speed there is more wind power at a coastal site (Los Angeles or Boston) than at Denver, Colorado (elevation 5,000 feet above sea level). There is 14 percent less wind power at Denver (see table 4-1). For our calculations, we are going to assume conditions at sea level. Keep this in mind. If you live at an elevation of more than 3,000 feet, use the values in table 4-1 to adjust any wind power estimates we produce.

TABLE 4-1
Wind Power vs. Elevation

Elevation (feet)	Percentage of Sea Level (%)	Elevation (feet)	Percentage of Sea Level (%)
Sea Level	100	Sea Level	100
500	99	6,000	83
1,000	97	7,000	80
2,000	94	8,000	77
3,000	91	9,000	74
4,000	88	10,000	70
5,000	86		

Area. Unlike changes in air density, changes in the area intercepting the wind influence power significantly. Wind turbines with large rotors intercept more wind than those with smaller rotors and, consequently, capture more power. Doubling the area of a wind turbine rotor, for example, will double the power available to it. This principle forms the foundation for most wind turbine design, and with it you can quickly size up any wind machine by noting the dimensions of the rotor.

Consider a conventional wind turbine whose blades spin about a horizontal axis. The rotor sweeps a disc the area of which can be found from the equation for the area of a circle, where A is the area, and R is the radius of the rotor (approximately the length of one blade).

$$A = \pi R^2$$

For conventional wind turbines, this tells us that the area swept by the rotor is proportional to the square of the rotor's radius, or the rotor's diameter (double the radius). Translated this means that relatively small increases in blade length or rotor diameter produce a substantial increase in swept area. Doubling the length of each blade—doubling the rotor's diameter—increases the area of the wind intercepted by the turbine four times.

$$A/\pi = R^2 = (2/1)^2 = \frac{2 \times 2}{1 \times 1} = 4$$

To illustrate this further, we will use three commercial wind turbines and examine the area of the wind swept by each (see table 4-2). The 4-meter wind turbine in table 4-2 sweeps almost twice as much area as the 3-meter turbine even though the diameter of the rotor is only 33 percent greater. The 6-meter turbine sweeps four times the area of the 3-meter machine and a little more than twice that of the 4-meter. If the efficiency with which each wind turbine captures the power in the wind and converts it to a useful form, say to electricity, is equivalent, then the 6-meter turbine is capable of producing four times the power of the 3-meter turbine, and twice the power of the 4-meter turbine as reflected in the last column in table 4-2. These values are roughly the actual generator capacities given by the manufacturers. This principle and how it can be used to estimate annual energy production will be expanded upon in chapter 5.

TABLE 4-2
Swept Area of Three Commercial Wind Machines

| Rotor Diameter | | | Swept Area | | Approximate Generator Size |
Meters	(Feet)	Model	Meters^2	(Feet)^2	(kW)
3	(10)	Bergey 1000	7.1	(79)	1
4	(13)	Enertech 1800	13	(130)	2
6	(21)	Enertech 21/5	28	(350)	4

Speed. Because the power in the wind is a cubic function of wind speed, changes in speed produce the most profound effect on power of the three parameters in the power equation: air density, intercept area, and wind speed. Consider an example where the wind speed at two sites differs by only 10 percent. At one site the wind speed is 10 mph, at the other, 11 mph. How much difference does this cause in the power available?

$$P/dA = (11/10)^3 (1.1)^3 = 1.1 \times 1.1 \times 1.1 = 1.33$$

By increasing speed only 10 percent (in this example, one mph) we have increased the power available by 33 percent. Let's try another example. Say the windier site has a wind speed of 12 mph.

$$P/dA = (12/10)^3 = (1.2)^3 = 1.2 \times 1.2 \times 1.2 = 1.73$$

This is why there is such a fuss concerning the proper siting of a wind machine. A small difference in wind speed caused by bordering trees or buildings of only 10 to 20 percent can reduce the power available by 30 to 70 percent. Unlike swept area, doubling wind speed does not simply double the power available. Instead, power increases a whopping eight times.

$$P/dA = (20/10)^3 = (2)^3 = 2 \times 2 \times 2 = 8$$

We can summarize the power equation with these general rules:

1. Power is not significantly affected by changes in air density except for sites in arctic and desert environments, or sites above 3,000 feet in elevation.
2. Power is proportional to the area intercepted by the wind. Double the area intercepting the wind and you double the power available.
3. Power is a cubic function of wind speed. Double the speed, and power increases eight times ($2 \times 2 \times 2 = 8$).

Now that you understand the significance of each term in the power equation, we are going to rearrange them slightly. The rate at which energy passes through a unit of area is called *power density* (P/A).

$$P/A = 1/2\ dS^3$$

The power available in the wind is presented as power density, or power flowing in the wind per unit area of the wind stream, more often than as wind speed. Power density is normally given in terms of watts/square meter (W/m²). It can just as easily be given in other units (e.g., watts/

80 • *Wind Energy*

square foot (W/ft²)).² The results of nearly all wind power research in this country are given in W/m², so these are the units we will use as well. (As you will see in a subsequent section, the use of W/m² doesn't create any undue hardships.) The beauty of power density becomes apparent when you know the size of the area intercepting the wind, say, the area swept by a wind turbine rotor. Estimating the power available to the wind turbine becomes simply the product of power density and area where A is the turbine's swept area and P/A the power density at the site.

$$P/A \times A = P$$

More on this later.

So, let us take the value for air density at sea level and a temperature of 60°F and plug it into the equation and see what we have.

$P/A = 0.05472\ S^3$, where S is in mph
$P/A = 0.08355\ S^3$, where S is in knots³
$P/A = 0.6125\ S^3$, where S is in meters per second⁴

Previously we learned that power is a cubic function of wind speed and that slight changes in speed produce significant changes in power. At this point an important question arises. What wind speed are we talking about? The average wind speed? If so, what average wind speed? The annual average?

Grab a cup of coffee, sit back, and ponder this statement. Using the average annual wind speed — alone — in the power equation will not give us the right numbers; our calculation could be off the actual power in the wind by a factor of two or more.

To understand why remember that wind speeds vary over time. Now let us calculate the power density for an average wind speed of 15 mph.

$$\text{At 15 mph;}\ P/A = 0.05472\ (15)^3$$
$$= 0.05472\ (3375)$$
$$= 185\ \text{W/m}^2$$

What happens if the wind blows half the time at 10 mph, and half the time at 20 mph? The average speed is still 15 mph

$$\frac{10 + 20}{2} = 15.$$

Right? Let's calculate the average power using these two wind speeds.

At 10 mph; $P/A = 0.05472\ (10)^3 = 55$ W/m^2
At 20 mph; $P/A = 0.05472\ (20)^3 = 438$ W/m^2
Average Power; $P/A = (55\ \text{W/m}^2 + 438\ \text{W/m}^2)/2 = 247$ W/m^2

How can this be? Both have the same average speed. The answer rests with the cubic relationship between power and speed.

The average of the cube of many different wind speeds will always be greater than the cube of the average speed. Or stated another way, the average of the cubes is greater than the cube of the average. In this case the average of the cube for two classes of wind speed (10 and 20 mph) is 1½ times the cube of the average.

The reason for this paradox is that the single number representing the average speed (see fig. 4-6) ignores the distribution of winds above and below the average. And it is those above the average that contribute most of the power.

Figure 4-6 is a speed histogram or bar chart. The vertical axis represents the number of occurrences per year or the percentage of yearly occurrences of winds at different speeds as shown on the horizontal axis. The speeds are arranged in speed classes, for example, from 8 to 12 mph, from 13 to 18 mph, and so on. Also shown is the *speed distribution*. This

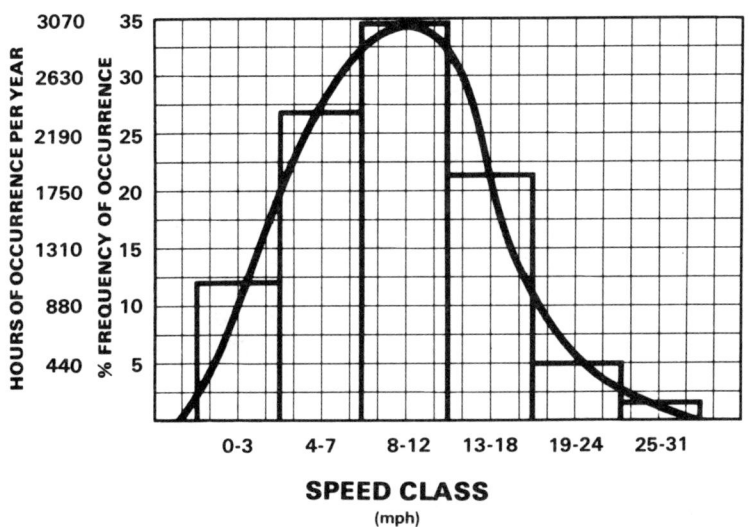

Fig. 4-6. Wind speed distribution and histogram for Greater Pittsburgh Airport.

82 • Wind Energy

ANNUAL
43,848 Obs.

B PERCENTAGE FREQUENCIES
OF WIND DIRECTION AND SPEED:

DIRECTION	HOURLY OBSERVATIONS OF WIND SPEED IN MILES PER HOUR												TOTAL	AV SPEED
	0-3	4-7	8-12	13-18	19-24	25-31	32-38	39-46	47 OVER					
N	.2	1.3	1.7	.5									3.7	8.7
NNE	.3	1.7	1.6	.3									3.8	7.7
NE	.3	1.7	1.1	.2	+								3.3	7.3
ENE	.2	1.6	1.8	.6	+								4.3	8.7
E	.3	1.6	1.8	.6	.1								4.4	8.5
ESE	.3	1.5	1.7	.8	.1	+							4.3	9.1
SE	.3	1.8	1.7	.4	+	+							4.2	8.1
SSE	.2	1.6	1.4	.5	+								3.7	8.2
S	.2	1.6	1.5	.5	.1	+							3.9	8.4
SSW	.2	1.6	2.1	1.4	.2	+							5.6	10.2
SW	.2	1.7	3.0	2.5	.5	.1	+	+					8.2	11.4
WSW	.3	2.4	5.4	5.8	1.3	.3	+	+					15.5	12.5
W	.3	2.0	3.0	2.8	.7	.1	+	+					8.9	11.6
WNW	.2	1.4	2.3	2.5	.6	.1	+						7.1	11.9
NW	.2	1.5	2.4	1.9	.3	+	+						6.4	10.9
NNW	.1	1.1	1.9	1.0	.1	+							4.3	10.1
CALM	8.3												8.3	
TOTAL	12.2	26.2	34.4	22.3	4.1	.7	.1	+					100	9.4

Fig. 4-7. Summary of hourly observations for the Greater Pittsburgh Airport. (NCC)

is what the distribution of wind speeds over time would look like if not summarized in the form of the histogram.

The speed distribution or histogram is compiled from long-term meteorological records. To calculate the power in the wind at a particular site, we need to average the power for each speed bin using the percentage of the time the wind blows within that interval (see the bottom of the table in fig. 4-7).

Let us work through the power calculation for the speed bin, from 8 to 12 mph (see table 4-3). First we need to find the midpoint of the speed class. In this case it is 10 mph $(8 + 12 = 20/2 = 10)$.[5] Next we find the percentage of occurrence for this speed class—what percent of the year the wind blows within this speed range (34.4 percent). Now we can calculate this bin's contribution to the average power.

$$\begin{aligned} 8\text{-}12 \text{ speed class; } P/A &= 0.05472 \, (10)^3 \times 34.4\% \\ &= 0.05472 \, (1000) \times 34.4\% \\ &= 54.72 \times 34.4\% \\ &= 19 \text{ W/m}^2 \end{aligned}$$

We then sum the contributions from all the speed classes to arrive at the average power for the year.

In this example for the Pittsburgh Airport, let us see what would have happened if we had used just the average wind speed of 9.4 mph shown in the lower right-hand corner of figure 4-7.

TABLE 4-3
Average Annual Power in the Wind
Greater Pittsburgh Airport

	Speed Class (mph)							
	0–3	4–7	8–12	13–18	19–24	25–31	32–38	39+
Midpoint of speed class	1.5	5.5	10	15.5	21.5	28	35	—
Percentage of occurrence	12.2	26.2	34.4	22.3	4.1	0.7	0.1	—
Power density (W/m²)	0	2	19	45	19	8	2	—

NOTE: Total power density—95 W/m² @ anemometer height of 20 ft.

$$P/A = 0.05472 \ (9.4)^3$$
$$= 0.05472 \ (831)$$
$$= 45 \ W/m^2$$

The actual power in the wind is 2.11 times greater than that calculated from the average wind speed for the same data. This difference, the ratio of the power calculated from the speed distribution to that calculated from the average speed is called the *energy pattern factor* (EPF). It is also called the cube factor or correction factor. I sometimes resort to labeling it a fudge factor.

$$EPF = \frac{\text{Power from Speed Distribution}}{\text{Power from Average Speed}} = \frac{95 \ W/m^2}{45 \ W/m^2} = 2.11$$

The energy pattern factor is a measure of the speed distribution's shape. It gives the proper weight to the contribution of winds at higher speeds. The EPF is one when the wind blows continuously at the same speed. Normally it ranges from somewhat less than two to five. For the EPF of the three sites whose average speeds were plotted in figure 4-3, see table 4-4. Generally, EPF decreases with increasing average speed, but terrain also appears to have an influence. In Pennsylvania, I have found that EPF varies from a high around three for airports sheltered within steep-sided valleys, to approximately two at airports on broad plateaus. These effects can be seen in table 4-4 for Amarillo, Texas, which experiences not only a higher average speed than the other stations, but also is located on the High Plains of the Texas Panhandle. The other two sites in the table are located at the bottom of valleys.

To calculate the power in the wind from the average wind speed then, you need to incorporate an estimate of the EPF for your area. Two pioneers in the field have taken the historical data for many of the long-

TABLE 4-4
Energy Pattern Factor for Three Stations

Station	A Avg. Speed (mph)	B Power from Avg. Speed (W/m²)	C Avg. Power from Speed Distribution (W/m²)	D EPF (C/B)
Harrisburg, PA	7.7	25	66	2.64
San Francisco, CA	10.5	63	176	2.79
Amarillo, TX	13.7	141	238	1.69

term recording stations and determined the EPF. Steve Blake's results differ somewhat from those derived by Jack Park for Neilson Engineering. Their work, nevertheless, is similar. Because of greater uncertainty in the distribution of wind speeds, this approach becomes more prone to error at low average speeds. To be on the safe side, use the more conservative values (see table 4-5).

If you want a fudge factor to get you in the ballpark for quick estimates, use an EPF of 2 for average annual wind speeds above 10 mph, and an EPF of 3 below 10 mph.

From table 4-5 you can see that average speed is the least you can use to estimate the power in the wind. It is always preferable to use an actual speed distribution where possible. If the average speed is all you have to work with there is a lot of room for error. Consider a site with an average wind speed of 8 mph. The values for EPF range from a low of 2 calculated from the Rayleigh distribution to a high of 3 from Jack Park. If you used the lower value, but the site's distribution of wind speeds produces an EPF closer to Park's value, you would underestimate the power in the wind by 30 percent.

TABLE 4-5
Energy Pattern Factors (Correction Factors) for Use with Average Speed

Average Annual Speed (mph)	Rayleigh Distribution	Steve Blake[a]	Park-Schwind[b]
8	1.9	2.4	3.2
10	1.9	2.1	2.7
12	1.9	1.9	2.4
14	1.9	1.8	2.1

[a] Steve Blake, Sunflower Power Company.
[b] Jack Park, Dick Schwind, Neilson Engineering.

What is the *Rayleigh distribution*? It is an imaginary distribution of wind speeds based on a mathematical formula. The average wind speed is all that is needed to define the Rayleigh distribution's shape. The EPF of the Rayleigh distribution for any average speed is approximately 1.9 (see table 4-5). The most you need to know about the Rayleigh distribution is that you should never have to use it. If you must make an assumption about the speed distribution in your area to calculate the average power in the wind, there is no need to construct the Rayleigh distribution and then break it down by speed class. Use the EPF for the Rayleigh distribution in table 4-5 and calculate the average power using the average wind speed. It is no more or no less accurate doing it this way than the method in table 4-3.

Let's work through a few examples using average speed and an assumed EPF.

Site 1. Average speed of 8 mph, an EPF of 3.
$$P/A = 0.05472 \ (8)^3 \ (3)$$
$$= 84 \ W/m^2$$

Site 2. Average speed of 10 mph, an EPF of 2.
$$P/A = 0.05472 \ (10)^3 \ (2)$$
$$= 109 \ W/m^2$$

Site 3. Average speed of 20 mph, an EPF of 2.
$$P/A = 0.05472 \ (20)^3 \ (2)$$
$$= 875 \ W/m^2$$

Note that as we learned before, if we double the wind speed, we increase the power in the wind eight times as shown by the eightfold increase in power from Site 2 to Site 3.

Where do we now stand? We have learned two ways to calculate the power in the wind.

1. By using the speed distribution in figure 4-6, or the speed summary in figure 4-7.
2. By using the average wind speed and an assumed energy pattern factor.

If you have an aversion for the mathematics involved, you can find the average annual power density from table 4-6 if you know only the average wind speed. A range of values is given for each average speed. The low end represents an assumed Rayleigh distribution, and the upper end of the range represents the power density using the EPFs suggested

TABLE 4-6
Average Annual Power in the Wind (W/m²)

Average Wind Speed (mph)	Range of Power in the Wind[a]	
	Low[b] EPF	High[c] EPF
8	53	90
10	104	148
12	180	227
14	285	315
16	426	448
18	606	638
20	832	876

[a] $P/A = 0.05472 \, (S)^3$ (EPF).
[b] EPF = 1.9 from Rayleigh distribution.
[c] EPF from Park-Schwind, table 4-5.

by Jack Park. The results give us power density at the height of the anemometer where the wind speeds were measured.

We are not quite ready to estimate what a wind machine is capable of doing with this power, because it will be on a tower tens of feet higher than the anemometer. How wind speed and power change with height is the subject of the next section.

WIND SPEED, POWER, AND HEIGHT

Wind speed, and hence power, vary directly with height above the ground. Wind moving over the earth's surface encounters friction caused by the turbulent flow over and around the mountains, hills, trees, buildings, and other obstructions in its path. Friction slows the wind down. These effects decrease with increasing height above the surface until unhindered air flow is restored. Consequently, as friction and turbulence decrease, wind speed increases.

As you can imagine, frictional effects differ from one surface to another depending upon its roughness. Friction is higher around trees and buildings than it is over the smooth surface of a lake. In the same manner, the rate at which wind speed increases with height varies with the degree of surface roughness. Wind speeds increase with height at the greatest rate over rough terrain as found in the Rocky and Appalachian mountains, and at the least rate over smooth terrain like that of the Great Plains. Because of this, it is more important to use a tall tower when siting in hilly terrain than it is in the Texas Panhandle.

At low wind speeds, the change in speed with height is less pronounced and more erratic. In light or calm winds, as may be encountered during a temperature inversion, wind speeds may increase slightly between the ground and a certain height, and then begin to decrease. Real world experience has shown that changes in wind speed with height are not constant. Note that in figure 4-8 the short-term wind speed at 100 feet is sometimes less than that at 30 feet. But—on an average—speed increases with height. The extrapolation techniques that researchers have derived are general rules that work best with average wind speed and in areas of moderate terrain. Hills, mountains, and numerous buildings often influence wind speed in an unpredictable manner.

With this caveat in mind, we can use the following formula to estimate how fast wind speed will increase with height.

$$\frac{S_o}{S} = \left(\frac{H_o}{H}\right)^a$$

where S_o is the wind speed at the original height (the height of the anemometer, for example), S is the wind speed at the new height, H_o is the original height, and H is the new height. The exponent a varies with surface roughness.

Graphically we represent the change in wind speed with height by what is called the *speed profile*. The shape of the profile is determined principally by roughness though temperature can play a role. Normally tem-

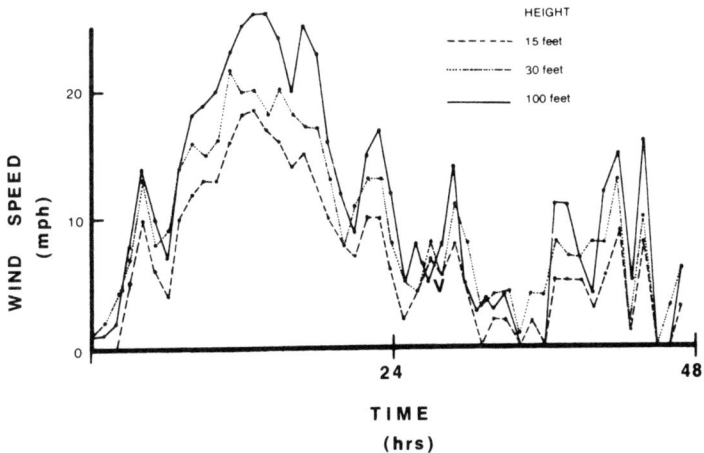

Fig. 4-8. Change in wind speed with height. (Park and Schwind, DOE)

perature decreases with height and speed increases. But during a temperature inversion, the opposite can be true, and the inversion may alter the wind speed profile.

Over smooth, level terrain covered by short grass the ⅐ power law applies; that is, the exponent a equals 0.14. Over tall row crops such as a field of corn, or over hedges and a few scattered trees, a equals ⅕ or 0.20. This relationship only holds true for the height above the corn or hedge row. The wind rushing over a field of corn, for example, sees the top of the corn as the effective ground level, not the soil on which it grows. When the surface is rougher, say more trees and a few buildings, a equals ¼, or 0.25. The speed profile is even steeper over woods and numerous buildings. Here again you are seeking the height above the "effective" ground level. For a wood lot this is the uppermost point where the branches of the trees are touching, not necessarily the top of the trees.

I have measured an increase in wind speed over an apple orchard (the trees were all about 10 feet tall) to a height of 60 feet that corresponds to an a of ⅓ or 0.33. This is the highest value I ever personally observed and I regard it as an upper limit. This rapid increase in speed with height leads to a big jump in the power available.

Let us see what all this means by examining the effect doubling the height above the ground has on speed and power at two sites, one over a sea of grass in Kansas where the ⅐ power law is appropriate, and the other, the apple orchard where a equals ⅓.

To double the height let us say that we plan to use a 60-foot tower ($H=60$) and that the average speed was recorded at 30 feet ($H_o=30$).

Grass: $\quad \dfrac{S_o}{S} = \left(\dfrac{H_o}{H}\right)^{\frac{1}{7}} = \left(\dfrac{30}{60}\right)^{\frac{1}{7}} = \left(\dfrac{1}{2}\right)^{\frac{1}{7}}$

$$S = \dfrac{S_o}{0.91} = 1.10 \; S_o$$

Apple Orchard: $\quad \dfrac{S_o}{S} = \left(\dfrac{30}{60}\right)^{\frac{1}{3}} = \left(\dfrac{1}{2}\right)^{\frac{1}{3}}$

$$S = \dfrac{S_o}{0.80} = 1.26 \; S_o$$

Doubling the height increased speed over the smooth site by 10 percent and by 26 percent over the apple orchard. But we cannot stop here. Power is a cubic function of speed. Remember from an earlier discussion that a 10 percent increase in speed causes a 33 percent increase in power. Where P_o is the power at the original height of 30 feet and P is the power at the new height, the formula for the increase in power with height is given as:

$$\frac{P_o}{P} = \left(\frac{H_o}{H}\right)^{3a}$$

$$P = \frac{P_o}{\left(\frac{H_o}{H}\right)^{3a}} = \frac{P_o}{0.74} = 1.33\, P_o$$

Or we can find the same result by cubing the increase in speed of 10 percent.

$$1.10 \times 1.10 \times 1.10 = 1.33\, P_o$$

What happens over the apple orchard?

$$1.26 \times 1.26 \times 1.26 = 2.0\, P_o$$

Double the height of a tower over rough terrain and you may be able to double the power available to a wind machine. This is the justification for installing wind machines on tall towers. It is also the reason why the height of the anemometer is always mentioned when wind data is given. The two go together. Wind speed means nothing without noting the height at which it has been measured. In general, when people speak of average wind speed they are referring to the speed at the height of the anemometer (from 20 to 40 feet). The confusion arises when you start talking to wind system dealers or reading manufacturer's brochures. The wind speeds they use may be either at the hub-height of the wind machine or at an anemometer height of 20 to 40 feet. It makes a big difference. Most rate their units at the hub; some don't and they can catch you off guard. Reputable manufacturers and their dealers will clearly state which they use.

All right, so you do not have a calculator that can compute fractional exponents and, besides, you never liked working with formulas anyway. All is not lost. Table 4-7 is for you. Here is how to use it. Say we want to find wind speed at 80 feet when the average speed is 10 mph.

Look down the left column of table 4-7 for the height you are interested in. Next, determine the roughness exponent a appropriate to your site. (Be conservative, the chance for error is greatest when using either right-hand column where a equals $\frac{1}{4}$ and $\frac{1}{3}$. As a rule, I don't use anything over $\frac{1}{5}$ unless I have measured it.) To find the wind speed at 80 feet when the roughness exponent is $\frac{1}{5}$, move along the 80-foot line until you reach

TABLE 4-7
Increase in Wind Speed (and Power) with Height

Height	Roughness Exponent (*a*)			
	1/7 0.14	1/5 0.20	1/4 0.25	1/3 0.33
30[a]	1.00 (1.00)	1.00 (1.00)	1.00 (1.00)	1.00 (1.00)
60	1.10 (1.34)	1.15 (1.52)	1.19 (1.68)	1.26 (1.99)
80	1.15 (1.51)	1.22 (1.80)	1.28 (2.09)	1.38 (2.64)
100	1.18 (1.66)	1.27 (2.00)	1.35 (2.47)	1.49 (3.29)
120	1.21 (1.79)	1.31 (2.30)	1.41 (2.83)	1.58 (3.94)

[a]Equivalent to anemometer height in most cases.

the column under 1/5. Take this value, 1.22, and multiply it times the original speed.

$$10 \text{ mph} \times 1.22 = 12.2 \text{ mph at } 80 \text{ feet}$$

To do the same with power, use the value in parentheses—1.80 under these conditions. For example, if the power at the original height was 100 W/m², then at 80 feet:

$$100 \text{ W/m}^2 \times 1.80 = 180 \text{ W/m}^2$$

This table assumes that the anemometer's original height is 30 feet. For anemometer heights less than 20 feet or greater than 40 feet use the formulas.

SOURCES OF WIND DATA—WHAT IS AVAILABLE

More than one new owner of a wind system has learned an expensive lesson—one that may seem patently obvious. Wind generators with little wind are like dams with little water.

Everyone has wind, but not everyone has enough. Ample wind is a prerequisite for successfully siting a wind machine—the more the better. But just how much is enough? How do you know whether the wind over your site is sufficient or not?

I get a lot of phone calls from people wanting to know if a wind machine will work in their backyard. I stutter and stammer for a few minutes trying to figure just where to begin. Often I start with the basics by asking a seemingly simple question, "Well, how much wind do you have?" I have heard all kinds of replies, but they generally fall into the category of "Why, it is always windy here." If they are living on the coast of Alaska or in the Panhandle of Texas, I take their word for it. Unfortunately, few people live where the wind has such a well-deserved reputation for being fierce and fairly constant. For most of us, we need a little better description of the wind than an "It's always windy here."

There are many sources of wind data. This section covers who has the data, the kind of data they have and its format. We will also discuss how this information has been updated and presented in a form easy for us to use.

The most extensive records have been collected by the National Weather Service (NWS) and the Federal Aviation Administration (FAA). These records are stored at the National Climatic Center (NCC) in Asheville, North Carolina. Wind data from other federal agencies (Civil Aeronautics Administration, Forest Service, and the Department of Defense) is also stored by NCC.

At some of the stations, wind speed and direction have been recorded for over thirty years. Periodically, this information has been tabulated into a Summary of Hourly Observations that provides the long-term average wind speed, as well as the speed distribution. An example for the Greater Pittsburgh Airport is shown in figure 4-7. The data is presented as the Percentage of Wind Direction and Speed. In this format, the percentage of time that the wind blows from one of sixteen compass points at one of nine speed classes is given. The percentage of wind in each speed class (e.g., 8 to 12 mph) from all directions is shown at the bottom of the chart, and the average speed at the lower right. Sometimes the Frequency of Occurrences (the actual number of observations) is given instead. Both are used in the same way except that the latter requires more time on the calculator. These summaries can be obtained by writing the NCC. Later we will explain how these summaries can be helpful when measuring the winds at your site.

The wind data from the NCC, however, is not always accurate. Much of the data has been collected at major airports, which are often sited such that they are sheltered from the wind. In general, wind data has been gathered near centers of population. People congregate in areas sheltered from storms and severe weather. (We live in valleys more than we live on windswept mountaintops.) Consequently, the data from our airports, military bases, and weather stations that serve us, may not reflect the winds that exist at more exposed—windier—sites. Using the data from airports alone may lead us to underestimate potential wind power. On the other hand, airports offer a vast clearing with minimal trees and buildings. In heavily wooded or developed areas, the open expanse at an airport may offset any sheltering effects due to its location.

The data may be unreliable for other reasons as well. The wind measuring instruments may have been inaccurate or poorly placed. At many sites the instruments were not properly maintained, and frequently they were located on or adjacent to the terminal building. At these stations, the data better reflects the turbulence around the building than the true wind conditions. At more remote sites, wind data was collected only during daylight hours or for a few hours each day during the fire season (by the Forest Service). In either case, the data does not represent what could be expected throughout the day. At some stations, wind speed was recorded in so few wind speed classes that the observer was jotting down wind speeds, in effect, as high, low, and medium winds. In Pennsylvania, for example, data from several military bases was collected during World War II. Taking their summaries at face value makes the area around these bases look attractive for wind development. Power densities over 200 W/m² are common with some as high as 300 W/m² to 400 W/m². The truth, unfortunately, is less encouraging. (Most of the data was collected during the day and recorded in only a few speed classes.) Average power densities in the range of 100 W/m² to 150 W/m² are more realistic for these areas.

Historical average speeds are also available from air quality monitoring stations at both conventional and nuclear power plants and from some industries. The limits on data from these sources are the same as those on the data from the airports. We measure air quality where wind speeds are low and where pollutants concentrate, such as in urban canyons and narrow mountain valleys. These are less than ideal locations for a wind machine, and the wind speeds do not represent what you can expect at better sites.

There are also numerous sources of short-term wind data: state energy offices, universities, nonprofit organizations, and wind system dealers. One or more of them might have collected wind data in your area. The winds could have been monitored from four months to four years during special programs. Check around and find out what work has been done and who has the results.

While looking about for local sources of wind data, do not be smitten by the discovery of an abandoned airfield outside town and dash off to Asheville for a search through NCC's archives. First, much of the data from your find will not be of any value to you. Second, the work has already been done for all federal sources of data, and for many private ones, too.

There have been several national studies of the data housed at NCC. Battelle Pacific Northwest Laboratory (PNL), under contract to DOE, conducted the most recent and most thorough review of existing wind data in their national assessment of wind energy resources. The results have been published in twelve regional atlases and are summarized in appendix C. Battelle not only used the summarized data, but also data from sources never before summarized, as well as data from short-term recording stations. Battelle carefully reviewed the records for accuracy. Their results are presented in two forms: a table of average annual wind speed and power, and a map of average annual wind power.

Battelle also adjusted the wind speeds for anemometer height. Prior to the sixties the anemometer could have been mounted anywhere. Since then, the height has been standardized at roughly 20 feet (6 meters) above ground, and the anemometer placed in the open near the end of a runway. What this means is that you do not need to track down the history of the anemometer for the airport nearest you. Battelle has done that and presented the data at a standard height of 10 meters (33 feet) for all stations. Now you can see how they really stack up. In Pennsylvania, for example, the Allentown Airport showed a promising annual power of 150 W/m^2; but, the anemometer was 60 feet above the ground. Battelle corrected the wind power to 110 W/m^2 (at 33 feet), a more realistic value considering the terrain near the airport.

The tables in Battelle's atlases are derived from the wind summaries for those NWS and FAA stations with reliable records. If your site is similar to the terrain around an airport and you are only a few miles away, the values in these tables may be representative of what you can expect. When not, use the wind power maps.

The maps of average annual wind power do not give wind power directly. Instead they identify portions of the state or region with a numerical rating that corresponds to one of seven Wind Power Classes. Each class represents a range of wind power. Allentown Airport, for instance, would fall into Class 2 with a wind power density of 100 W/m^2 to 150 W/m^2.

These maps are derived from computer modeling of the historical wind data, the terrain, and regional weather patterns. The values shown represent only those sites well exposed to strong prevailing winds. Good exposure is equated with hilltops, ridge crests, mountain summits, and large clearings free of obstructions. Estimates for ridge crests or mountain summits are shaded to indicate that the wind power there could be significantly greater if the terrain increased local wind speeds.

The feature I like best about these maps is their use of Wind Power Classes. By giving a range of possible wind speeds (or power) rather than a single number, they do not lure you into the mistaken notion that we can estimate wind speed with such precision. We cannot. Talk to a farmer, or anyone who works with natural resources, and he will tell you it is hard to project what will happen one year to the next, even one decade to the next. When evaluating your site always present the results as a range of wind speeds (or power). It is more realistic than pretending precision.

SITE SURVEYS — CLASSIFYING THE WIND AT YOUR SITE

To evaluate the potential wind power at a site begin by asking yourself a few questions. What is it that I want? What am I looking for? Is it the instantaneous wind speed in some fraction of a second, the average speed over a year, or the distribution of wind speeds? The answers are determined by what you plan to do with the data.

If you want to know when it is too windy to go sailing, instantaneous wind speed will suffice. If you want to estimate the annual energy output of several different wind machines for supplemental power, then at least average wind speed (or power) is necessary, and preferably the wind speed distribution. If you plan to use the wind turbine as your sole source of power at a mountaintop retreat, for example, then more detailed information may be required, such as the number of calm days and the time between them.

In most cases you are interested in how much energy a wind machine can produce in your area; more specifically, at your site. This is the bottom line. To estimate annual energy output, the speed distribution is preferred because it gives you more accurate results, but it is not necessary. Average speed or — better yet — average power will suffice.

Now that you know what is needed, find out if someone has already done the work for you. Start by locating anyone nearby who has installed a wind machine or anemometer. Pay them a visit. What have they found? How much energy have they generated? What is their estimate of the average speed or power? Before you take any of their data to heart, though, determine if your site and theirs are comparable. Is the data typical of what you could expect? If so, you have saved a lot of time and effort. Normally you will not be lucky enough to get the answers to these questions. You will probably be the first in the county to give the matter serious thought.

If that is the situation, you will have to come up with your own data and make your own estimates. The maps of wind power in appendix C are the first place to look. Next, consult the table of average wind speed and power for the airport or weather station with reliable records nearest you. How does the data compare? Are the two estimates of wind power reasonably close, that is, within the same power class?

Find out what you can about the airport. Is it representative of the winds that you can expect at a well-exposed site as shown on the wind power maps? Is it similar to your site? For example, if the airport is sheltered at the bottom of a steep-sided valley its wind data will not be the same as what you could expect on a plateau several hundred feet higher. While you are at it, take a good hard look at the trees in your area.

Trees and shrubs are frequently touted as a good qualitative indicator of wind speed. High winds and a harsh environment of ice and snow will deform them. The severity of the deformation, whether the tree is slightly flagged or completely bent to the ground, can be used as a rough gauge of wind speed. The types of deformity are:

- Brushing: branches and twigs sweep downwind. This can be observed on both conifers and deciduous trees.
- Flagging: branches sweep downwind, upwind branches are cropped short.
- Throwing: trunk sweeps downwind.
- Carpeting: trunk bends to the ground. Found frequently in Alpine or severe environments where trees do not grow more than a few feet above the ground.

Use the scale of deformity in figure 4-9 with table 4-8 to find the range of wind speeds represented.

There are, though, a few limitations to this technique. First, do not get excited by one or two odd-shaped trees. One flagged tree is insufficient to make a judgment. You need to note several of the same species with an equal amount of deformation. (Each species varies in its susceptibility to flagging, so you can't mix them.)

The use of deformation works best on conifers, especially pine and fir trees. Deciduous trees (those that lose their leaves in the fall) can also be used, but they are less reliable indicators. Moreover, deformation is more obvious where freezing salt spray or ice frequently accompany high winds, such as along the Atlantic or Pacific coasts and in mountainous areas.

There are many other causes for a tree to look flagged besides the wind. The absence of flagging, moreover, does not necessarily indicate a low speed.

Too often the value of trees as a wind speed indicator is overplayed. At best it gives only a crude range of possible wind speeds, and even then, it cannot be employed with any degree of certainty without being in an area where conifers dominate.

TABLE 4-8
Average Annual Wind Speed by the Griggs-Putnam Index

I	II	III	IV	V	VI
7–9	9–11	11–13	13–16	15–18	16–21

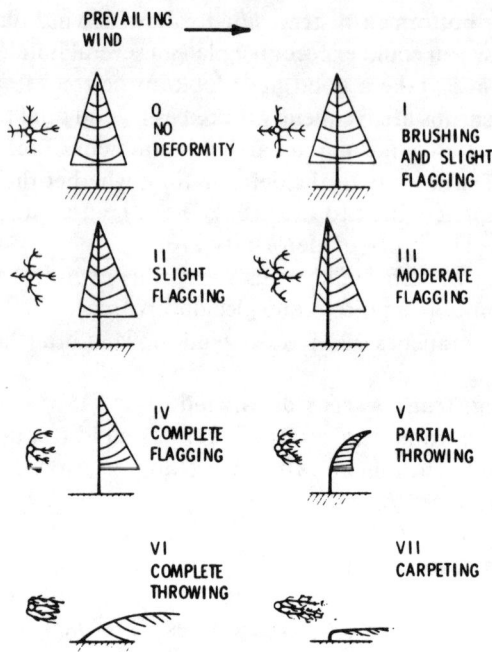

Fig. 4-9. Griggs-Putnam Index of Tree Deformity. (Battelle PNL, DOE)

Now, put all the pieces together. Estimate what you think is your average wind speed and power. Give yourself some room for error. Avoid pinning your hopes on one number alone. Instead of saying 10 mph give a range of 9 to 10 mph (or 10 to 11 mph, if you think the site might be a little windier). In other words, present your estimate as a Wind Power Class. A 9-mph to 10-mph range falls within Class 2 for example.

Be reasonable. Most people overestimate the amount of wind available. I frequently carry a hand-held wind speed meter that I use to play a guessing game with my clients. We both estimate the wind's speed then check it with the meter. Try it yourself. After a while you will develop a feel for the wind and your guesses will be more on the mark; but, at first, you are likely to be too high.

With these numbers in hand, use the techniques in the next section to estimate the output from several different size wind machines. Pick the one that delivers the amount of energy you need and determine its cost. Call the local wind system dealer for realistic estimates. (Most published information where prices are included is outdated before it hits the streets.) Take a look at the economics. If the numbers seem reasonable, your next step is to decide whether you want to spend the time and money to monitor the winds at your site. If the costs are much more than you anticipated, if they exceed your budget, or if the economics don't appear even marginally attractive, there is no need to go any further.

Measuring the wind at your site is useful when the wind system is a marginal investment and you do not want to take the risk that the winds are less than needed to give you the desired return. Normally you do not need to measure the wind yourself if there is a good chance the winds are strong enough to make the project pay, or when the winds are so poor there is nothing you can do to make the investment lucrative. In table 4-9 are some general rules. This table is to be applied to the wind speed and power from a nearby airport or the speed and power from the wind power maps in appendix C.

TABLE 4-9
Need for Site Survey with Anemometer

No	When a nearby airport has an average wind speed of 12 mph or greater (200 W/m^2) and your site is within 10 miles over smooth terrain.
No	When a nearby airport has an average wind speed of 12 mph or greater (200 W/m^2 and your site is more exposed, e.g., on a mountaintop.
No	When the average speed is less than 8 mph (75 W/m^2).
No	If there are numerous trees and buildings over 20 feet tall within 150 feet of the site, and the anemometer will be less than 50 feet above the ground.
Yes	When the average speed is between 8 to 12 mph (75 to 200 W/m^2).

Table 4-9 assumes that wind machines are generally economic where wind speeds are 12 mph or above and uneconomic where wind speeds are less than 8 mph. Your situation may be different. To find out if it is, try this. Select the wind machine you prefer and determine what wind speed is necessary for it to be a marginal, a good, and an excellent investment. Let us say you learn that an average wind speed of 10 mph will give you a good investment and that an average speed of 12 mph an excellent investment. From your careful review of the published wind data you estimate that your site has an average speed of 12 mph. Installing an anemometer, then, is an unneeded and a time-consuming expense because you can afford to be off one or two mph and still make the investment pay.

Table 4-9 also assumes that reliable measurements cannot be made where there are trees and buildings near the anemometer. Ideally, measurements, if they are to be made at all, should be taken at the intended location of the wind machine and at the proposed height. If not, they should be taken as near as possible. This is particularly true in rough terrain or where there are trees and buildings. Most wind machines, however, are installed on towers 80 to 100 feet tall. Site surveys, though, are often performed with masts 30 to 40 feet tall because they are cheap and easily erected.

The 30- to 40-foot mast height is also comparable to the 20-foot anemometer height at most airports and is close to the World Meteorological Organization's standard anemometer height of 10 meters (33 feet). Data from a 30-foot mast, for example, can be directly compared to airport data without adjusting for height; from a 40-foot mast, data can be compared directly to any site with data at 10 meters.

It does not make any sense to install an anemometer on a 30- to 40-foot mast if the site is sheltered by trees, buildings, or the terrain. I have found that any trees at all (within 150 feet), or nearby buildings, obstruct the wind to such a degree that the anemometer gives misleadingly low results. Even tall corn and shrubs raise the effective ground level and reduce wind speed at the anemometer. If you need to measure the winds at your site, take a survey of the trees and buildings and estimate their heights. You may find that, contrary to your original intent, the anemometer should be erected elsewhere.

Estimating Tree Height

There are three simple methods for estimating the height of an object, whether it is a tree or tall building, without a transit. The easiest is a method learned by art students. Remember those comic pictures of a ragged artist thrusting his thumb at the world. There is a good reason for it. Artists use their thumbs to gauge proportions. A pencil works better though. Here's how to use it.

Identify an object of known height at about the same distance from you as the tree that concerns you. TV antennas, telephone poles, and houses work well. Hold the pencil at arm's length and sight along it. Line up the top of the pencil with the top of the tree. Slide your thumb down the pencil until it lines up with the bottom of the tree. Now turn to the object of known height and again sight along the pencil. While keeping the pencil at arm's length, move your arm up or down until your thumb lines up with the bottom of the object. Judge the proportions by noting how much of the pencil extends above the object. Is it twice the height, one-third greater, or is it the same height?

Say you are using a TV antenna for comparison. The antenna is 20 feet tall (two 10 foot sections are common). The tree appears to be twice its height or 40 feet tall. Reverse the procedure and see if you get the same result. This is the method I use whenever I visit a site. It is quick, easy, and good enough for a first cut.

When you need to be more accurate, a similar approach is better. Find a pole or post between you and the tree. The side of a building will also do. Position yourself such that your line of sight to the top and bottom of the tree falls across the post. Note your position and measure the distance to the tree and to the post. Measure the distance between the upper and lower lines of sight where they strike the post. The height of the tree to its distance from you is the same as the span between the upper and lower lines of sight to your distance to the post. This is expressed as:

$$H = h/d \times D$$

where H is the height of the tree, h is the span between the upper and lower lines of sight, d is the distance from the post to you, and D is the distance from you to the tree. For example, the tree is 100 feet away. At 20 feet stands a post. The spread between the upper and lower line of sight is 12 feet. How tall is the tree?

$$H = 12/20 \times 100$$
$$= 0.6 \times 100$$
$$= 60 \text{ feet}$$

You can do this by yourself, but another person is helpful for marking the lines of sight.

Try this technique on a sunny day. You will not need any help at all. Measure the shadow cast by the tree. Then plant a yardstick or ruler in the ground and measure its shadow. The height of the yardstick to the length of its shadow is the same as that of the tree to its shadow. Use the same formula as before. Say the tree casts a shadow 100 feet long (D) and a three-foot yardstick (h) casts a five foot shadow (d). How tall is the tree?

$$H = h/d \times D$$
$$= 3/5 \times 100$$
$$= 0.60 \times 100$$
$$= 60 \text{ feet}$$

Let us assume that you want to measure the wind at your site to get a better picture of what is there. (The wind data in appendix C is unconvincing for such a large investment.) You have a clear site. There are no trees or buildings other than your house a few hundred feet away. What next?

Anemometers measure wind speed. Wind vanes indicate direction. That seems simple enough. What you need and how you use it is less clear cut. In the next section we will go over the equipment that is on the market and discuss what probably meets your needs best.

Equipment

An anemometer is composed of two parts: the *sensor* (called the anemometer head) and a means for displaying the data it measures. Both the sensors and the data displays are varied. In each case the sensor generates an electrical signal that is proportional to wind speed. The National Weather Service's sensor, for example, uses a three-bladed propeller to drive a DC generator (a miniature wind machine). Changes in wind speed cause a change in DC voltage that is then read by a calibrated voltmeter. The data is not recorded. The needle swings to and fro like the speedometer on your car. To record the data, the anemometer must be read manually.

Cup anemometers are more common today. The spinning cups drive either a DC generator or AC alternator, or they produce an electrical pulse. Many low-cost anemometers (such as those sold in electronics supply houses for less than $100) use a cup anemometer, DC generator, and voltmeter. With the advent of pocket calculators, digital displays have become more popular. (The data is the same, it just looks different.) Better quality instruments, however, use AC alternators and measure output frequency. The response of these anemometers is more accurate than those of the DC systems. Another type produces an intermittent electrical signal by opening and closing a small switch. The number of on-off signals corresponds to the number of anemometer revolutions. These signals are then displayed as instantaneous wind speed, accumulated, or both.

Whether DC, AC, or an electrical pulse, the sensor can be used to drive a recorder. Unlike displays of instantaneous wind speed—what I call wind speed meters—recorders store information for future retrieval. They use mechanical or electrical counters, strip charts, magnetic tape, or silicon chips. Only a few years ago nearly all meteorological data was recorded on strip charts, but wind prospecting and the boom in electronics have revolutionized wind speed measurement.

Before we go any further, let us clear up a common misconception. Wind speed meters (whether using a meter or flashing neon lights) are next to useless for finding the average wind speed at a site. They are cheap and fun to watch, but that is it. To be of value in a site survey you would have to check them every hour, twenty-four hours a day, every day for months on end. If you have that kind of time, fine. They are helpful, I have found, for developing a better understanding of the wind. Use them only to calibrate your wind sense or to monitor the performance of a wind machine.

Strip chart recorders, though still in widespread use, are now almost obsolete. They get their name from a long strip of paper wound between two spools. Wind speed is recorded on this paper as a neat tracing on the fancier models and as a mass of dots on the cheaper ones. Usually a roll of chart paper will last a month. I say usually, because the chart drives do not always run at the correct speed. (When using a strip chart recorder you must always mark the beginning and end of the recording period on the chart paper.) Their chief drawback is the tedious steps necessary to analyze the data. Moreover, the older models (the ones you are most likely to find used) can easily be up to 10 percent off. This amounts to 1 mph in an area with a 10 mph average speed, or a 30 percent error in estimating power. One advantage is that they display wind speed immediately, as well as record it. You can use them as wind speed meters if need be. If you can find one cheap (less than $100), have plenty of time and patience, and can assure accuracy, then a strip chart recorder may be for you. But an accumulator is an overall better choice.

Accumulators are a simpler form of recorder (see fig. 4-10). Like the odometer on your dashboard they count miles, in this case the miles of wind—the *wind-run*—that pass the anemometer. These wind-run odometers use either mechanical or electronic counters. Both work well. To estimate average wind speed all that is necessary is to determine the time that has elapsed between readings. The miles of wind, or fractions thereof, are divided by the recording time in hours to give the miles per hour for that period. For example, if 240 miles of wind have passed within a twenty-four-hour period, the average wind speed for that day is 10 mph (240/24 = 10).

Wind-run odometers register miles, 1/100 of a mile, or 1/60 of a mile of wind. When using 1/60 of a mile recorders, the accumulators can be used to indicate wind speed directly like a wind speed meter by recording the wind-run over one minute (60 seconds). The resulting display is the one minute average wind speed in mph. (If you cannot wait 60 seconds, take a 30-second count and double the value, or take a 15-second count and multiply it by 4, and so on.)

Accumulators are being used wherever cost is a major concern and average wind speed will do the job. With anemometer, batteries, and signal wire they can be purchased for under $300 (1983). Their accuracy is good, too. In wind tunnel tests, I have measured error rates consistently less than

Fig. 4-10. Typical wind-run odometer. (Courtesy Aeolian Kinetics, Inc.)

2 percent on new equipment.

Accumulators are easy to use. You merely walk up, note the date and time, and press a button. You then jot down the numbers displayed on the register. The difference between the current and the previous reading is the number of fractional miles of wind.

A slightly more sophisticated accumulator records elapsed time, wind-run, and a parameter called V^3. This parameter is used to derive the energy pattern factor. Though this recorder costs twice as much as the simple accumulators, it is the only low-cost way to calculate average power density accurately.

Accumulators have evolved into more sophisticated data loggers. In essence, *data loggers* consist of multiple accumulators that tally the data falling into each accumulator's domain. For example, each accumulator could represent a given wind speed range. Winds from 0 to 1 mph would fall into the first accumulator, winds from 2 to 3 mph would fall into the second, and so on. At the end of the observation period, the contents of each register can be used to plot the speed distribution. This distribution can then be used to calculate power density or a wind machine's potential power output using its power curve, an approach that is much more accurate than that using the average wind speed and an assumed speed distribution. More complex data loggers can record wind speed by time of day (see fig. 4-11).

The accuracy of these recorders has sometimes been questioned. Occasionally a counter, for instance, would not register properly. When monitored closely, however, recorders should give reasonably good results. Still, their cost prevents their use by the average person. The degree of sophistication required to produce a speed distribution from a data logger is unwarranted in most cases and can produce a false notion that estimates based on the data are more reliable than those from simple recorders. (The potential for error in estimating wind turbine performance is considerable, and it is not limited to the wind measurements, but also encompasses the reliability of the wind turbine's power curve.)

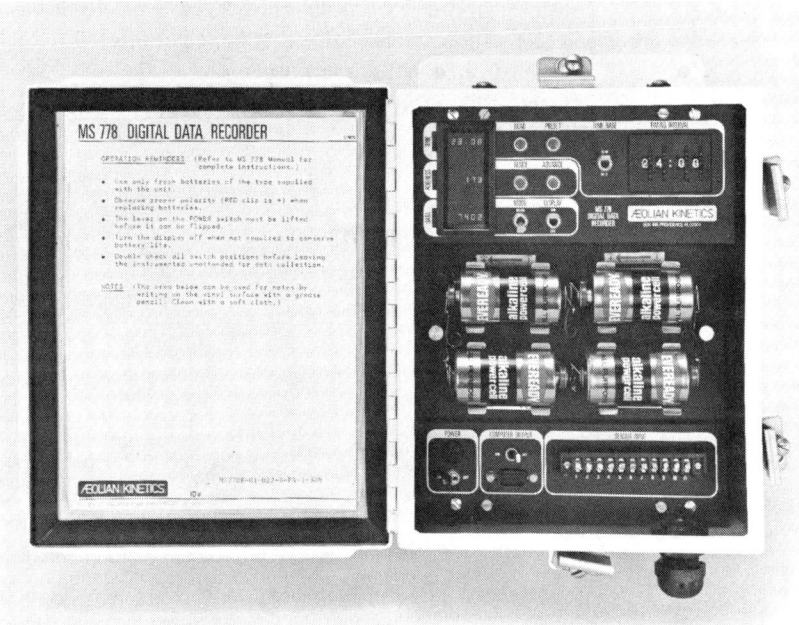

Fig. 4-11. Typical data logger. These are more sophisticated instruments than wind-run odometers and can record significantly more data useful in characterizing the wind at a site. (Courtesy Aeolian Kinetics, Inc.)

104 • Wind Energy

Data loggers can be tailored to meet a customer's specific needs. For example, one model converts wind speed into the power output that would be produced by a specific wind turbine by using the machine's actual performance characteristics. At the end of the measurement period, the Wind Energy Monitor indicates the number of kilowatt-hours that would have been generated had a wind turbine been in place. The data from such recorders is easy to interpret, but it has limited application unless you are interested in only one manufacturer's product.

Some data loggers are designed to preprocess the data and, using data compression techniques, record the results. These "smart" recorders significantly reduce the time and expense of analyzing the data because they, as Jack Park says, "crunch" the raw data. These recorders are capable of storing much more data within the data logger than when the recorder stores raw data.

As a rule, equipment designed for the wind turbine market is less expensive, easier to use, and gives you more of the information that you seek than do standard meteorological instruments. This includes wind vanes. For most siting studies, wind direction is unnecessary. We perform a site survey as an aid in estimating potential wind energy. We don't really care what direction the wind comes from, just so it is there. Whether it is or is not will be measured by the anemometer.

Table 4-10 summarizes the equipment available for site surveys. My preference is a wind-run odometer with V^3 or, when money is tight, a simple wind-run odometer.

TABLE 4-10
Equipment Summary

Type of Recorder	Data Produced	Comments
Wind speed meter	Instantaneous wind speed	Cheap, good only for developing wind sense, satisfying curiosity or performance monitoring
Wind-run odometer	Average wind speed	Least costly, easiest to use, best choice for most
Wind-run odometer with V^3	Average speed with EPF	More costly than simple odometer but more useful
Data logger	Average speed, EPF, speed distribution, maximum speed, degree of calms	Most costly, provides most data, unneeded complexity for most applications
Strip-chart recorder	Average speed, EPF, speed distribution, maximum speed, degree of calms	Difficult to use, data difficult to read

Measuring the Wind • 105

You can either buy your own or rent an anemometer. If you buy, you may be able to sell the anemometer later to a neighbor wanting to do his own survey. Or you could mount it on the wind machine tower should you install one. Check with your state energy office or utility and ask them if they rent anemometers. If not, most wind system dealers do. You could also hire a professional consultant who specializes in this kind of work. He would rent you the anemometer, install it, and analyze the data you collect. Fees vary considerably, so it is a good idea to shop around. No matter who you rent from or hire, look at their record and determine what kind of reputation they have. This is important because, though few in number, there are some unscrupulous salesmen out there who do not know what they are doing when it comes to measuring the wind.

Building Your Own Anemometer

You can build your own low-cost recording anemometer for less than $100. Gil Masters of Stanford University describes in a recent edition of *Other Homes and Garbage*[6] how to construct a wind-run odometer using inexpensive pocket calculators. Most cheap anemometers merely indicate current wind speed with a swinging needle or ever-changing digital display. They cannot be used to record wind speed. This one can.

Masters suggests building the anemometer head from scratch using L'Eggs (plastic eggs used to package panty hose) cups. Unfortunately, such an anemometer requires expensive calibration before you can put it to use. A better approach is to use an off-the-shelf sensor and leave calibration to the experts. The sensor may be a little more costly than the one you build yourself, but it will work — and reliably, too.

The best choice is M. C. Stewart's (see appendix D for the address) contact anemometer. It provides an on-off signal from a reed switch for every $\frac{1}{60}$ of a mile of wind that passes the anemometer's cups. This signal is then used to drive the pocket calculator as a counter.

An alternative is a cup anemometer by Maximum. It also uses a reed switch, but gives an on-off signal four times per revolution. Pocket calculators incorporate an anti-skip feature that prevents them from counting the signals fast enough when the wind gets above 10 to 20 mph. You could build a signal conditioner to deal with the problem, but that increases the complexity and cost. It is not worth it. The Stewart anemometer works just fine and costs only $50.

First find a calculator that can do the job. Not all will. The calculator must have the logic allowing it to count when the "1", "+," and "=" keys are pushed in sequence. Subsequently, striking the "=" key again will cause the calculator to count and register the count on the display.

Each time you push the "=" key one (1) should be added to the total displayed. Calculators using both LED (light-emitting diodes) and LC (liquid crystal) displays with the proper logic can be found for less than $10 (1983). LED calculators are preferred because you can buy power supplies for extended operation at less than $5. (Make sure the calculator has a jack for remote power.) The small, thin calculators using LC displays lack the remote power option. Moreover, most automatically turn themselves off when not in use to conserve power. During light winds or calms you are likely to find the calculator has switched itself off, clearing the accumulated data from its memory.

Once you have picked your calculator, carefully pry it open. Next, determine what happens when you push the "=" key (see fig. 4-12). Your objective is to insert the anemometer's on-off action for that of the "=" key. On an LED calculator look for two sides of the switch. You will probably find these on the circuit board. Delicately solder two leads at appropriate nodes on the board to bypass the "=" key. Test by programming the calculator with "1," "+," and "=" key strokes, and then short the leads a few times. It should begin counting. (If not, review your work.) Now connect the leads to the anemometer and spin the cups. After several revolutions of the Stewart anemometer, the display on the calculator should begin registering counts. Wiring an LCD calculator is even easier

Fig. 4-12. Inexpensive pocket calculators can be modified to perform as wind-run odometers.

because the circuitry is clearer. Find the metal fingers representing the " = " sign and trace them on the circuit board. Solder two leads to the circuit and test as before. Close the case — being careful not to crimp the leads — and you are done. It takes just minutes.

Neither calculator is suited for outdoor use. (For remote sites you will need professional equipment.) Nevertheless, try to place the anemometer head as far away from your house, other buildings, and any trees as possible. And get it up as high as you can without putting it on a building. Keep the calculator (now a recorder) indoors. Bell wire works well as the signal wire between the anemometer head and the calculator. Read the calculator at least once a week and jot down the data. Divide the difference from one reading to the next by sixty to find the miles of wind, and divide that by the number of hours since the last reading to find the average speed. That is all there is to it.

Towers

Most site surveys use telescoping masts or TV antennas. Unarco-Rohn's telescoping towers are a frequent choice. Each bay is approximately 10 feet long. The total height of a four-bay tower, for example, is 38 feet. A five-bay tower is also available. Taller towers can be assembled from a Rohn 25-G guyed tower. This is an expensive approach, though, and can be justified only on commercial installations.

I have found that a tower made from 16-gauge TV antenna mast is less costly and easier to handle. The 10-foot sections can be found in any building supply or hardware store. Using three sections, a tower 29 feet tall can be erected with two guy levels (see fig. 4-13). At some stores a 5-foot section is available. This can be added to the tower to bring the total height to 34 feet — the limit to this type of tower (the sections are too weak to support anything taller). If this tower will meet your needs, then read on for some of the pertinent details.

The tower should be guyed at two levels: one at the midpoint of the second section and one at the midpoint of the third section. Use four guy cables and anchors at each level. This insures maximum protection for the anemometer head if any one cable or stake fails — and they occasionally do. Four guys also allow the tower to be raised by tipping it into place as will be described.

Multistrand galvanized steel wire (18- or 20-gauge six strand, e.g., 20/6) can be used for the guy cables. The cable comes in 50-foot bundles, is difficult to work with, and can be dangerous. (Always wear glasses or goggles when working around steel cable.) Polypropylene rope is also suit-

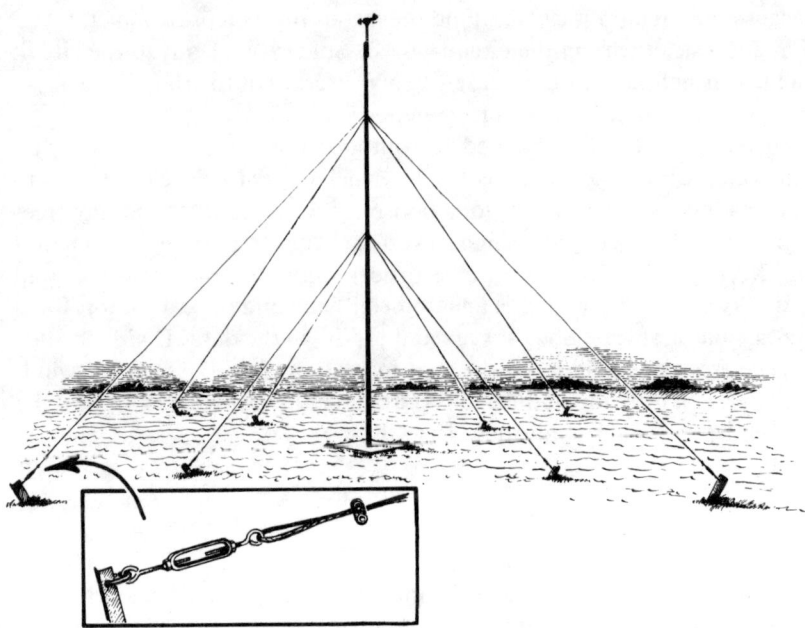

Fig. 4-13. Low-cost anemometer tower with stake detail. (Center for Alternative Resources)

able. Guys of steel wire are attached with cable clamps (Crosby clamps) at each end.

Inexpensive stakes can be cut from one-inch angle iron. After drilling a ¼-inch hole in one end of a two-foot long stake, insert a chain repair link (monkey link) and attach a turnbuckle (see fig. 4-13).

Mount the anemometer support tube to the tower with a pair of U-bolts or hose clamps. The guy brackets can be made from a pair of eyebolt clamps. Each clamp has two eyes for attaching the guy cables. The recorder should be strapped to the tower with nylon cable ties or U-bolts.

Many homeowners try to install the anemometer on the top of their house because trees and buildings may prevent the use of a ground-mounted anemometer tower, or simply because they do not want to clutter up their yard. Unfortunately, turbulence around the building lowers wind speed dramatically. Avoid it if you can. You will never get an accurate reading unless the tower is 1½ times the height of the building.

Let us say you want to install an anemometer on the roof of your house, which is 30 feet above the ground. You will need an anemometer tower 45 feet tall (1½ × 30 = 45) to clear the turbulence. Most of the time, mounting an anemometer on a building is impractical. Where do you attach the guy cables, for example? Will it be necessary to drill holes in the roof for anchors? If that does not stop you, imagine trying to erect a 30-foot mast on a steeply pitched roof. It is an accident waiting to happen.

A farm silo is a better bet if you have one, and if you can extend the anemometer well above the silo cap. I have had mixed results with silos. It depends on the site. From my experience, the silo should be tens of feet taller than any other silo, trees, or buildings nearby for it to be of any value.

Installing an Anemometer

The following sequence was developed for erecting a tower from TV antenna mast (see fig. 4-14). It is a three-step process that takes two people. First, the rigging is laid out and attached. Second, the mast is raised without the anemometer. Third, the anemometer is attached and the tower is raised into place.

Begin by siting the base for the tower. Lay out the stakes by measuring from the base with either a 50-foot tape or a premeasured cord. Drive a stake at 15 feet and one at 25 feet. Run the tape or line at right angles to the first set and drive the next two stakes. Continue until all eight stakes are in place.

Decide in which direction you want to raise the tower. Drive two stakes at the base, keeping in mind the direction across the ground the tower will slide when being raised. Place a restraining board upright against these

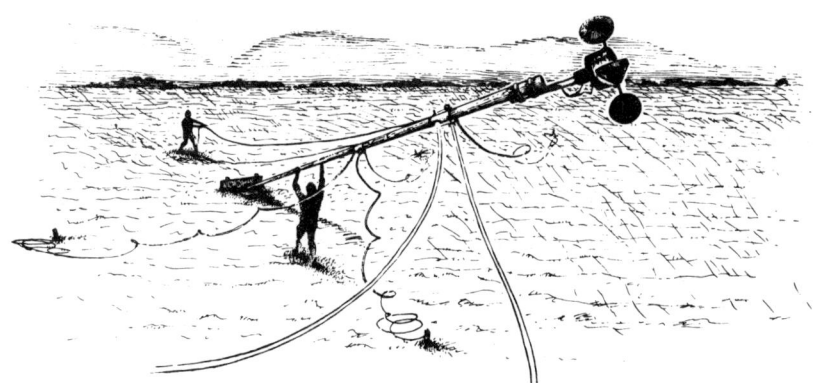

Fig. 4-14. Erecting anemometer. (Center for Alternative Resources)

stakes. Next, lay a second board flat on the ground next to the first. This will act as a pad or platform for the tower.

Lay out the tower sections in the direction you selected. Slide the sections together (if you ever plan to disassemble the tower, lightly grease the mating surfaces), place the bottom end atop the tower pad, and butt the end against the restraining board. Place the assembled mast as near the line of stakes as practical.

Connect the guy cables. One cable clamp at each attachment point is sufficient. You can attach both ends of six cables while the tower is on the ground. The remaining two cables can be attached only to the tower; their opposite ends will remain free until the tower is upright. The uppermost free guy will be used to pull the tower into the upright position.

Check the guy cables, stakes, and mast. You are now ready to raise the tower. Grab the uppermost free guy cable at the free end and begin pulling the tower into the air as someone else lifts it off the ground. They should then continue to push the tower upright by carefully walking the tower towards the vertical, being careful to avoid tripping over the stakes. When the tower is upright, note any guys that need gross adjustments. If the cables are properly cut to length and the stakes laid out accurately, the tower should reach the vertical without toppling in any direction. Reverse the procedure and carefully lower the tower. Adjust the cables as required.

This test assures that all the cables are attached securely. If anything goes wrong at this stage, the anemometer head will not be damaged.

Next, lift the top of the mast off the ground and rest it on a box so the anemometer will not be damaged by supporting the weight of the tower. Attach the anemometer head and connect the signal wire leads. Do not overtighten the terminal nuts (they'll pull out of their plastic base). If you plan to leave the anemometer up for an extended period or if the atmosphere is particularly corrosive (i.e., near the coast or in areas of acid rain) then coat the terminals with silicone rubber.

Run the signal wire down the tower securing it to the mast frequently with electrical tape or nylon cable-ties. This prevents the signal wire from flapping in the wind and pulling on the leads at the anemometer terminals. Connect the signal wire to the recorder and insure that the recorder is picking up the signal from the anemometer (that the counter is registering). Either attach the recorder to the tower or set it out of the way.

Raise the tower as before and attach the free guy cables to their respective stakes. Adjust the guy wires as needed until the tower is vertical. Use the turnbuckles for minor adjustments. Do not overtighten the guy cables or the tower will buckle. Attach flagging to each guy cable so no one will trip over the stakes or tangle with the guys. Remove the restraining board and stakes.

A different procedure is used for erecting a tower of telescoping tubes. Lay out the stakes as before. Attach the guy cables and the anemometer to the tower. (On the Unarco-Rohn masts, cable rings with holes are provided for this purpose.)

While the sections are still nested together, tip the tower into place and attach the first set of guy cables. When secured, lean a ladder against the mast, climb the ladder, and position yourself so you can extend the remaining nested sections. It will take some muscle and it is helpful to have someone give you a hand to keep the cables untangled, hold the ladder, and shout encouragement. Remember that as you are raising each section you must attach the signal cable from the anemometer to the mast. It may seem like three hands are required, but you can do it with two.

The Rohn telescoping towers have a screw clamp that can hold the topmost sections in place while you take a break to strap the signal wire to the mast. When a bay is fully extended, thrust a cotter pin through the holes provided. The bays rest on these pins. Do not depend on the screw clamps. They are not strong enough to support the tubes when the cables are tensioned.

Survey Duration

"Yeah, you got a fine site here," said the dealer as he installed the anemometer George had ordered. "I bet you've got 12 mph."

Two days later the dealer returned. After examining the anemometer he said to George, "Just as I thought, an easy 12 mph average." The dealer then persuaded George to buy his wind machine.

A wet finger in the air on the first visit would have been just as accurate. Maybe the dealer did not know how to measure wind speed, or maybe he was a con artist. It is hard to tell. The site was indeed a good one and the wind turbine (made by a reputable manufacturer) was installed in a workmanlike manner. The site could have had a 12 mph average wind speed, but the dealer or the buyer would not have known that after two days of measurements.

How long is enough? That is another tough one to answer. Average wind speeds can vary as much as 25 percent from year to year. But it is obviously impractical to gather ten years of data from a site before you decide whether or not to install a wind machine.

Battelle suggests gathering one year of data. Even so, your site's average speed will be dependent on how normal the year has been with respect to the long-term average. Was it a typical year, or was it windier? To answer that question, you must examine the wind data from the nearest airport for the same year and compare it with the airport's historical average. To do so, you will need copies of Local Climatological Data for your airport.

Monthly summaries of Local Climatological Data[7] are published by NCC for most airports and weather stations (see fig. 4-15). NCC issues an annual summary at the end of the year (the annual summary also includes the history of the anemometer, where it has been located, and at what height). These summaries not only present the average daily wind speeds (as shown in fig. 4-15), but also the average three-hour speeds.

If you need these summaries to determine if it is a normal year, you might as well try to establish a correlation between your site and the airport as the data is being collected. If you are lucky, you may find that a full year of measurements is not necessary. But four months is a minimum. Anything less is guesswork (see fig. 4-16).

Measurements, ideally, should be made in all seasons with an em-

SEP 1981
ERIE, PENNSYLVANIA
NATIONAL WEATHER SERVICE OFC
ERIE INTERNATIONAL AIRPORT

Local Climatological Data

MONTHLY SUMMARY

LATITUDE 42° 05' N LONGITUDE 80° 11' W ELEVATION (GROUND) 731 FT. STANDARD TIME USED: EASTERN WBAN #14860

SEP 1981 ERIE, PENNSYLVANIA

	TEMPERATURE °F					DEGREE DAYS BASE 65°		WEATHER TYPES ON DATES OF OCCURRENCE 1 FOG 2 HEAVY FOG 3 THUNDERSTORM 4 ICE PELLETS 5 HAIL 6 GLAZE 7 DUSTSTORM 8 SMOKE, HAZE 9 BLOWING SNOW	SNOW, ICE PELLETS OR ICE ON GROUND AT 07AM IN.	PRECIPITATION		AVG. STATION PRESSURE IN. ELEV. 737 FEET M.S.L.	WIND					SUNSHINE		SKY COVER TENTHS		
DATE	MAXIMUM	MINIMUM	AVERAGE	DEPARTURE FROM NORMAL	AVERAGE DEW POINT	HEATING (SEASON BEGINS WITH JULY)	COOLING (SEASON BEGINS WITH JAN.)			WATER EQUIVALENT IN	SNOW, ICE PELLETS IN.		RESULTANT DIR.	RESULTANT SPEED M.P.H.	AVERAGE SPEED M.P.H.	FASTEST MILE SPEED M.P.H.	FASTEST MILE DIRECTION	MINUTES	PERCENT OF POSSIBLE	SUNRISE TO SUNSET	MIDNIGHT TO MIDNIGHT	DATE
1	2	3	4	5	6	7A	7B	8	9	10	11	12	13	14	15	16	17	18	19	20	21	22
1	85*	69	77*	12	64	0	12	3 8	0	.20	0	29.10	18	10.8	11.8	17	24			8	9	1
2	77	69	73	8	68	0	8	1 3 8	0	.34	0	29.13	20	5.9	7.2	15	17			10	10	2
3	76	69	73	8	67	0	8	1 8	0	.30	0	29.21	16	8.8	8.9	13	14			10	10	3
4	77	65	71	7	64	0	6	8	0	.02	0	29.16	20	8.3	10.1	16	23			9	10	4
5	74	65	70	6	64	0	5	1	0	.38	0	29.24	24	6.7	8.6	12	29			10	10	5
6	71	63	67	3	64	0	2	1	0	.35	0	29.32	24	.8	1.3	6	28			10	9	6
7	75	60	68	4	61	0	3	8	0	T	0	29.21	24	2.4	4.3	8	31			5	6	7
8	70	60	65	2	55	0	0	1	0	.41	0	28.99	27	6.7	10.6	16	29			9	8	8
9	64	50	57	-6	45	8	0		0	0	0	29.14	29	8.6	11.2	16	31			6	5	9
10	75	52	64	1	51	1	0		0	0	0	29.09	21	12.2	12.7	21	20			8	8	10
11	78	64	71	8	62	0	6	8	0	0	0	29.17	22	6.4	7.2	10	21			9	9	11
12	75	55	65	2	60	0	0	8	0	0	0	29.22	24	6.7	7.6	13	30			5	4	12
13	79	55	67	5	59	0	2	8	0	0	0	29.18	21	4.7	6.8	13	21			1	2	13
14	77	67	72	10	64	0	7	1 3 8	0	.19	0	29.07	24	6.1	9.5	15	22			9	9	14
15	69	55	62	0	49	3	0		0	0	0	29.17	02	5.2	6.0	13	07			9	9	15
16	67	53	60	-1	47	5	0		0	.08	0	29.21	07	7.7	8.2	15	07			9	9	16
17	65	50	58	-3	48	7	0		0	T	0	29.27	19	6.7	8.2	14	19			6	8	17
18	63	44	54	-7	44	11	0	1 8	0	T	0	29.32	19	6.3	6.6	12	18			7	5	18
19	67	43	55	-6	46	10	0	1 3	0	.32	0	29.02	21	9.5	12.1	17	20			9	8	19
20	60	50	55	-5	42	10	0		0	0	0	29.08	28	8.9	9.9	15	29			8	8	20
21	60	46	53	-7	44	12	0	1	0	.28	0	29.22	13	3.4	8.2	17	07			9	9	21
22	55	51	53	-7	45	12	0	1	0	.45	0	29.25	02	10.7	11.4	20	01			10	9	22
23	58	49	54	-5	39	11	0		0	0	0	29.48	36	12.4	13.1	17	02			7	7	23
24	62	48	55	-4	41	10	0		0	0	0	29.51	31	7.0	10.1	16	27			2	3	24
25	66	47	57	-2	47	8	0		0	T	0	29.49	18	5.4	5.8	8	23			10	9	25
26	81	57	69	11	55	0	4	8	0	T	0	29.27	18	12.6	13.1	20	19			9	9	26
27	74	59	67	9	52	0	2	8	0	.15	0	29.13	23	13.0	14.1	28	28			5	6	27
28	61	46	54	-4	37	11	0		0	.01	0	29.35	30	16.1	16.4	28	28			9	8	28
29	56	42	49	-9	40	16	0		0	0	0	29.42	28	1.4	4.6	9	24			10	7	29
30	54	39*	47*	-10	41	10	0	3	0	.78	0	29.37	11	2.4	2.5	15	15			10	8	30
	SUM	SUM				TOTAL	TOTAL			TOTAL	TOTAL	FOR THE MONTH:						TOTAL	% FOR MONTH	SUM	SUM	
	2071	1642				153	65	NUMBER OF DAYS		4.26	0	29.23	23	3.4	8.9	28	28			238	231	
	AVG.	AVG.	AVG.	DEP.	AVG.	DEP.	DEP.	PRECIPITATION		DEP.						DATE: 28+		POSSIBLE		AVG.	AVG.	
	69.0	54.7	61.9	0.5	52	12	32	≥ .01 INCH 15		0.70										7.9	7.7	

NUMBER OF DAYS				SEASON TO DATE				GREATEST IN 24 HOURS AND DATES			GREATEST DEPTH ON GROUND OF SNOW, ICE PELLETS OR ICE AND DATE	
MAXIMUM TEMP.		MINIMUM TEMP.		TOTAL	TOTAL	SNOW, ICE PELLETS ≥ 1.0 INCH	0	PRECIPITATION		SNOW, ICE PELLETS		
≥ 90°	≤ 32°	≤ 32°	≤ 0°	162	573	THUNDERSTORMS	5	.78	30	0	0	
0	0	0	0	DEP.	DEP.	HEAVY FOG	0					
				-46	200	CLEAR 2	PARTLY CLOUDY 7	CLOUDY 21				

Fig. 4-15. Local Climatological Data, Monthly Summary. Data such as this is useful for comparing wind speeds from an airport or other long-term recording station with those recorded for the same time period from a site survey. (NCC)

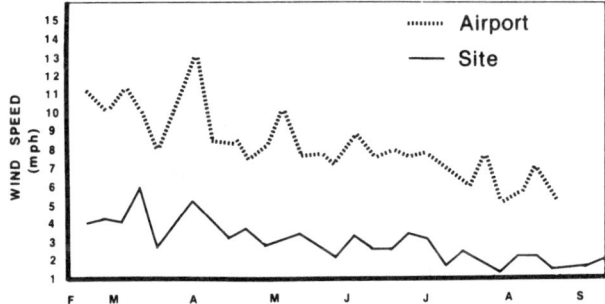

Fig. 4-16. Comparison of wind speeds from an airport presented on a summary of Local Climatological Data, with those measured at a proposed site for a wind machine. Wind speeds at the two sites correlated well; the site had a consistently lower average weekly speed than the airport. (Center for Alternative Resources)

phasis on the winter and spring months when winds are strongest. The windiest months for much of the country are December, January, February, March, and April. (This isn't true everywhere.) The reason for this is that the correlation may change with average speed. For example, during the summer months at one mountaintop site in Pennsylvania the correlation was good, but the site was only slightly windier than the Pittsburgh Airport. During the spring, however, the correlation was less clear but the site had consistently higher average speeds. We are interested in the relationship between the site and the airport when wind speeds at the site are highest. It is best to start your measurements during the fall and continue through the windy months of winter and spring (or whenever the windy season begins).

Data Analysis

Making sense out of the data from a site survey is as much alchemy as science. Much is left to the judgment of the observer: whether the data is representative of the site and not surrounding obstructions, whether the year is normal, and whether there can be any direct relationship between the site and a nearby station with long-term records. The simplest method, and that used most, is comparing your site's average speed to that of a nearby airport for the same time period. You then:
1. Calculate the ratio between your site's average speed and the airport's, and
2. Adjust the airport's historical average by this ratio.

114 • Wind Energy

Step one established whether your site is windier or less windy than the airport. The second step normalizes the results for a typical year.

This approach assumes that a correlation exists between your site and the airport. There may not be one, particularly in rough or mountainous terrain. Try to use an airport in terrain similar to yours and with a similar exposure to the wind. The nearest airport may not always be the best choice.

You will find that, as a rule, hourly or daily wind speeds are too erratic to establish a correlation between the two sites. Average weekly speeds are more stable. They tend to smooth out the passage of weather systems and local diurnal variations.

Another factor may be the way NWS reports wind speed. They take a one minute reading every hour, often using a dial wind meter. The observer watches the swinging needle and then picks a number. Three hours of such observations are then combined. Eight, three-hour averages are tallied to give an average daily wind speed. Wind-run odometers (the recorder you'll most likely be using), on the other hand, count every mile of wind that passes. There is less chance of error in the resulting average speed.

Let us assume you are installing a wind-run odometer. Record the data from the odometer at least weekly, more often if you like (see fig. 4-17). Note the date, time, and the count on the register (see table 4-11). The key to reliability is frequent inspection. Site visits can reveal a failure of the anemometer head or the recorder. The more frequently you read the odometer, the less data you are likely to lose in the event of a malfunction or vandalism.

Order the Local Climatological Data summaries for the airport you are using as a comparison site. (It takes four to six weeks for NCC to

TABLE 4-11
Sample Wind Speed Data and Analysis

Date	Time	Reading	Increment	Wind-run Miles	Elapsed Time Hours	Site	Average Speed (mph) Comparison Site #1	#2
6/29	1930	1428460	112110	1869	196	9.5	8.1	
6/21	1545	1316350	81640	1361	169	8.0	7.6	
6/14	1500	1234710	90943	1516	167	9.0	6.5	
6/07	1600	1143767	100951	1683	165	10.2	6.5	
5/31	2100	1042816	87820	1464	171	8.6	5.8	
5/25	1745	954996						
							9.1	6.9

Fig. 4-17. Measuring wind speed. Reading a wind-run odometer merely entails pushing a button and writing down the numbers flashed onto the screen. (Center for Alternative Resources)

process the summaries, so they will not be sent immediately.) Compare the average weekly wind speeds between your site and the airport (see the last two columns of table 4-11). The comparison site in this example is the Greater Pittsburgh Airport. The comparison will not always be consistent. You are looking for general trends. Find the ratio between your site and that of the airport. In our one-month sample

$$\frac{\text{site at 9.1 mph}}{\text{airport at 6.9 mph}} = 1.32.$$

Now locate the historical average for the airport in appendix C, in this case 10.3 mph.[8] Multiply it by this ratio to find the corrected long-term average.

$$1.32 \times 10.3 \text{ mph} = 13.6 \text{ mph}$$

Another method is to use linear regression analysis. It not only gives a measure of the ratio between the two sites, but also tests the degree of correlation between them, as well as projecting the site average speed. Mathematically, it is drawing the best fitting line between two sets of data. Pocket calculators with engineering or statistical functions make the job a cinch. Consult the calculator handbook for details.

You are not finished yet. Remember that average speed is only part of the picture. We need to know the average power in the wind before we can estimate how much energy a wind machine will produce. There are three ways to derive the average power. One is to use an odometer such as Helion's which records V^3. We can then calculate the energy pattern factor for the specific wind data at your site. Or we can use historical wind data and assume that the speed distribution at your site is the same as that at the airport.

The easiest historical data to use is the average speed and power for the airport as given in appendix C (all the data has been adjusted to a height of 10 meters). Calculate the average power using the average speed, then find the ratio between it and the power density given in the table to derive the EPF. For the Pittsburgh Airport, the average power calculated by Battelle is 107 W/m².

$$P/A = 0.05472 \ (S^3)$$
$$P/A = 0.05472 \ (10.3 \text{ mph}^3) = 59.8 \text{ W/m}^2$$
$$\frac{107 \text{ W/m}^2}{59.8 \text{ W/m}^2} = 1.79$$

You now use this EPF to calculate the average power with the corrected wind speed for your site. We found earlier that the site had a 13 to 14 mph average speed (13.6 mph).

$$P/A = 0.05472 \ (S^3) \ (EPF)$$
13 mph; $P/A = 0.05472 \ (13^3) \ (1.79) = 215 \ W/m^2$
14 mph; $P/A = 0.05472 \ (14^3) \ (1.79) = 269 \ W/m^2$

From this analysis we could say that the site in table 4-11 falls within Wind Power Class 4 having an average power from 200 W/m² to 250 W/m². This seems reasonable, even though we used only a one-month sample, because the wind power maps for this mountaintop site indicate a Class 4 potential.

Battelle used the speed distribution for calculating average power, not the average speed. If you had the speed distribution as shown in Figure 4-6, or the speed summary shown in figure 4-7, you could calculate average power directly. You would simply adjust each speed class by the ratio between your site speed and the airport's (1.32 in this example).

Another Option

After all this you may have had enough. If you cannot install an anemometer (too many trees for example) or you do not want to (too much money, too much trouble) there is one avenue left; install a wind system. Yes, go ahead and install a small wind machine (with or without an anemometer). Sounds crazy doesn't it? That is what I thought the first time I heard it. Agreed, this is not a low-cost way to test the wind, but it works. What you get is an operating wind system. You gain hands-on experience and you learn exactly what you want to know: how well it will perform—how much energy it will generate—at your site.

For this method to be worthwhile, though, you must be interested in a larger wind turbine, and the product you are considering must be capable of being upgraded. Let us say you found that a 7-meter wind turbine would meet your needs best (it would fulfill your energy and cost-effectiveness goals), but you are not certain you have the wind to make it pay and you cannot install an anemometer. You decide to minimize the risk of installing a wind system by spending only part of the money you set aside and install the wind system in stages. Phase one is a small machine. If it is successful, you go to phase two: the 7-meter wind turbine. If it is unsuccessful, that is, it does not produce as much energy as you hoped it would, you are stuck with a wind machine, but at least you have not invested all your money.

There are a couple of ways to do this. You could install a foundation and anchors suitable for the 7-meter machine and erect the small turbine on a light-duty tower. Or you could install the small turbine on a heavy-duty foundation and tower. In the first case, the tower and turbine would be traded in for a heavy-duty tower and the 7-meter machine. In the second scenario you would trade in only the small turbine for the bigger one. This will not work with all wind systems, as tower and turbines may not be interchangeable.

5

Estimating Output and Payback

Once we have defined the site where we want to put a wind machine and determined how much wind is available to it we can proceed to the next step: estimating the amount of energy that can be produced. With an estimate of the annual energy output in hand we can examine the economics of various sizes and brands of wind machines to find the one that offers the most for our money.

OUTPUT – HOW MUCH ENERGY CAN YOU EXPECT

There are three methods that we will use. The first is a technique using the swept area of the wind turbine. This is what I call the back-of-the-envelope approach. With it you can quickly evaluate the potential output of any wind machine by finding the average speed or power and simply sizing up its rotor. If you do it often enough the technique will become so familiar you will be able to do it in your head. The second method is more involved and requires a speed distribution for your site and a power curve for each wind turbine you would like to evaluate. The third approach uses the published annual energy output.

As in the preceding sections, formulas are presented to describe precisely what we are doing (so you know where the numbers come from). They also aid in the explanation of important concepts. As before, where formulas are used a table will be included summarizing the results of calculations for a standard set of conditions.

Swept Area Method

Our first step is to find the power in the wind—the power density in W/m². Once we have determined power density (by whatever means necessary) we can easily estimate the potential power output from a wind machine. All we need to do is calculate the area swept by the turbine's rotor.

Think about it. What captures the wind in a wind machine? Is it the tower, the transmission, the generator? No, of course not. It is the spinning rotor. Yet this is a concept difficult for most to understand and that is why I presented it this way with the tower first. The tower is important as we have learned, and so is the transmission and the generator. But they do not have anything to do with capturing the wind. Invariably people look at the size of the generator first. Yet the generator tells you very little about the size of the wind machine. It is the size of the rotor that tells the story.

As an example, say we have a wind turbine whose rotor sweeps 100 square meters and intercepts winds with 100 W/m² of annual average power. From an earlier equation:

$$\text{Power} = P/A \times A$$

where P/A is power density in W/m² and A is the area swept by the rotor in square meters.

$$\text{Power} = 100 \text{ W/m}^2 \times 100 \text{ m}^2$$
$$= 10{,}000 \text{ watts}$$

The wind machine in our example intercepts 10,000 watts of power or 10 kilowatts (1,000 watts equals 1 kilowatt).

Now you can see why the metric system is used to describe rotor size. It is not because I have a special affinity for meters. Using meters, though, makes the job easier when wind power is in units of W/m², and it usually is.

Power, unfortunately, is not what we are seeking. Energy is. When you pay your utility bill you are not paying for the power you used, you are paying for energy.[1] The amount of energy consumed is the product of power and time—how long, the length of time, the power was used. In the case of a wind machine the energy it intercepts is a function of average power and how long it is available.

We are using an average annual wind power of 100 W/m². This means that on average throughout the year our wind turbine intercepts 10 kW of power. There are 8,760 hours in a year. Consequently, there are 87,600 kilowatt-hours (kWh) of energy annually that pass through the rotor.

$$\begin{aligned} \text{Energy} &= \text{Power} \times \text{Time} \\ &= 10 \text{ kW} \times 8,760 \text{ hrs/yr} \\ &= 87,600 \text{ kWh/yr} \end{aligned}$$

That is a lot of energy. Alas, we cannot capture all of it. If we did, the wind would come to a halt at the rotor and nothing further would happen. The maximum that we can capture at the rotor, the theoretical limit, was derived by a German scientist, A. Betz. The "Betz limit" as it is called is 59.3 percent of the energy available to the rotor. Betz reached this conclusion using commonly held theories of momentum. He said in essence that a portion of the wind must keep moving through the rotor, and that there is a balance between what can be extracted by the rotor and what must flow through the rotor. In practice the conversion efficiencies of wind turbine rotors are much less than the Betz limit. Optimally designed rotors reach levels slightly above 40 percent. Usable energy is even less.

Energy is lost in transmission, generators, and power conditioning (the equipment necessary to convert the energy into a form we can use). There are also losses due to rapid changes in wind speed and direction that are not accounted for in our simple formulas. Well-designed drive trains operate consistently above 90 percent efficiency. The efficiency of generators, on the other hand, varies significantly depending on how they are loaded. When running at their rated output—where they were designed to operate best—generator efficiency can also be above 90 percent. But wind turbines drive their generators at the rated output infrequently. Much of the time the generator is partially loaded and its efficiency suffers as a result. Power conditioning for an interconnected wind system usually is not a source of important losses. Unlike an anemometer, which measures gusts, a wind turbine does not respond to all gusts because of its inertia. The energy available in a gust as registered by an anemometer may not be used by the wind turbine. It may not "see" the gust, particularly if the speed of the gust goes above the turbine's operating limits. Yawing or changing the direction of the wind turbine as the wind changes direction causes a similar problem for conventional wind machines. Cup anemometers capture wind from all directions. But a wind turbine takes time to change its position, thus it misses a portion of the wind recorded by the anemometer. This is not a problem with vertical axis wind turbines because, like the anemometer, they are omnidirectional. When you put all this together with a well-designed rotor that delivers 35 to 40 percent of the energy in the wind

Rotor		Transmission		Generator		Power Conditioning Yawing and Gusts		Overall Conversion Efficiency
35% to 40%	×	90%	×	90%	×	90%	=	26% to 29%

This is what you can get out of it and can put to use. Most wind turbines perform at the lower end of this range. They can reach higher levels of conversion efficiency at certain instantaneous wind speeds (see table 5-1) but the overall conversion efficiency (see table 5-2) on a yearly average approximates 25 percent below an average speed of 12 mph and less at an average speed of 12 mph or above.[2]

TABLE 5-1
Instantaneous Conversion Efficiency

Rotor Diameter (m)	Wind Speed (mph)							
	16	18	20	22	24	26	28	30
3	27	28	29	29	25	23	20	16
4	27	30	25	22	19	17	14	11
6	24	28	25	20	19	16	13	11
10	29	30	26	20	15	12	10	8
10	23	24	29	28	28	28	27	24

TABLE 5-2
Overall Conversion Efficiency

Rotor Diameter (m)	Average Wind Speed at Hub Height[a] (mph)		
	10	12	14
3	22-33	23-29	20-22
4	25	22	19
6	27	22	18
10	24	22	18
10	20	22	22

[a]Rayleigh distribution, data from manufacturers' published estimates of annual energy output corrected for height.

Estimating Output and Payback • 123

So let us assume our hypothetical wind turbine converts 25 percent of the energy in the wind to usable electricity and see what we get.

$$25\% \times 87{,}600 \text{ kWh/yr} = 21{,}900 \text{ kWh/yr}$$

In our example we assumed that a wind turbine's rotor swept an area of 100 m². When we look at a wind turbine or at the product literature describing it, the swept area is not always apparent. But what is obvious, or clearly stated in the literature, is the rotor's size. For conventional wind machines the rotor's diameter is given. For vertical axis wind turbines (those like the Darrieus or eggbeater that rotate about a vertical axis) the height of the blades and the rotor's diameter are both provided.

Because the rotor on a conventional wind turbine (see fig. 5-1) sweeps a disc, use the formula for the area of a circle to find the swept area. Use the formula for the area of a rectangle when calculating the swept area

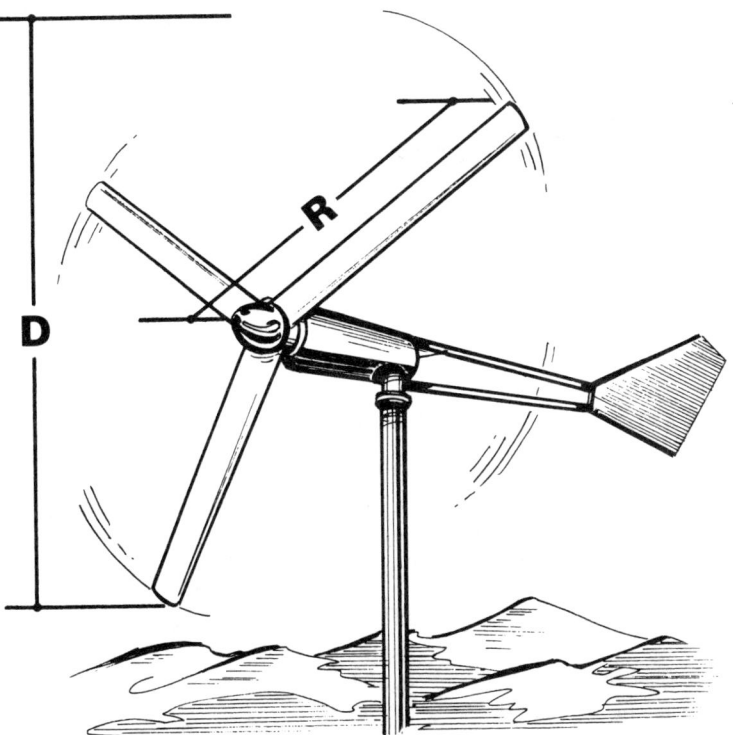

Fig. 5-1. Swept area, conventional rotor.

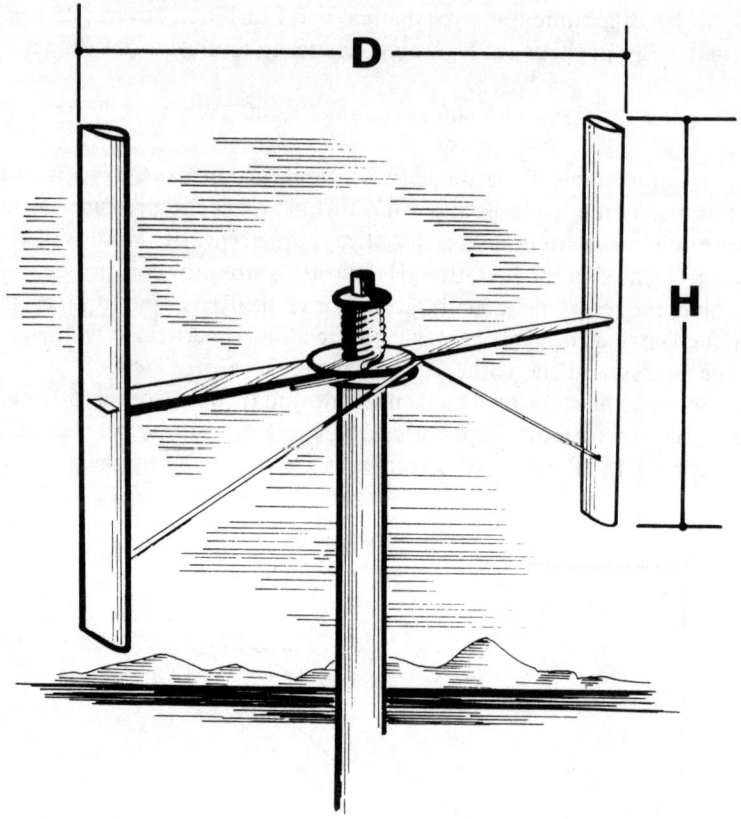

Fig. 5-2. Swept area, H-rotor.

of an H-rotor (see fig. 5-2). For Darrieus rotors (see fig. 5-3), because of their curved blades, use the formula approximating the area of an ellipse.

Type of Wind Turbine	Formula for Swept Area
Conventional rotor	$A = \pi \times R^2$
H-rotor	$A = D \times H$
Darrieus rotor	$A = 0.65\ D \times H$

where R is the radius of the rotor (½ the diameter), D is the diameter, and H is the height of the blades.

Using the swept area and an assumed conversion efficiency of 25 percent (this seems reasonable in light of experience) we can calculate the

Estimating Output and Payback • 125

annual energy output at the height where the wind data was recorded. Usually this will be around 30 feet. To estimate what we can expect at greater heights where speed and power are correspondingly greater we need to incorporate the height of the tower into our calculation.

Assume that our hypothetical wind turbine will be installed on an 80-foot tower and that the wind power was measured at 30 feet. Our site has a few hedge rows so a roughness exponent of $\frac{1}{5}$ is justified. In chapter 4 for the same conditions we found that power increased 80 percent. Therefore, our wind turbine will produce

$$21,900 \text{ kWh/yr} \times 1.80 = 39,400 \text{ kWh/yr}.$$

This is enough for a small, all-electric home.

To summarize, here are the steps necessary to estimate the annual energy output of any wind turbine.

1. Find the average annual power in the wind.

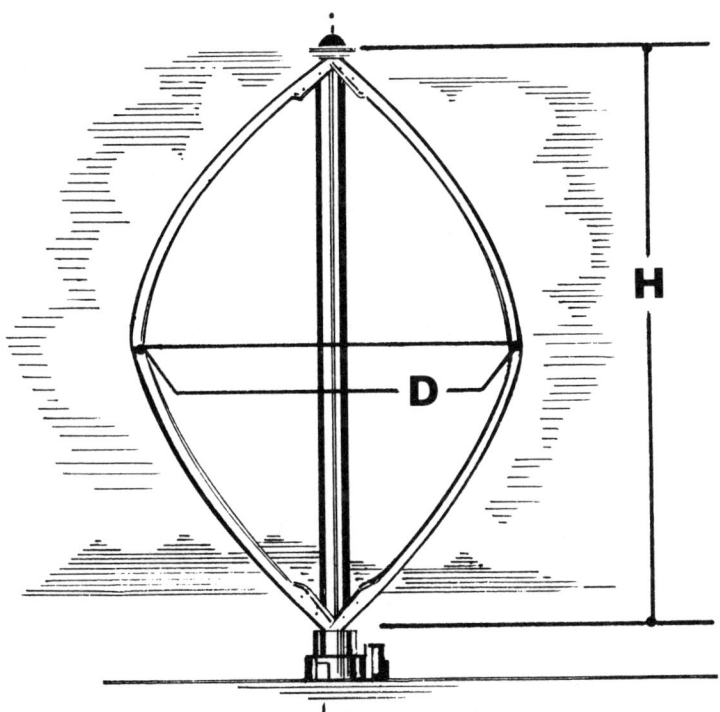

Fig. 5-3. Swept area, Darrieus rotor.

126 • *Wind Energy*

   a. You can do this by looking up the value for your area on the wind power maps in appendix C.
   b. You can calculate the average power from the speed distribution.
   c. You can calculate the average power from the average speed and an assumed energy pattern factor.
2. Adjust the power in the wind by the appropriate factor for the height of the tower (table 4-7 in chapter 4), or use the formula for the increase in power with height.
3. Calculate the area swept by the wind turbine you are considering.
4. Assume an overall conversion efficiency. (Do not use any value higher than 25 percent unless there is proof.)
5. Find the product of the average annual power, the height adjustment, swept area, conversion efficiency, and the hours in a year.

$$AEO = P \times \text{Height Adjustment} \times A \times \% \text{ efficiency} \times 8{,}760 \text{ hrs/yr}$$

You can get the feel of this technique by working the following examples.

Example 1: You have identified the area where you live on the maps in appendix C and found a Class 3 wind power (150–200 W/m^2). You want to know what is the maximum energy output you can expect from a conventional wind turbine 3 meters in diameter that is 25 percent efficient and will be installed on an 80-foot tower. The roughness exponent is $\frac{1}{5}$.

Solution: Power Density = 200 W/m^2
Height Adjustment = 1.80 (from table 4-7)
Swept Area = π (1.5^2) = 7 m^2
AEO = 200 W/m^2 × 1.80 × 7 m^2 × 25% × 8,760 hrs/yr
     = 5,500 kWh/yr

This power density is roughly that you would find in an area with a 12 mph average speed. The 3-meter turbine is the size a homeowner without electric heat would seek to make a dent in his utility bill.

Let us try another wind turbine at a site with a 12 mph average speed.

Example 2: You know the average wind speed is 12 mph and would like to determine how much energy a wind turbine would produce that uses an H-shaped rotor 10 meters in diameter with straight blades 5 meters long (see fig. 5-1). It will be installed on a 60-foot tower in the Texas Panhandle where the 1/7 power law fits conditions best.

Solution: Power Density = 0.05472 (S³) (EPF)
= 0.0542 (12³) (2)
= 190 W/m²
Height Adjustment = 1.34 (from table 4-7)
Swept Area = $D \times A = 10 \times 5$
= 50 m²
AEO = 190 W/m² × 1.34 × 50 m² × 25% × 8,760 hrs/yr
= 27,900 kWh/yr

If such a wind machine were available, it could be used by homeowners with electric heat to reduce their bill substantially.

Example 3: You own a dairy farm along the shores of Lake Erie and the local airport has an average power of 200 W/m². You are considering a larger Darrieus turbine with a rotor 123 feet tall and 82 feet wide. The center of the rotor is 80 feet above the ground and you feel the increase in speed with height would correspond to a roughness exponent of 1/5. How much energy could the turbine generate?

Solution: Height Adjustment = 1.80 (from table 4-7)
Swept Area = 123 ft × (1 ft/3.28 m) × 82 ft × (1 ft/3.28 m) × 0.65
= 610 m²
AEO = 200 W/m² × 1.80 × 610² × 25% × 8,760 hrs/yr
= 481,000 kWh/yr

That is enough electricity to milk quite a few cows.

If you would like to avoid the aggravation of making repetitive calculations of annual energy output for conventional wind machines use table 5-3. The output of wind machines from 3 meters to 15 meters in diameter has been estimated for average wind power from 50 W/m² to 1,000 W/m². I've assumed that the overall conversion efficiency of each wind turbine in the table is 25 percent for Wind Power Class 1 through Class 5, 20 percent for Class 6, and 10 percent for Class 7.

The values shown are for the output at hub-height. You must take into account an increase in power that accompanies an increase in height. There are two ways to do this: adjust average power to the height where

TABLE 5-3
Annual Energy Output[a]
(Thousand kWh)

Power Class	Power Density	Avg.[b] Speed	(m)[c] (ft.)	Rotor Diameter								
				3 / 10	4 / 13	5 / 16	6 / 20	7 / 23	10 / 33	12 / 40	13 / 44	15 / 50
1	50	8		0.8	1.4	2.2	3.1	4.2	8.6	12	15	19
2	100	9.8		1.6	2.8	4.3	6.2	8.4	17	25	29	39
3	150	11.5		2.3	4.1	6.5	9.3	13	26	37	44	58
4	200	12.5		3.1	5.5	8.6	12	17	34	50	58	77
5	250	13.4		3.9	6.9	11	15	21	43	62	73	97
6	300	14.3		4.6	8.3	13	19	25	52	74	87	120
7	400	15.7		5.0	8.8	14	20	27	55	79	93	120
	1000	21.1		6.2	11	17	24	34	69	99	120	150

[a] Assumed conversion efficiency of 25 percent in Class 1 through Class 5, 20 percent in Class 6, and 10 percent in Class 7.
[b] Derived from power density and Rayleigh distribution.
[c] In meters and approximate feet equivalent.

the wind turbine will be operating, or adjust the annual energy output in the same way. It does not matter which you do, so long as it is done.

Here is an example of how to use table 5-3. Say we want to know the output from a 3-meter turbine on an 80-foot tower in an area with an average power of 100 W/m² (an average speed of around 9 mph to 10 mph). We will use a roughness exponent of ⅕ to find the increase in power on an 80-foot tower from table 4-7 in chapter 4 (1.80).

In the left-hand column find the line for 100 W/m². Locate the column for a wind machine with a rotor 3 meters in diameter. Find the value where the line and column intersect (1,600 kWh/yr). Adjust this value for the power on an 80-foot tower.

$$AEO = 1{,}600 \text{ kWh/yr} \times 1.80 = 2{,}900 \text{ kWh/yr}$$

When you examine the table the concepts expounded on earlier should become clearer. If you double the rotor diameter from 3 meters to 6 meters the amount of energy output increases four times.

Power Curve Method

Where you have access to the speed distribution or speed summary for your site or at least for the nearest airport you can use the turbine manufacturer's curve of power output to calculate annual energy production. You match the speed distribution to the power curve and determine how many hours per year the wind generator operates at different power levels.

The power curve in figure 5-4 is for the Bergey 1000, a wind turbine whose rotor is just shy of 3 meters in diameter. We will work through an example that we used previously using this wind machine and the speed distribution for the Greater Pittsburgh Airport. First, here are some points of reference on the power curve.

Start-up Speed. The wind speed at which the rotor first begins to turn.

Cut-in Speed. The wind speed at which the generator begins to produce power. In this example, the start-up and cut-in speeds are the same: 9 mph. The start-up and cut-in speeds of another popular brand, Enertech, are also the same. At a wind speed of 10 mph the brake locking the Enertech rotor in place is released, and the rotor is motored up to run-

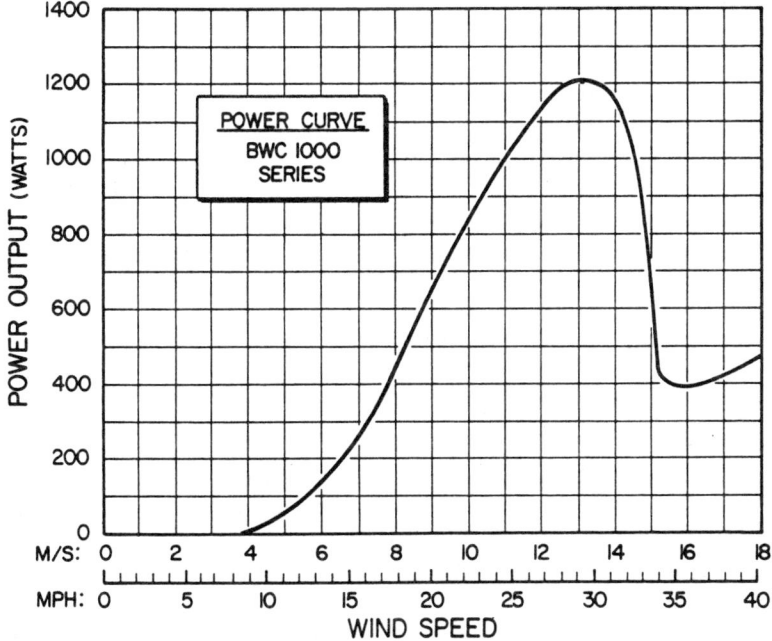

Fig. 5-4. Power curve for the Bergey 1000, an upwind turbine using a rotor approximately 3 meters (9 feet) in diameter. (Courtesy Bergey Windpower Co.)

ning speed. The Bergey on the other hand has a free-spinning rotor. On other brands, the rotor may also be free-wheeling and begin to spin at, say, 8 mph but because they use a generator that must be excited the rotor does not produce power until the generator's magnetic field is energized at 10 mph.

Rated Speed and Power. The wind speed at which the generator produces the advertised power. For the Bergey 1000 the generator produces 1,000 watts at 11 meters per second or approximately 25 mph. Though frequently used as a reference to the size of the wind machine, rated speed and power have little utility. The Bergey 1000, like most other wind machines, will produce more than its rated output.

Maximum Power. Simply the maximum power the generator is capable of producing (1,200 watts in this example).

Cut-out Speed. The wind speed at which the generator stops producing power. This is accomplished by applying a brake or other mechanism for physically stopping the rotor. The Bergey 1000 does not use a brake and continues to produce power at all wind speeds.

Furling Speed. The wind speed at which a wind machine using a tail vane begins to fold or furl towards the tail. The Bergey wind turbine does not use a brake to lock the rotor in place to protect itself in high winds. Instead, the wind machine folds towards the tail at a wind speed of 30 mph to decrease the area of the rotor intercepting the wind.

Before we use the wind speed distribution we need to adjust it for the height where the wind machine will operate. As in previous examples let us use an 80-foot tower and a roughness exponent of $\frac{1}{5}$. Consequently, speed increases 1.22 times (from table 4-7). Adjust the midpoint of each speed class (10 mph in the 8–12 mph speed class) by this value.

$$(10 \text{ mph} \times 1.22 = 12.2 \text{ mph})$$

To estimate the annual energy output take the adjusted midpoint of each speed class and from the power curve determine how much power the wind turbine should generate at that speed. Multiply this power by the number of hours the winds occur within this speed range. Then sum the energy production for all speed bins. At 12.2 mph, for example, the Bergey 1000 will generate 100 watts or 0.1 kW. Winds occur within this range 3,000 hours per year. This speed class will contribute 300 kWh per year

(0.1 × 3,000 hrs = 300 kWh)

to the total production of 1,870 kWh (see table 5-4). Output estimates from the power curves for the Enertech 1800 (a 4-meter turbine) and the Enertech 21/5 (a 6-meter turbine) are included for comparison.

Flip back to table 5-1 and note that overall conversion efficiency drops off at higher average speeds. Because winds above the cut-out or furling speed occur a greater percentage of the time, the generator is running less often. The same effect is produced from increased speed due to a taller tower. Though the total amount of energy is greater the turbine is less efficient at extracting it. This can be seen in the last columns of table 5-4 for the 3- and 4-meter turbines. Some wind machines are designed to continue producing power at greater wind speeds than others. The Enertech 21/5, for example, the 6-meter turbine in table 5-4, cuts out at 45 mph. It is capable of generating 3.5 kW of power in the 32–38 mph speed class (adjusted for height of 43 mph). The 3- and 4-meter turbines in contrast are not generating any power in this speed range.

TABLE 5-4
Average Annual Energy Output[a]

	8–12	13–18	19–24	25–31	32–38	39+
Midpoint of speed class (mph)	10.1	15.5	20.5	28.0	35.0	—
Percentage occurrence	34.4	22.3	4.1	0.7	0.1	—
Hours of occurrence	3000	1950	360	60	10	—
Midpoint adjusted for height[b]	12.2	18.9	25.0	34.2	42.7	—
Wind Turbine 3 Meters in Diameter, 1 kW Rated						
Power output (kW)	0.1	0.6	1.0	0.7	0.0	
Energy output (kWh)	300	1170	360	40	0.0	
Annual Energy Output = 1,870 kWh/year						
Wind Turbine 4 Meters in Diameter, 1.8 kW Rated						
Power output (kW)	0.4	1.0	1.8	1.6	0.0	
Energy output (kW)	1200	1950	650	100	0.0	
Annual Energy Output = 3,900 kWh/year						
Wind Turbine 6 Meters in Diameter, 4 kW Rated						
Power output (kW)	0.8	2.6	4.0	4.0	3.5	
Energy output (kWh)	2400	5070	1440	240	40	
Annual Energy Output = 9,190 kWh/year						

[a]Speed distribution for Greater Pittsburgh Airport
[b]80-foot tower, $a = 1/5$

132 • Wind Energy

Power curves are usually presented as smooth lines. In the real world, however, the power output for a particular machine is a range of values at each wind speed depending on whether the rotor was coasting from a previous gust at the time the measurement was made (in which case power would be greater than average) or was coming up to speed and the anemometer registered the gust but the rotor did not (power would be less than the norm). Power curves, like many other aspects of wind energy, are just approximations of what happens in a complex environment. They should not be taken literally.

Performance Ratings

Practice in this country has been to describe a wind machine's size by referring to its generator capacity in kW at some rated wind speed. This power "rating" is then used extensively in product promotion. For a moment, though, let us take an excursion. Put yourself in Denmark — Roskilde to be specific — at the Danish Test Plant for Small Windmills.

"Helge, that's a beaut, what size is it?"
"10 meters."
"What? No, I meant how big is it, my Danish isn't too good."
"It's a 10-meter, but there's a bigger 12-meter down the road. Would you like to see it?"
"We're not going another step until you tell me how big that is!"
"You Americans are so demanding."
"All right, one more time, how big a generator does it have?"
"Which one?" Helge asked quizzically.
"Boy, what a case of jet lag," the American said to himself.
"I want to know how big the generator is. You know, the thing that generates the electricity, the guts of the machine."
Helge's patience was beginning to wear thin. "It has two generators, the largest of which is 30 kW, but it has a 10-meter rotor driving it."
"Phew, I thought I'd never get it out of you," said the exasperated Yankee.

Such an exchange has probably taken place more than once. We find ourselves in this predicament — identifying wind machines by their generator size — because many of the early pioneers in wind development came from the electric utility industry or were designing wind turbines for the utilities. In common parlance we refer to power plants by the combined size of their generators. For example, an Eskimo village runs a 50 kW diesel generator, it is a 500 megawatt (MW) coal-fired power plant that clouds the valley, or it is the 900 MW Unit 1 nuclear reactor that was damaged at Three Mile Island.

Utilities try to run their generators at a fairly constant output so the generators perform the most efficiently and they can make the most money. As a result, a 50 kW generator normally produces 50 kW, and engineers understandably use this power rating when they talk to one another.

Wind machines are different. Because of fluctuations in wind speed, generator output is not constant. Seldom does the generator produce its rated output for any extended period. Moreover, optimum rotor and generator combinations depend on the wind regime. A wind turbine with a 10-meter rotor may, for instance, perform most efficiently (deliver the most energy) matched with a 9 kW generator in one part of the country, but with a 25 kW generator in another where winds are strong enough to frequently drive the generator at or near its rated power. One 10-meter wind machine that uses a 9-kW generator will outproduce a turbine the same size with a 25-kW generator in areas with average speeds up to 11 mph. But put the two side-by-side on a California wind farm, say in the Altamount Pass, where there is an 18 mph average speed, and the 10-meter turbine with the 25-kW generator will deliver more energy.

Fortunately, our rating practice began falling out of favor several years ago. Windworks led the way with their "Windworker 10" signifying that the wind machine uses a rotor 10 meters in diameter. Since then Aerolite has introduced their 7-meter turbine and Enertech has begun marketing their E-44 (the designation stands for a 44-foot rotor). More manufacturers are expected to follow suit.

Rotor diameter is a much more practical measure since it is the rotor and not the generator that captures the wind and converts it to a form we can use. The generator comes later in the conversion process. Northern Europeans, especially the Danes, invariably refer to the size of their machines by rotor diameter.

Increased awareness of rotor diameter's importance in the conversion process has led to the adoption of a hybrid designation using rotor diameter *and* generator capacity. Enertech's model 21/5, for example, uses a rotor 21 feet in diameter to drive a 5 kW generator. Such designations are particularly helpful to wind farm developers for whom generator capacity is important.

The rated power at speed system is not only confusing because there is no reference speed to compare one turbine with another (rated speeds range from 20 mph to over 30 mph) but also because some manufacturers rate their machines at peak power output and others do not. Enertech, for example, rates their machines at 25 mph. Yet the generator will produce more power than rated. Their model 1800 will actually produce 2200 watts.

A few, shall we say, less than reputable manufacturers have taken advantage of our emphasis on generator size by adding large generators to relatively small rotors. In essence it is possible to slap a 6-foot two-by-four on a 25-kW generator and call it a 25-kW wind turbine using this rating system. And it will produce 25-kW—in a hurricane!

Because of questionable power ratings by some manufacturers and general confusion by consumers as to what the numbers mean, the American Wind Energy Association (AWEA) has attempted to shed light on machine designations and performance ratings by calling for a standard list of parameters. These include maximum power (not rated) and, most importantly, the annual energy output at various average speeds. This is intended by AWEA to eliminate the rated power at rated speed nomenclature with values that make more sense. It represents the wind industry's equivalent to the Environmental Protection Agency's mileage sticker on new cars. And like the mileage sticker, the AEO is based on assumed conditions. "Your performance may vary." The AEO may be given as a table of outputs at several average wind speeds or as a graph.

To standardize the estimates from one company to the next AWEA recommends that the Rayleigh distribution be used to project the AEO. A heartening development has been the presentation of the AEO as a range of outputs (which include those produced by the Rayleigh distribution) rather than as a single line (see fig. 5-5). This band better reflects reality. Output estimates are not precise. They should not be displayed that way.

If you are adverse to the calculations described in this chapter and you are not satisfied with table 5-3 use the AEO estimates. Find your average speed, and adjust it to the height of your tower (if the AEO is for hub-height). Check the table or graph first; some are presented for anemometer height in which case no adjustment is needed. But, it is always wise to examine some of the values against the swept area of the rotor to see if they are within reason. This has been done (see table 5-5) for the three turbines used earlier in table 5-4.

In order of importance there are three ways to increase output:

1. increase wind speed; that is, find a better site or use a taller tower
2. increase the swept area of the rotor
3. improve the wind turbine's conversion efficiency

Efficiency is placed last because output is so sensitive to the other factors: speed and swept area.

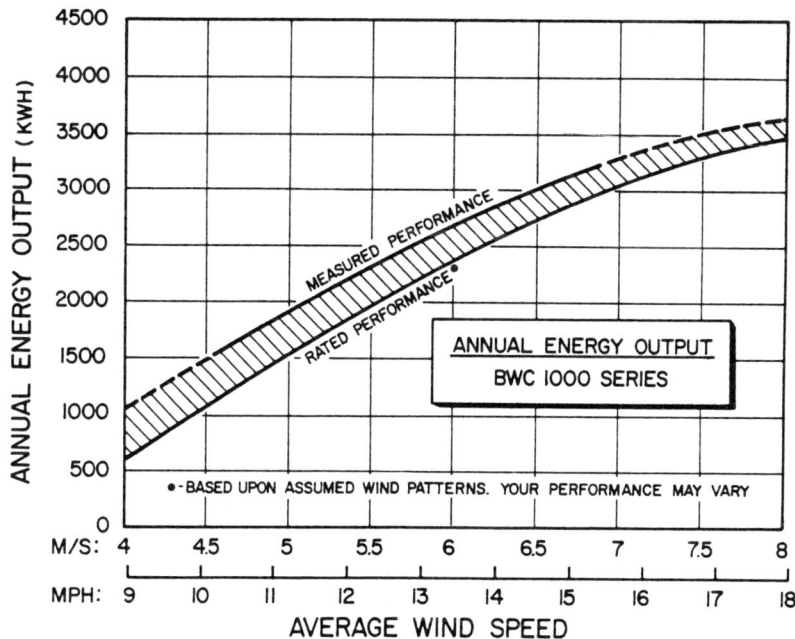

Fig. 5-5. Performance curve for the Bergey 1000. Note that the annual energy output is presented as a range of values. Graphs such as this, or tables of the annual energy output, are the wind industry's equivalent to EPA's mileage labels on new cars. (Courtesy Bergey Windpower Co.)

TABLE 5-5
Comparison of Annual Energy Output

Rotor Diameter	AEO in kWh per Year		
	From Table	From Power Curve	From Manufacturer
3-meter	2900	1900	1000–2000
4-meter	5000	3900	4400
6-meter	11000	9200	9700

NOTE: In all three examples the manufacturer's AEO estimate is less than that computed by the swept area method indicating that the estimates are legitimate. The discrepancy between the columns is due to several factors: the Bergey 1000 is less than the assumed 3 meters in diameter, and the conversion from average power to average wind speed introduces considerable error. When making such a comparison always use the lower values. If the wind turbine performs better than you expect, great; it will pay for itself that much sooner.

To gauge the size of a wind machine, ignore the size of the generator and get right to what matters most: the annual energy output for your average speed or power. But if the AEO is not available, whether you are in Europe or the States nothing outside of the wind itself, no other single parameter, is more important in determining a wind machine's capability of producing energy than rotor diameter.

Investors have lost staggering amounts of money (not just homeowners but savvy businessmen who should know better) because they did not grasp this concept. Even after years of experience with operating wind machines, reams of technical documents and studies, there is always someone who comes along that has made a startling discovery of a way to beat the Betz limit. (Most of them have never heard of Betz, unfortunately.) Their wonderful new device not only is more efficient than any wind machine preceding it, but it produces more energy than theoretically possible. That is not to say that it cannot be done—the Betz limit is only a theory—but that no one has done it yet and plenty have said they could. And as sure as the rain will fall there will be someone willing to part with their savings before punching a few numbers into a calculator.

ECONOMICS—WILL IT PAY

Wind machines are not a luxury item. They are a far cry from a hot tub, swimming pool, or a rack of snowmobiles taking up space in the backyard. These frills only take from your paycheck. Wind machines, in contrast, give something back. They save or even earn their owners money. Wind machines are as useful as any other energy system (oil, gas, electric) now serving your home and business. Mike Bergey, of Bergey Windpower Company, likes to point out that you are paying for a wind machine whether you want to or not each time you mail your monthly contribution to the local utility. Though, like death and taxes, you may not be able to avoid dealing with the utility entirely, you can at least reduce your payments. But will a wind machine be a sound investment? Will it pay for itself? Or more simply, is it worth the trouble?

These are questions frequently asked. The answers, though, can be elusive. They are dependent upon a number of speculative variables not subject to precise calculation (inflation and interest rates are two that quickly come to mind). Moreover, there are several different approaches, each with its own merits as well as bias, to analyzing the economics. No one method tells you everything you need—or want—to know.

All too often, for example, consumers look only at the initial cost of installing a wind system. They contrast this with what they are accustomed to paying their utility and throw up their hands in despair. There is no contest! "The wind may be free," they might be overheard saying, "but it sure takes a lot to catch it." There are always two sides to every equation and the other side of the wind system's cost is the revenue it earns or the money it saves. You cannot look at one without examining the other.

Another common misconception is that to justify a wind system investment it must generate nearly all of the energy currently supplied by the utility. In other words, it must eliminate your present electric bill. Sometimes this may be best. But a wind machine may be a good buy even when it provides only a small proportion of the energy demand. In many applications it is better to select a wind system that will produce less—not more—energy than is currently used because often utility buy-back rates are less than retail rates. (It does not make any sense to sell the utility energy for $0.06/kWh when they will turn around and sell it back to you for $0.08/kWh. As an investment the wind system should stand on its own merits: the economic generation of energy. And it should be viewed in that way.

Determining the *cost of energy* (COE for those with a bent for acronyms) produced by a wind system is an approach popular with government agencies and academic institutions. The COE takes into account the fact that the wind system will be generating energy for many years to come. It also includes the cost of money and maintenance over the same period. This "life-cycle cost" method produces a figure in cents/kWh for the wind system over its life span. The results can be easily compared with today's cost of conventional energy. And, unfortunately, that is just what most academics and bureaucrats do. They fail to note that conventional energy costs will escalate faster than that due to inflation alone. Rather than comparing the COE to today's electric rates it should be compared to the average future cost of electricity. Moreover, the COE only reveals whether or not the wind system will produce energy at less cost than that supplied by the utility. It does not tell you how much of a bargain—or cost—the wind system might be. The COE, consequently, cannot be used to judge whether or not your money would be more productive in an interest-bearing account at the local bank or in some other investment. Don't bother with the cost of energy; payback is more useful.

Finding the payback, or the time it will take for the investment to pay for itself, is a popular and fairly simple way to gauge an investment's worth because payback is related to *return on investment* (ROI). A short payback offers a high return on investment. Banks and businesses try to maximize their ROI—to get the most for their money. Nevertheless, the payback approach also has its limits.

Payback gives no indication of the future earnings a wind turbine will produce after it has paid for itself. Since wind generators are designed to last twenty years or more, much of the earnings take place in later years after payback has occurred. Some wind machines will pay for themselves many times over.

To get a good idea on what a wind system will do, it is necessary to look at both payback and future earnings.

Simple payback can be found from the formula:

$$\text{Payback} = \frac{\text{Adjusted Total Cost}}{\text{Adjusted Annual Value}}$$

where *adjusted total cost* is the installed cost of the wind system less any tax credits or tax deductions, and *adjusted annual value* is the annual value of the energy produced adjusted for the user's tax bracket. Let us take a look at the factors used in this equation.

Machine costs not only include the cost of the wind machine, tower, and installation, but also the cost of capital to buy the machine, and the cost of maintenance, insurance, and depreciation. The cost for the wind system itself is self-explanatory. That is what you end up paying when you buy one and have it installed. Capital costs are a little more mysterious. You have paid for the use of money before if you have ever bought a car or appliance "on time." Most potential buyers are aware of interest costs if they have considered using a loan to buy their wind system. But you cannot avoid these costs by paying with cash. The money spent could have been earning interest at the bank, and that lost interest on a cash purchase is another cost that must be attributed to using wind energy—a cost that can be significant.

Many are also aware that a wind system can add other costs by increasing their property tax burden (except in those areas where they are exempted). And, it is possible that at some time the income derived from a wind system could be taxable as well.

Maintenance costs, on the other hand, are almost an entirely unknown factor. Though much experience had been gained on the maintenance required of the small machines used during the 1930s, little information is available on the costs to maintain modern wind machines. For some brands the cost is minimal; for others the sky is the limit. The industry consensus is that maintenance will cost one to two percent of the total wind system investment annually.

Insurance is a cost often overlooked. Not only should the owner have insurance on the wind system itself but also liability insurance for any accidents due to the wind machine. Since most wind systems require an extended period to pay for themselves, an owner would be taking a great risk operating an uninsured wind machine. Unexpected events that designers cannot anticipate, such as a horse or car entangling itself in a guy cable and toppling the tower, may wipe out an expensive wind system before you can recoup your investment. In the same vein, wind systems can be dangerous and are a potential hazard. Personal injuries and property damage can occur in a multitude of ways: a blade can be thrown, the tower can fall over, or (the most likely form of accident) someone can fall off the tower. Insurance companies have little or no experience with wind systems, and their rates vary from a minimal addition to homeowners' coverage to a more specific policy with a higher cost.

Depreciation, on the other hand, does not necessarily affect small system users. It is a cost of more concern to utilities who must invest continually in new equipment.

We invest in a wind system on the premise that at some time the income derived or money saved from the energy produced will offset its cost. Consequently, the other side to the economic equation, once costs are known, is to determine the value of the energy produced.

How much is the energy worth? Like costs, the answer depends upon a number of factors, the most important of which is the retail cost of electricity from the utility. If we assume that whatever energy the wind machine generates will displace electricity that otherwise would be bought from the utility, the value of this energy (per kilowatt-hour) would equal the retail cost of electricity.

If you are a homeowner, the cost of electricity per kilowatt-hour (kWh) is easy enough to estimate. Simply pull out all the old electric bills, sum them, and divide by the total amount of electricity consumed to get an average cost in cents/kWh. Bear in mind, though, that your rates might actually go up after installing a wind system.

Utilities charge for electricity based on both the amount of energy consumed and on the rate at which the energy is used. As a rule, those who consume more electricity pay a lower unit rate (their total bill may, of course, be higher) in cents/kWh. In addition, commercial customers pay a *demand charge* based on the maximum power consumed (in kW). This is done to compensate the utility for maintaining *spinning reserve* or generators on line to provide power at the demand of the customer. The demand for most residential customers is insufficient to warrant the utility to add a demand charge to a homeowner's monthly electric bill. But a homeowner who uses a wind generator consumes less energy from the utility (that is the whole intent—right?). This may cause the utility to place the homeowner into a higher rate-paying category. The increased cost of this backup energy, if any, should be acknowledged and deducted from the value of the energy from the wind system. In many states being bumped by the utility into a higher rate-paying bracket is considered discriminatory and would be prohibited.

Once the value of the wind generated energy in cents/kWh has been found, the annual value of the electricity produced can be calculated by multiplying the retail rate ($/kWh) by the amount of energy (kWh) produced yearly. The result is in $/year.

One of wind energy's chief advantages over generating electricity by conventional means is that the fuel (the wind) is free. The bulk of the cost for a wind system occurs all at once. When paid for, the energy produced costs little over the remaining life of the wind system. Conventional generation, on the other hand, consumes nonrenewable fuels whose costs continue to escalate. Thus, our payback analysis would be incomplete if we did not account for the rising price of electricity. Like other aspects of energy production the rate at which utility prices will rise over the next twenty years is hotly debated. Estimates by the utility industry differ sharply from those of almost everyone else. Projected price escalation ranges from a low of 5 percent by the utilities to more than 15 percent per year by others.

Utility prices rose during the seventies and early eighties due in part to the rapid rise in the cost of oil caused by the oil embargo and OPEC's pricing strategy. Oil-fired utilities were hit first and their rates jumped dramatically. Coal and gas-fired utilities, though, were not immune as the cost of these fuels generally tracks the price of oil. Their rates rose more gradually rather than in fits and spurts. But they still rose.

Many utilities in the early seventies committed themselves to massive construction programs expecting continued growth in electricity consumption. These plants, both coal-fired and nuclear-powered, are only now being completed. They are enormously expensive, particularly the nuclear reactors. Consumption, moreover, has not kept pace with the projections of a decade ago, so a portion of the capacity in these plants will be idle. Someone has to foot the bill. As a consequence, consumers will be seeing higher electric rates for years to come. Rates for one electric utility are expected to increase 20 percent per year for at least the next two years as they attempt to pay for the completion of a nuclear reactor.

Electric utility rates are projected to continue rising through the next decade even though the world price of oil—the benchmark of energy costs—is now falling. The international oil market remains as volatile as it has ever been. Future conflicts in the Middle East could once again cause shortages and higher prices here. It would not be prudent to assume that the price of oil and in turn other sources of energy will return to pre-OPEC levels.

With the passage of the Federal Energy Act in 1978, Congress offered tax credits to those who purchase wind systems.[3] Because the credits reduce an individual's tax liability (instead of merely reducing taxable income) directly, they in effect reduce a wind system's initial cost and can be easily incorporated into this method of estimating payback.

A credit of 40 percent is earned on purchases up to $10,000 for residential use. For a wind system costing $10,000, a $4,000 credit is gained. This credit would, as a result, cut the cost of the $10,000 wind system to $6,000. If the credits exceed the tax liability they can be carried forward for several years.

The energy credits work differently in commercial applications. In a strictly commercial venture, only 15 percent of the purchase price can be credited to a business tax payment. But, there is an additional 10 percent investment tax credit. Total credits amount to 25 percent, and there is no limit on the size of the credit.

Both residential and commercial credits are effective through 1985 when they will expire unless renewed by Congress. The investment tax credit for businesses will remain in effect after 1985.

The credits can be prorated between home and business. For example, if you are a farmer, you might buy a wind machine large enough to meet both your domestic and business consumption. The Internal Revenue Service requires that you apportion the tax credits so that they reflect how much of the wind system will be used for your home and how much for your business.

Many states have their own tax credit programs. Check with your state energy office or any wind system dealer; they will be able to tell you which credits you qualify for and which you do not.

Let us try to make some sense out of all this with an example. Assume that we are installing a wind turbine 6 meters (20 feet) in diameter atop an 80-foot tower in an area with a 12 mph average annual wind speed (200 W/m^2). Under such conditions this wind turbine will produce approximately 20,000 kWh per year. We will assume that this electricity is worth seven cents per kWh, and that the wind turbine can be installed for a total cost of $15,000 (1983).

Residential Payback

This wind system qualifies for the maximum credit of $4,000. Therefore,

$$\text{Adjusted Total Cost} = \$15,000 - \$4,000 = \$11,000.$$

The value of the energy produced is:

$$20{,}000 \text{ kWh/yr} \times \$0.07/\text{kWh} = \$1{,}400.$$

Energy savings are tax free, that is, one dollar saved is worth considerably more than one dollar earned. If you are in the 50 percent tax bracket you receive only fifty cents for every dollar earned. One dollar saved, thus, is actually worth two dollars before taxes ($1/50%). For those in lesser tax brackets the benefits are not as great, but they are still considerable (see table 5-6).

Assume that you had a good year and you find yourself in the 40 percent tax bracket. Then the:

$$\text{Adjusted Annual Value} = \$1{,}400 \times 1.67 = \$2{,}300$$

We can now calculate the payback using today's cost of electricity.

$$\begin{aligned} \text{Payback} &= \$11{,}000/\$2{,}300 \\ &= 4.8 \\ &= 5 \text{ years} \end{aligned}$$

This payback is equivalent to a 20 percent return on the investment. Is this realistic? Yes, even when adding the cost of money, insurance, and maintenance to the cost of the wind system. Why? Because we have also left out an important factor on the income side of the ledger—the escalating cost of energy. Here are the details.

If a buyer makes a down payment on the wind system equivalent to the tax credit and borrows the remaining $11,000 from a bank or credit union, the first year out-of-pocket expenses are nil because the credits are refunded. The cost of the loan can be found by using interest tables for the *capital recovery factor* (see appendix B). A five-year loan for our 6-meter turbine at 15 percent interest will cost:

$$\begin{aligned} \text{Cost} &= \text{Capital Recovery Factor (Loan Amount) (Loan Period)} \\ &= 0.3\ (\$11{,}000)\ (5 \text{ years}) \\ &= \$16{,}500 \end{aligned}$$

$$\text{Cost of Loan} = \$16{,}500 - \$11{,}000 = \$5{,}500$$

TABLE 5-6
Value of Savings

Tax Bracket		Increased Value of Savings
50%	(1/50%) =	2.00
40%	(1/60%) =	1.67
30%	(1/70%) =	1.43

Inflation, though, erodes the value of future dollars used to repay the loan. The loan is repaid with less valuable money. If we assume that the inflation rate is 10 percent the loan costs considerably less when we discount it for inflation.

15% interest rate − 10% inflation rate = 5% true loan interest

Let us apply this interest rate to the cost of the loan. The capital recovery factor for 5 percent over a 5-year period is 0.23.

$$\text{Cost} = 0.23 \, (\$11{,}000) \, (5 \text{ years})$$
$$= \$12{,}700$$
$$\text{Cost of Loan} = \$12{,}700 - \$11{,}000 = \$1{,}700.$$

By discounting the cost of the loan for inflation we have reduced its true amount by 70 percent.

Interest compounding has a similar effect on the cost of electricity. If utility rates escalate at 14 percent annually over the twenty-year life of the wind system, the value of the energy it generates in the twentieth year is:

$$\text{Future Value} = (\text{Compound Amount Factor}) \, (\text{Initial Value})$$
$$= 13.7 \times (\$0.07/\text{kWh})$$
$$= \$0.96/\text{kWh}$$

But *future value*, like the interest cost on a loan, should be discounted for inflation. This gives a more realistic picture of future earnings. With an inflation rate of 10 percent the *discounted future value* of energy grows at a rate of:

14% escalation rate − 10% inflation rate = 4% true escalation rate

From appendix B, the *compound amount factor* for 4 percent at twenty years is 2.19.

$$\text{Discounted Future Value} = (2.19) \times (\$0.07/\text{kWh})$$
$$= \$0.15/\text{kWh}$$

We now have the discounted future value of electricity ($0.15/kWh) and its present value (7 cents/kWh). To approximate the discounted average future value—the *levelized value*—of electricity over the life of the machine we need to find the average between them.

$$\text{Levelized Value} = (7+15)/2 = 11 \text{ cents/kWh}$$

The discounted average future value, we will call this the levelized future value, tells you exactly how much the energy you produce is worth. It takes into account both the escalating value of the energy you generate as well as inflation, yet still gives us a single number we can use in our payback calculation.

The assumptions used in the preceding example may not reflect the conditions at a particular time and place. The inflation rate may be lower or the projected rate increases by the local utility may be lower than those used in the example. It is the relationship between interest rates, energy price escalation, and inflation that counts, not the specific numbers. Utility rates could be rising at 10 percent annually, for example. But with an inflation rate of 6 percent the difference between the two is the same as that in the example: the cost of electricity will rise 4 percent faster than inflation.

Let us refine our payback calculation by incorporating the additional costs of both a loan on the wind system and of its maintenance. But let us balance these costs with the additional revenues generated by using the levelized future value of electricity.

For a loan at 15 percent interest over five years, the discounted cost (the real cost) is $1,700. Assume that annual maintenance costs will amount to 2 percent of the initial cost (an estimate commonly used in the industry) or $300 per year. Over the life of the machine (twenty years) $6,000 will be spent on maintenance. (On some products this estimate is too low; on others it is too high.)

$$\text{Adjusted Total Cost} = \$11,000 + \$1,700 + \$6,000$$
$$= \$18,700$$
$$\text{Adjusted Total Value} = \text{Energy/yr} \times \text{Tax Adjustment} \times \text{Levelized Value}$$
$$= 20,000 \text{ kWh} \times (1.67) \times \$0.11/\text{kWh}$$
$$= \$3,700/\text{yr}$$
$$\text{Payback} = \$18,700/\$3,700 = 5 \text{ years}$$

Though we used a more sophisticated approach we still conclude that this wind machine will pay for itself in five years. Over its twenty-year life, it will pay for itself four times over and earn its owners — in real dollars — $33,000.

Commercial Payback

Businesses can take advantage of favorable depreciation deductions that, in combination with the commercial tax credits, may reduce the cost of a wind system up to 75 percent. In fact businesses are in a much better position than homeowners to put the tax system to work for them. Because there is no limit on the amount of tax credits, a business can buy the most economical size wind machine for their application. For a homeowner, though, the tax credits take a smaller and smaller bite out of the cost of a wind system as the cost goes above $10,000. Homeowners consequently reap the most benefits when they keep the total cost of the wind system under $20,000. Businesses do not suffer this restriction. If they buy a million dollar wind turbine they qualify for a quarter million dollar tax credit.

Thanks to recent changes in the law, wind systems can be depreciated over a five-year period. Depreciation deductions offer substantial tax benefits to users of capital intensive equipment such as wind systems. For example, a business in the 50 percent marginal tax bracket gains fifty cents in value for every dollar in deductions. Because of the tax rate, for every dollar deducted from gross income the business no longer owes Uncle Sam fifty cents. They can keep it. Thus, you can "write-off" the cost of the wind system. At the end of five years you have obtained a piece of capital equipment for 50 percent of its original cost. Deductions for interest payments, for maintenance, and for other costs of doing business are also available. But their impact on the "bottom line" is not nearly as great as depreciation.

These tax benefits are somewhat tempered by the fact that energy costs are a deductible expense. The actual cost of electricity to a business, as a result, is less than the amount paid to the utility. Savings in energy costs must be taken at face value—or reduced—to account for this. Where residential users gain value in excess of the amount saved, businesses lose value.

Let us work through the example of the 6-meter turbine being installed in a commercial application. All other conditions will remain the same.

Adjusted Cost = Cost − Tax Credit − Value of Deductions
Tax Credits = 25% ($15,000) = $3,800
Value of Deductions = $15,000 × 50% five-year depreciation
= $7,500
Adjusted Cost = $15,000−$3,800−$7,500
= $3,700

Deductions are calculated on initial cost, not on the cost after the tax credits are applied. It is also no longer necessary to reduce the initial cost by the salvage value remaining at the end of the depreciation period. The wind machine, of course, is not worn out at the end of five years—it should last another fifteen—but it is as far as the Internal Revenue Service is concerned.

To be fair we should include the cost of a loan or the "opportunity cost" of money used to buy the wind system. The cost of a discounted loan over five years on the initial cost less the tax credits ($11,200) is $1,700. (This assumes a loan at 15 percent interest with a 10 percent inflation rate.) And we cannot forget maintenance. It adds another $6,000 to the cost. But in both cases these costs are deductible expenses. Their true cost is only half that shown for a business in the 50 percent tax bracket.

$$\text{Total Cost of Loan and Maintenance} = (\$1{,}700 + \$6{,}000) \times 50\% \text{ five-year depreciation}$$
$$= (\$7{,}700) \times 50\%$$
$$= \$3{,}900$$
$$\text{Adjusted Total Cost} = \$3{,}700 + \$3{,}900 = \$7{,}600$$
$$\text{Annual Value} = 20{,}000 \text{ kWh} \times \$0.11/\text{kWh} = \$2{,}200$$
$$\text{Payback} = \$7{,}600/\$2{,}200 = 3 \text{ years}$$

This wind machine in a commercial application will earn approximately a 30 percent ($\frac{1}{3}$) return on investment, and over its lifetime will generate $37,000 more than its cost. Commercial users are not limited to a specific size (cost) of wind machine nor to a specific number as residential users are because tax credits and deductions can be claimed on any amount. This is the reason that there is such great interest in *wind farms*—arrays of multiple wind machines. Large installations gain economies-of-scale where the cost per unit declines as the number installed increases. Larger, more expensive machines (those from 7 to 15 meters in diameter) are also more cost-effective than the small turbines (3 to 6 meters in diameter) that most homeowners select. They generate a better return on their owner's investment.

Implicit in these payback calculations has been the assumption that all the energy generated was used on site offsetting energy bought from the utility. This may not be the case. Some of the energy may be generated at night when there is little or no need for it. The excess energy then will be sold back to the utility and, in all likelihood, at a price lower than the purchase rate. This reduces the overall savings produced by the wind system.

Using the 6-meter wind turbine and the same assumptions as before let us estimate how the sale of 30 percent of the turbine's output to the utility at $0.05/kWh affects payback. The levelized value of five cents is approximately eight cents.

Adjusted Total Value = (1.67)($0.11/kWh)(70%)(20,000 kWh) +
(1.67)($0.08/kWh)(30%)(20,000 kWh)
= 2,570 + 800
= $3,400
Payback = $18,700/$3,400 = 5.5 = 6 years

Thermal Application

Say you are going to use the 6-meter wind machine solely for heating air or hot water to reduce the operation of your oil-fired furnace. What do you do? First, convert the kWh of electricity generated by the wind machine to gallons of oil using table 5-7. Hold on, though; this is not the amount of oil saved. For every gallon of oil the furnace burns, 40 percent goes up the chimney. Only the remainder is put to work heating your house. On the other hand, the conversion of electricity to electric heat is about 90 percent efficient. You need to take all this into account because a gallon of oil generated by the wind machine has more heat value than a gallon of oil burned. Putting it together:

$$20,000 \text{ kWh} \times \frac{0.034 \text{ gal}}{1 \text{ kWh}} \times (90\% \text{ conversion to heat}) \times$$

$$(\frac{1}{60\%} \text{ adjustment for furnace efficiency}) = 1,000 \text{ gallons of oil}$$

For heating oil at $1.10 per gallon the value of the wind turbine's output in the first year is:

Annual Value (first year) = 1,000 gal/yr × $1.10/gal = $1,100

TABLE 5-7
Energy Equivalency of Common Fuels

1 kWh = 3,413 BTU
= 3.41 ft^3 of natural gas
= 0.034 gal of oil
= 0.00017 cord of wood

1 Therm = 10^5 BTU
= 100 ft^3 of natural gas
= 1 gal of oil
= 29.3 kWh of electricity
= 0.005 cord of wood

1 gal of oil = 1 × 10^5 BTU
1 cord of wood = 2 × 10^7 BTU
1000 ft^3 (Mcf) natural gas = 1 × 10^6 BTU

The escalation in the price of of oil (or any other fuel) should be incorporated in the payback estimates as in the example for electricity.

Here is a summary of the steps to estimating payback.
1. Determine the levelized value of the energy produced using the fuel cost escalation rate, the inflation rate, the life of the wind machine, and the initial value of energy.
2. Adjust the total annual value of the energy generated by the tax factor for residential users.
3. Reduce the cost of the wind system by applicable tax credits and the value of depreciation deductions.
4. Find the years to payback by dividing the adjusted cost by adjusted annual value.

COST-EFFECTIVENESS

Though sometimes used interchangeably, cost-effectiveness is not the same as efficiency. You could have the most inefficient wind machine ever built, the kind that brings a tear to an engineer's eye, but if it works, and is cheap enough, it could be more cost-effective than a modern engineering marvel costing much more. The reason for installing most wind machines is the generation of low-cost energy. (Some are intended for remote sites where cost is not a primary concern.) If you deliver lower cost energy with an inefficient wind machine than with an efficient one, so be it.

There are several measures of cost-effectiveness in use: cost per kW of installed capacity, cost per swept area, and cost per kWh of energy generated. (The high-tech boys occasionally throw around the cost per pound of materials used in the wind machine. The term is seldom used in reference to small wind systems.)

The most frequently used measure of cost-effectiveness is cost per kW. It is about as meaningful as the power rating in describing the size of a wind machine. This measure came into use the same way. Utility engineers were accustomed to using it. Like the power rating the cost per kW works well for power plants that run at constant output. But for wind machines, because they produce variable amounts of power, something more descriptive is needed.

A few manufacturers, moreover, were taking advantage of consumer confusion over what parameters really mattered. Because consumers were using cost per kW to compare one wind machine against another, these manufacturers offered wind machines with higher generator ratings relative to the size of the rotor than was the norm. Their products were not any better, nor would they produce more energy than their competitors, but their cost per kW was lower. In one case the manufacturer's power rating was based on the generator's intermittent capability not on its rating under continuous use. As a result, the nameplate rating on the generator, and the wind turbine, was nearly three times that of the same generator in use by others. Where most wind systems were costing well over $1,000 per kW this wind machine came in under $700.

If a truer measure of a wind machine's size is the area swept by the rotor why not use the $/ft^2 or $/m^2 instead of $/kW. It is preferable, but as with our energy estimates the limitation on using cost per swept area is the assumption that all wind machines are created equal when it comes to conversion efficiency. They are not.

Our payback calculation does not use either kW or swept area. Instead it uses the kWh of energy generated. This is what it is all about — energy production. Why bother with the other measures? They are just shorthand for the cost per kWh. Use what matters, the cost per kWh.

Note in table 5-8 that the cost per kWh continually decreases as the size of the wind machine increases. But the cost per kW is less consistent. The 13-meter wind turbine, for example, costs more per kW than the 10-meter turbine, yet its cost per kWh of energy is markedly less. Why? The 13-meter turbine (the Enertech E-44) uses a 20 kW generator;[4] the 10-meter turbine uses a 25 kW generator. This is a good example of how misleading generator size can be when judging cost-effectiveness.

TABLE 5-8
Installed Cost-Effectiveness of Typical Wind Machines in 1983, Before Taxes, in a 12 mph Wind Regime, 80-foot Tower[a]

Size	$/kW	$/kWh
3-meter	5,000–7,000	1.08–1.33
4-meter	5,000–6,000	1.01–1.21
6-meter	3,800–5,000	0.75–1.0
7-meter	1,300–2,100	0.50–0.82
10-meter	1,200–3,800	0.50–0.65
13-meter	2,000–2,500	0.38–0.48

[a]200 W/m^2, $a = 1/5$

The cost per kWh in table 5-8 is not equivalent to the cost per kWh you pay for electricity from the utility. They are two different animals. The cost per kWh measure of cost-effectiveness does not incorporate the life of the wind system, interest costs, fuel escalation, nor all the other factors affecting the economics of a wind system investment. It is merely a measure useful for comparison shopping — nothing more. Don't read into it what is not there.

6

Evaluating the Technology

Wind machines come in all shapes and sizes. They run the gamut from the familiar farm windmill that dots the range to NASA's sleek giants. Selecting the one that is right for you is much like buying a car. You don't base your decision solely on what is under the hood or the type of transmission it uses. You look at the complete automobile. The same is true of buying a wind machine. You weigh all the pluses and minuses of each component. You look at the whole (including such nontechnical aspects as the reputation of the manufacturer), not just the individual parts. But it helps to understand how the parts combine, how they function, and what their limits are to know the whole.

In this chapter we will discuss wind machine technology and the developments that have brought us to where we are today, the important difference between wind machines that use drag to drive their rotors and those that use lift, why modern wind machines use only two or three blades, what materials are used to make these blades and the advantages of each, the kinds of controls used to prevent the rotor from flying apart, and the types of transmissions and generators being used.

BACKGROUND

"Hey, that's a funny lookin' windmill you got there. What is it?"
"A VAWT."
"Don't get smart with me son. I asked you a simple question."
"Actually, it's an articulating, straight-bladed VAWT."
"Can't you just tell me what it is? I don't work for the government you know."
"Some call it a giromill."
"Well, that's better. Why didn't you say so in the first place? For a moment there I thought you were speaking in tongues."

In dealing with the wind industry you are going to encounter plenty of jargon. It is pervasive. You will find it in advertising brochures as well as technical reports. In its proper place jargon serves a purpose. It allows complex concepts to be conveyed in as few words as possible. Jargon is technical shorthand. As wind technology has expanded so have the words and terms used to describe it. In the following paragraphs I will introduce you to the jargon as I discuss the technology.

The recent surge of activity in wind power began in the early seventies. It received its major push after the 1973 oil embargo. Developments have continued apace since then but much of the groundwork for today's technologies had already been laid. Today's developments are improvements and refinements on techniques and schemes first explored years ago. No one has made any earthshaking new discoveries that have revolutionized the generation of electricity with wind power, though many profess to have done so. As today's automobile still resembles and still functions in the same way as automobiles of years past so do modern wind machines operate much like their predecessors. One change has been the switch to downwind rotors on many small wind machines; another is the continuing development of the Darrieus rotor.

Wind machines can rotate about either a vertical or horizontal axis. As its name implies, the rotor of a vertical axis wind turbine (VAWT, to those with a penchant for jargon) spins about a vertical axis like that of the simple wind machine in figure 6-1.

The rotors of more conventional wind machines, like that of the Dutch windmill found throughout northern Europe and the American farm windmill, spin about a horizontal axis (a line parallel to the horizon) as in figure 6-2. The rotors of conventional wind machines may be upwind of the tower or downwind. In figure 6-2 the tail vane keeps the rotor oriented upwind of the tower regardless of changes in the wind. The rotors of Dutch windmills were also oriented upwind of the tower, though the mechanism for doing so was often not evident. On most, the miller had to turn the wind-

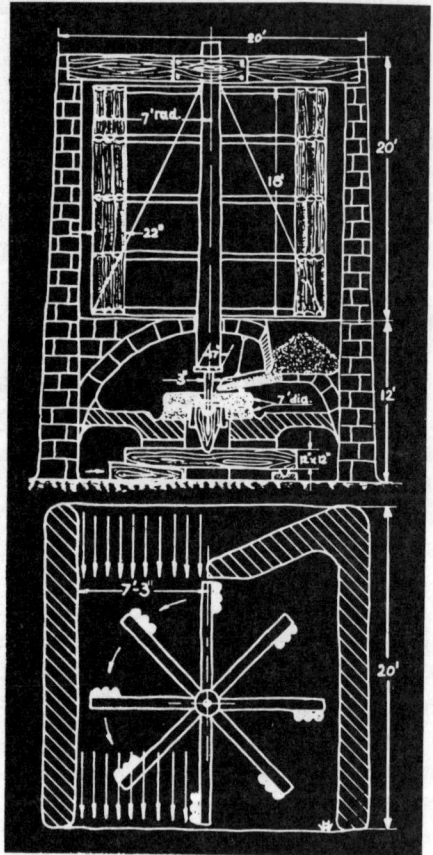

Fig. 6-1. Panemone. Bundles of reeds strapped to wooden posts power this simple Persian windmill. Note that the rotor spins about a vertical axis. (Mitre, DOE).

mill physically into the wind. Later versions used fan tails (see fig. 6-3) to turn the rotor mechanically into the wind.

Tail vanes and fan tails are means for passively changing the direction of the wind turbine. They use forces in the wind itself to orient the rotor upwind of the tower. These same forces can be used to orient the rotor downwind of the tower (see fig. 6-4). Downwind rotors do not need tail vanes or fan tails. Instead the blades are swept slightly downwind giving the spinning rotor the shape of a shallow cone with its apex at the tower. This "coning" causes the rotor to orient itself downwind.

Wind machines above 10 to 15 meters in diameter frequently use "active" yaw controls (i.e., either an electric or a hydraulic motor mechanically orients the rotor downwind of the tower). In figure 6-5 the rotor is mechanically driven downwind of the tower, though the downwind sweeping blades allow the rotor to orient downwind automatically when required. The Vestas turbine in figure 6-6 uses active yaw controls to orient the rotor upwind of the tower.

UPWIND VS. DOWNWIND

Downwind machines are thought to have a sleeker appearance than upwind machines using tail vanes, and some feel they have a more modern look. They do have one clear advantage. Downwind machines eliminate the cost of tail vanes.

Upwind proponents point out, however, that downwind machines occasionally get caught upwind and run in reverse when winds are light and variable. They add, moreover, that the tower creates a "shadow" that disrupts the air flow over the blades as they pass behind the tower. This decreases performance and increases wear, they charge.

Vaughn Nelson of West Texas State University notes, though, that upwind machines suffer a similar performance penalty because the wind piles up in front of the tower. Others disagree and the debate simmers.

Fig. 6-2. Early Halladay wind engine. In 1854, Daniel Halladay invented the first self-regulating windmill. In contrast to the Persian panemone, Halladay's wind machine, like most conventional wind machines, spins about a horizontal axis. The tail vane keeps the rotor pointed into the wind. (Courtesy T. Lindsay Baker, Panhandle Plains Historical Museum)

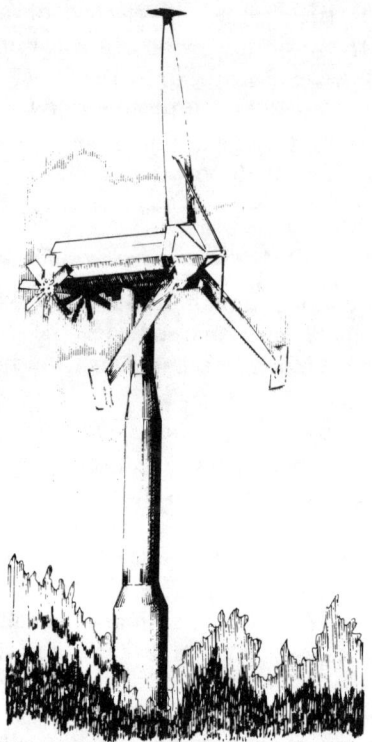

Fig. 6-3. Dutch upwind turbine with fan tails. Rotor on the Polenko is mechanically oriented into the wind by fan tails on either side of the nacelle — a common approach on wind machines in northern Europe. (Courtesy Holec, USA)

One disadvantage to downwind machines is yaw control in a high wind. For upwind machines with a tail vane the rotor can be furled (turned 90 degrees to the wind). Unless a downwind machine has a mechanical drive the rotor always stays downwind of the tower during high winds. Other mechanisms for controlling the rotor must be used.

Upwind rotors with tail vanes also seem limited to wind machines less than 10 meters in diameter. Above this size the tail vane becomes large and cumbersome.

You can avoid the whole upwind-downwind question by spinning the rotor about a vertical axis as in the Persian windmill at the beginning of the chapter. By doing so you also avoid any performance losses during the time it takes for the rotor to realign with the wind.

Evaluating the Technology • *155*

Fig. 6-4. Halladay rosette or umbrella mill. The rotor passively orients itself downwind of the tower. Segments of the rotor are shown furled here.

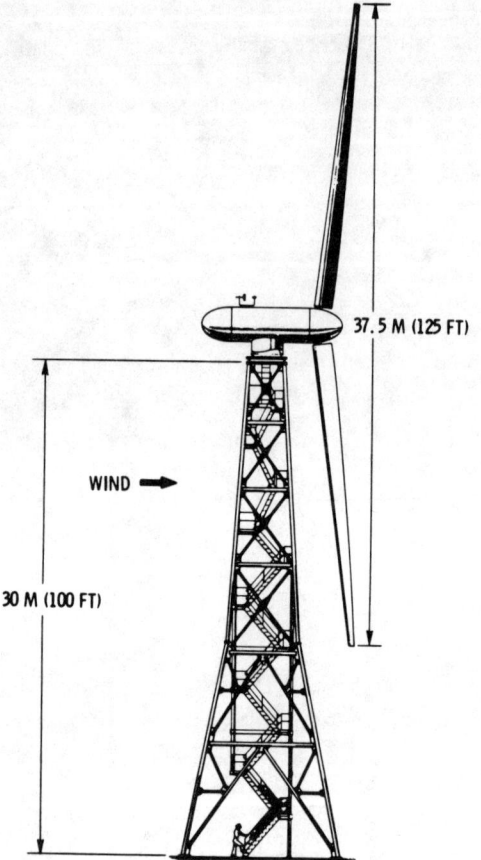

Fig. 6-5. Downwind rotor. Note the downwind sweeping blades on DOE/NASA's Mod OA built by Westinghouse. (DOE)

VERTICAL AXIS

Generically, the principal advantage of vertical axis wind machines over conventional wind machines is that they are omnidirectional—they can accept wind from any direction. This simplifies their design and eliminates the serious problems imposed by gyroscopic forces on the rotor blades of conventional turbines as they turn into the wind.

Another advantage to a vertical axis of rotation is that the turbine need not be placed on a tower to keep the blades from striking the earth (see fig. 6-7). (It is a weak argument, I agree, but vertical axis advocates

Fig. 6-6. Danish upwind turbine. The rotor is mechanically pointed into the wind with a motor controlled by a wind vane on the nacelle (not visible here). This wind machine uses a rotor 15 meters (50 feet) in diameter. (Courtesy Vestas a/s)

158 • Wind Energy

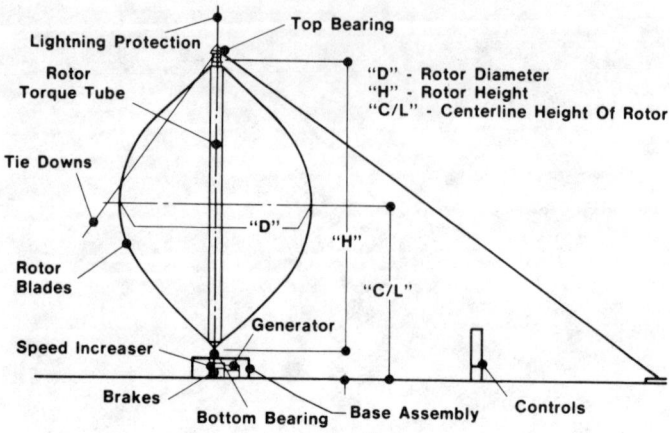

Fig. 6-7. Detail of Darrieus rotor with nomenclature. (Courtesy Alcoa)

are fond of using it. The main reason we put wind machines on towers, at least today, is to get more wind.) Nor do the generator and gear train need to be supported some distance off the ground. They can be mounted at ground level instead. Both features minimize the turbine's structure to only that needed to keep the rotor erect. Vertical axis turbines also simplify maintenance and repair of the generator, drive train, and brakes. (You appreciate this feature if you've ever had to repair a wind turbine 80 feet above the ground in sub-zero weather.)

Vertical axis turbines, like their conventional counterparts, can be divided into two major categories: those that use aerodynamic drag to extract power from the wind (for example, the cup anemometer) and those that use lift from an airfoil. We can subdivide those using airfoils further between wind machines with straight blades and those with curved blades.

The simplest configuration uses two or more straight blades attached to the ends of the rotor cross-arm (see fig. 6-8). From the side, this rotor appears like a large H—the blades being the uprights. The chief disadvantage of this configuration is that centrifugal forces throw the blades away from the axis inducing severe bending stresses in the blades at the point of attachment.

During the 1920s a Frenchman, D. G. M. Darrieus, patented a wind machine that cleverly dealt with this limitation. Instead of using straight blades in an H-shape he attached curved blades to the rotor in such a manner that it gave the turbine a more circular appearance. The curved blades (formed in the shape of a spinning rope or troposkein) directed centrifugal force to act through the blade's length. These forces cause tension in the blades rather than bending. Because materials are stronger in tension than in bending, the blades could be lighter for the same overall strength and operate at higher speeds than straight blades.

Darrieus' patent faded into obscurity until the National Research Council of Canada re-invented it in the mid-sixties. Though the phi (∅) configuration, or eggbeater as it is called, is the most common, Darrieus

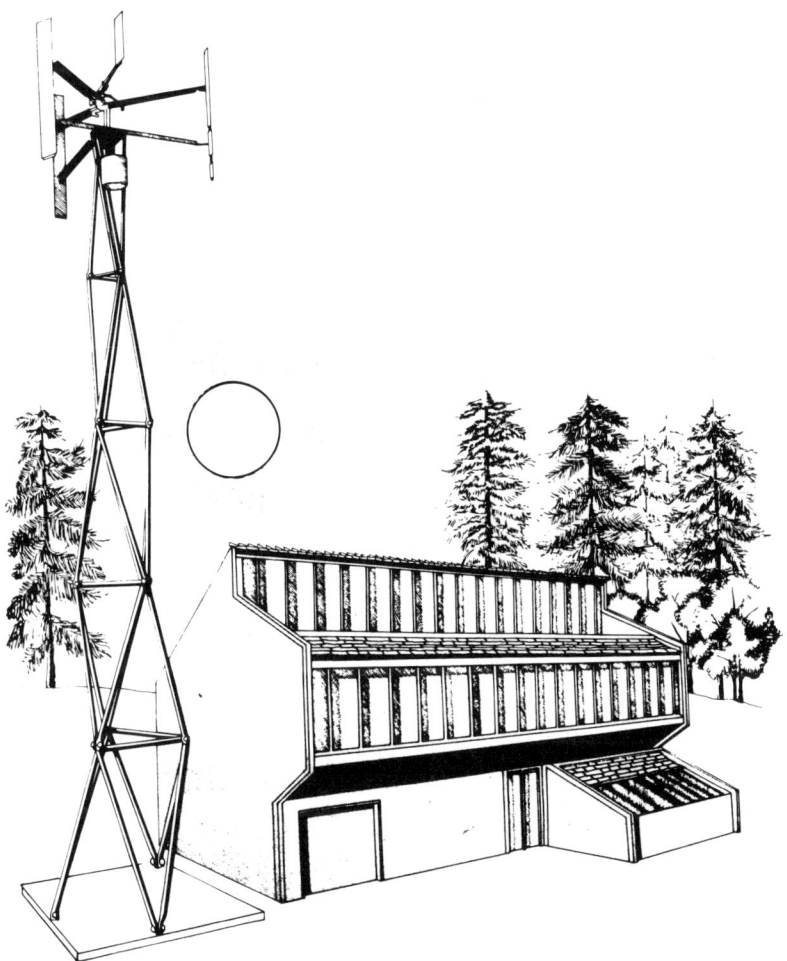

Fig. 6-8. Pinson Cycloturbine. Wind vane atop rotor changes blade pitch through a mechanical linkage keeping the blades at the optimum angle of attack. (Courtesy Natural Power, Inc.)

160 • *Wind Energy*

conceived several other versions. Delta, Diamond, and Y configurations have all been built.

Darrieus turbines are not reliably self-starting. Their fixed-pitch blades cannot be depended upon to drive the rotor from a standstill. To provide self-starting capability several researchers in this country and abroad reverted to the H-rotor. By using straight blades they can vary blade pitch as the blades orbit around the rotor's axis. (The blades on Darrieus rotors are attached rigidly to the torque tube.) Technocrats identify this kind of wind machine as an articulating, straight-bladed, vertical axis wind turbine—or *giromill* (see fig. 6-8).

The H-rotor has another advantage over the eggbeater Darrieus. It captures more wind. The intercept area of an H-rotor is a rectangle. For the same size wind machine—that is, the height and diameter are the same—the H-rotor will sweep more area than an ellipse.

LIFT VS. DRAG

There are two major classes of wind machines: drag devices and lift devices. *Drag devices* are simple wind machines that use flat, curved, or cup-shaped blades to turn the rotor. In figure 6-1 reeds form a flat paddle typical of many drag devices. Cup anemometers are also representative of drag devices as is the American farm windmill. In each the wind merely pushes on the cup or blade.

Drag devices characteristically produce high starting torques because much, if not all the area swept by the rotor, is covered by blades. (There is a lot of surface for the wind to push against.) They are good for pumping water in low volumes. But they have inherent drawbacks that limit their use for generating electricity economically. At best only one-third of the power in the wind can be captured by such machines. In comparison, the maximum possible for a lift device is 59 percent (the Betz limit). They also require more materials than comparable wind machines using lift.

Experimenters have tried numerous approaches to improving the performance of drag devices. In fact backyard tinkerers turn their attentions first to drag devices rather than lift devices because they are easier to understand. The wind pushes on a big wide blade and it moves. What could be simpler? The way it works is clear cut. (Lift devices are more complicated. We will see why in a moment.)

The farm windmill and the Savonius rotor are two drag devices where there has been some success. Early farm windmills used flat wooden slats for blades (a good example of a blade using drag). In 1888, Aermotor introduced its "scientific" windmill substituting sheet metal for wood. Aermotor stamped a broad curve into the blade to trap more air (see fig. 6-9).

Fig. 6-9. Multiblade farm windmill. The windmill that won the West. It performs better than a wind machine dependent upon drag alone, yet it is not a true lift device. Also note that the axis of the rotor is offset from the axis running through the top of the tower about which the wind machine pivots. The tail vane is aligned with the pivot axis.

162 • Wind Energy

It did, and in doing so directed the air to flow over the backside of the following blade. This cascade effect heightened the difference in pressure from one side of the blade to the other improving performance. Unfortunately, the farm windmill is still able to extract only about 15 percent of the power in the wind.

The Savonius or S-rotor was developed by the Finnish inventor Sigurd Savonius in 1924 (see fig. 6-10). It is sometimes called an S-rotor because of its shape from above. Principally, it is a drag device. But there is some recirculation of flow between the two halves of the rotor. Air striking one blade is directed through the separation and onto the other blade. Researchers have measured conversion efficiencies of almost 30 percent, considerably more than that extracted by other drag devices. In practice, however, S-rotors, like the farm windmill, extract less than 15 percent of the power in the wind.

There are hybrids as well, wind machines that don't fall neatly into either category. Anton Flettner, another Finn, built several such devices. Using the *Magnus effect* as a means of propulsion on two upright spinning cylinders, he sailed across the Atlantic in 1925. The following year he built a horizontal axis wind turbine 65 feet in diameter and used four spinning cylinders for the rotor's blades.

The Magnus effect is the lift or thrust produced by air moving over the surface of a spinning object (see fig. 6-11). It is what produces the curve in the curve ball. The pitcher imparts a spin to the ball as it leaves his hand. Air rushing over the spinning ball causes it to curve. In 1933 J. Madaras constructed a 90-foot cylinder 28 feet in diameter at Burlington, New Jersey in hopes of using the Magnus effect to drive cars around a track. Like similar attempts to harness the Magnus effect it was abandoned because the spinning cylinders used materials inefficiently. The same results could be obtained from true airfoils at less expense.

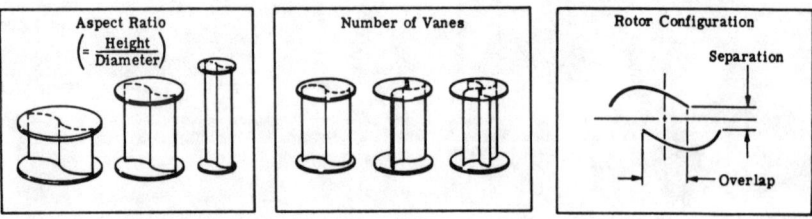

Fig. 6-10. Savonius rotor. Another hybrid device where performance exceeds that of a simple wind machine dependent upon drag alone. To achieve optimum performance the two blades must be separated to permit some recirculation of flow. (Mitre, DOE)

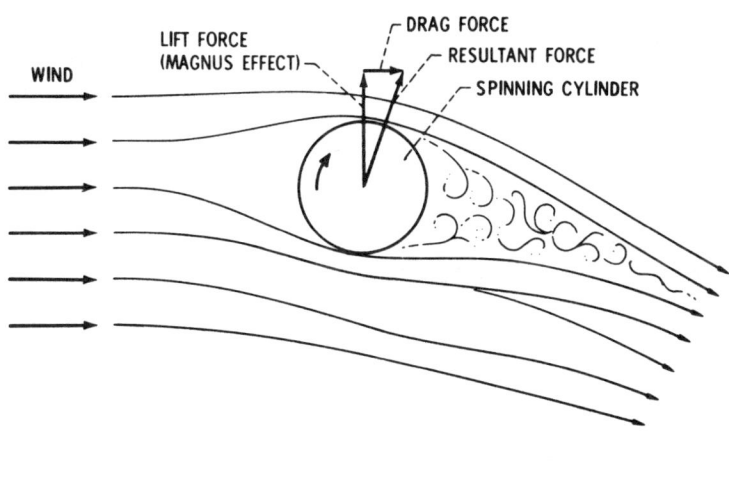

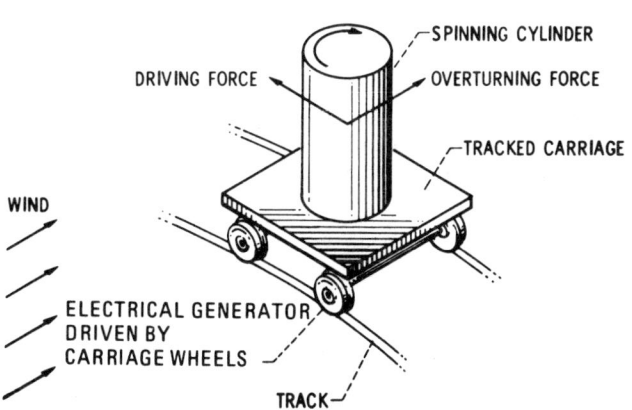

Fig. 6-11. Magnus effect. Hybrid device using the lift created by a spinning cylinder to drive a wind machine. In the Madaras concept, the spinning cylinder drives a car around a circular track. (Mitre, DOE)

Instead of paddles or cups, *lift devices* use airfoils like those in the wing of an airplane (see fig. 6-12). The limited number of blades on lift devices contrasts markedly with the multiple blades of drag devices. At first it seems mysterious that a windmill with only a few blades can operate more efficiently than one with a large number of blades. But a modern wind turbine, because it uses lift, can capture the same amount of power with a smaller rotor using fewer blades than a multi-blade drag device.

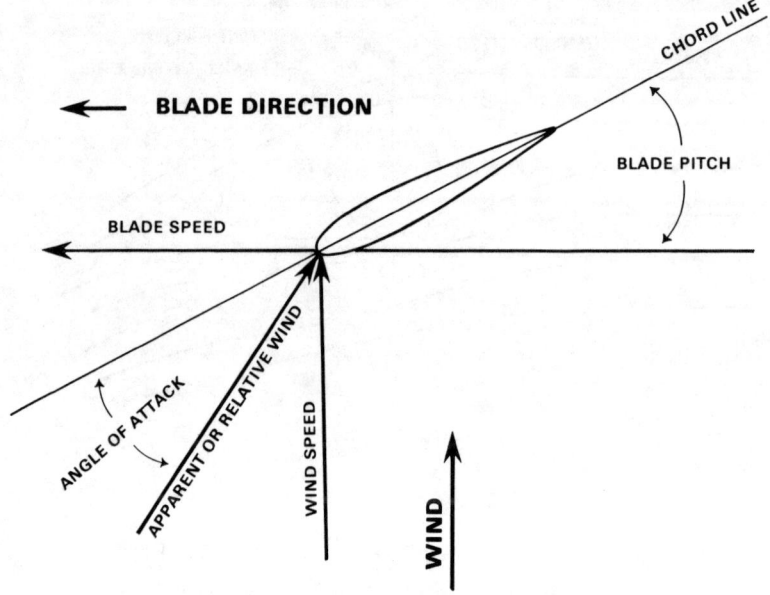

Fig. 6-12. Lift on an airfoil and the relative wind. Airfoil performance is gauged by the ratio of lift to drag. Lift is determined by the angle of attack. The pitch of the blade, the speed of the blade through the air, and the speed of the wind control the angle of attack and, consequently, lift.

On the basis of blade surface area, wind turbines using lift can extract one hundred times more power from the wind than a drag device. This is the reason why today's wind machines look so different from the farm windmill. Modern wind machines, those using lift, do more with less.

Why is this so? Why do some wind machines use multiple blades like the farm windmill where others use only a few? Why do some blades taper from the root to the tip where others taper from the tip to the root? To understand the answers to these questions we need to delve into wind turbine aerodynamics.

AERODYNAMICS

As a kid I was always intrigued by the fact that a sailboat could travel faster than the wind, and that it did so not when the wind was pushing from behind but when the wind was from the side. It wasn't until I began working with wind machines that I learned that a blade of a wind turbine is much like the sail of a sailboat and that lift propels them both.

To begin, let's look at the factors affecting the amount of lift from the airfoil in figure 6-12 representing the blade of a wind turbine. Air flowing over the blade causes both lift and drag. The sum of these two forces generates a thrust pulling the blade on its journey through the air. Engineers rate airfoil performance by the lift to drag ratio. Lift provides the forward motion to the blade; drag retards it. We want a high lift/drag ratio for best performance.

The lift/drag ratio is determined by the blade's *angle of attack*—the blade's angle with respect to the apparent wind. The lift/drag ratio increases with increasing angles of attack until a point is reached where airflow over the blade becomes turbulent. Lift then deteriorates rapidly, the ratio declines, and the airfoil is said to *stall*. The angle of attack at which this occurs varies from one airfoil to another. Stall is a dangerous condition in flight. But in a wind machine it can be put to good use. We'll see why in a moment. But first consider the angle of attack. It is a function of the blade's angle to the plane of rotation—its pitch—and the apparent wind. Assume the pitch is fixed, which it is on most small wind machines.

The *apparent wind* is the wind "seen" by the blade. It is the result of the wind due to the blade's motion through the air and the wind across the ground. On a sailboat the wind you feel on your face is the apparent wind. It is due to the wind blowing across the water and the wind created by the boat moving through the water. If you recall some of your high school physics you will note that the resultant of these two forces is dependent on their relative strengths. For example if both were equal (say the wind speed was 10 mph and the speed of the blade was also 10 mph) the apparent wind would be acting at a 45 degree angle between the two.

What would happen if wind speed increased while we kept the blade speed constant? The apparent wind would swing towards the wind across the ground because it is stronger. As it does, it increases the angle of attack. Let us reverse them and see what happens. Blade speed is now stronger and causes the apparent wind to shift towards the direction of motion decreasing the angle of attack. Because there is an optimum angle of attack, a point where the lift/drag ratio is a maximum, for each airfoil designers must decide how best to deal with this relationship.

For a fixed-pitch blade, to maintain an optimum angle of attack, blade speed must increase in proportion to wind speed. As wind speed increases the rotor must spin faster. On the other hand, if rotor speed must be constant then it will be necessary to change the pitch of the blade as wind speed varies. Or, designers must be willing to sacrifice performance if they choose a constant speed rotor with fixed-pitch blades. As wind speed increases the airfoil will begin to stall and performance will decline. And that is just what many do. Stall is desirable to designers because it reduces the rotor's power in high winds making it easier for them to design protective controls to keep the rotor in one piece.

The amount of thrust driving the rotor is not only a function of the lift/drag ratio and consequently the angle of attack, but also the area of the blade and the speed of the apparent wind over it. To increase the load an airplane can lift, you either increase the size of the wings, increase the plane's speed, or do both.

We have been considering an example of a blade as if the conditions it sees were constant along its entire length. They are not. Even when the pitch of the blade is fixed and rotor speed constant the speed of the blade through the air changes with distance from the hub. Blade speed is higher at the tip than near the hub because it has more space to cover in the same amount of time. (On larger wind machines the tip can approach the speed of sound.)

Because blade speed increases with distance from the hub, the apparent wind varies as well. It increases in strength and shifts towards the plane of rotation as you move out along the blade. If the blade designer wants to maintain the angle of attack (to optimize performance) at the same time blade speed is increasing, the pitch of the blade must decrease towards the tip. As a result, blades are twisted from root to tip with the greatest pitch at the root and the least at the tip. Glance up at the next wind turbine you see and note that the tip of the blade is almost parallel with its direction of travel.

Wind machine designers long ago learned that blade area (the number of blades as well as their length and width) governs the amount of *torque,* or turning force, a rotor can produce. The more blades the rotor has for the wind to push against, the more torque it will produce. Greater *solidity,* the ratio of the blade area to the area swept by the rotor, generates greater torque. If you were designing a wind machine for pumping water on the semi-arid Great Plains, you would want a rotor that provides high torque at low wind speeds. This would assure you that the windmill would be able to lift the pump's piston (and the water with it) during late summer when the need is greatest but the winds are lightest. During the rest of the year you would not really care how the rotor performed because there would be less demand for water and more wind than needed to pump it.

There is no better example of a high-solidity rotor than the American farm windmill. It uses multiple blades that taper from the tip to the root (see fig. 6-9) so that nearly the whole rotor disc is covered by blades (80 percent solidity). It was designed to deliver high torque at low wind speeds. It does its job well.

The region east of my home is renowned for the farms of the Pennsylvania Dutch. These descendants of German settlers are known for their simple way of life, wise husbandry, and prosperous farms. They shun power equipment, farm with horse-drawn teams, and use buggies to get about. Water-pumping windmills are a familiar sight on the Pennsylvania Dutch landscape so I receive a lot of calls commenting that wind machines are obviously a good idea if the thrifty Dutch are still using them. These people fail to realize that the Pennsylvania Dutch abhor modern conveniences such as electricity. Windmills are their only source of water. Cost is no object. But to these homeowners cost *is* the object.

The farm windmill, however, was not designed to deliver power economically. When you are dying of thirst you don't care how much the windmill costs or how much power it produces just so long as it pumps some water. But the demands on electricity or power-generating wind machines are different. We don't need the power. Our lives are not dependent on it. We want the power (are willing to pay for it) only when it is a better buy, when it is cheaper than power from a conventional source. To produce economical power we design the wind machine to extract as much power as we can from the wind in the most efficient and least costly manner possible. The farm windmill is not the way to do it.

Admittedly the multiblade farm windmill looks like it would capture more wind than a modern machine with two slender blades. Intuitively, we feel that the rotor should have more blades to capture more wind. If this were true, however, consider what would happen if we carried this belief to its logical extreme. The optimum rotor would cover the entire swept area with blades, in effect producing a solid disc. No air would pass through. The wind would pile up in front of the rotor and flow around rather than through it. The wind speed behind the rotor would be zero. Instead of capturing more wind we would not capture any. There must be air moving through the disc and it must retain enough kinetic energy so it can keep moving to make way for the air behind.

We must reach a balance between a rotor that completely stops the wind and one that allows the wind to pass through unimpeded, between the amount of wind striking the rotor and the amount flowing through. Our friend Albert Betz demonstrated mathematically that the optimum is achieved when the rotor reduces wind speed by one-third. (Academics argue that Frederick Lanchester should also be given credit for deriving the theoretical limits of a propeller turbine. He published his findings in 1915 several years prior to Betz's widely quoted paper in 1920. They both reached the same conclusions.)

When the wind flys into the wide vanes of the farm windmill it is deflected slighty as it continues downstream. The effect from all the blades causes the wind leaving the rotor to spin in a spiral like a corkscrew. To maximize the amount of work the wind can perform (to approach the theoretical limits as near as we can) we must minimize this deflection, minimize the rotation in the wake.

Swirling of the wake is greatest, and so is the power lost, in rotors producing high torque. High solidity rotors such as the farm windmill produce plenty of torque, but they also lose more energy in the wake than do lower solidity rotors delivering less torque. But, you may ask, if we were to lower the rotor's torque wouldn't we be lowering the power it can produce even if it is going to be more efficient at producing it? Yes, if we kept everything else the same. We don't. Power is a product of torque and rotor speed. For the same power output we need to increase rotor speed by an amount equivalent to the reduction of torque. What we give up in torque we gain in rotor speed—and improved conversion efficiency. We come out ahead.

Torque is the product of a force acting on a lever. In our case the lever is the blade of a wind turbine. The lever is longest at the tip and shortest near the hub. To reduce the torque produced by the blade the force acting on it must progressively decrease as the lever arm increases in length or as we proceed along the blade toward the tip.

Lift provides the force needed to produce torque. Lift is a function of blade area and speed. As we move toward the tip blade speed increases and so does lift. We now have two factors reinforcing each other: blade speed increases towards the tip as does the length of the lever through which it will act. To decrease torque we need to decrease blade area—solidity—more at the tip than near the hub. The blade, as a result, tapers from the root to the tip (see fig. 6-13). This also explains why so few blades are used. Designers want to keep solidity as low as possible.

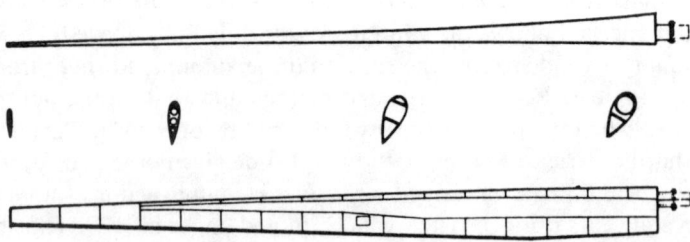

Fig. 6-13. Taper and twist. Rotor blades on conventional wind machines often taper from the root, where the blade attaches to the hub, to the tip. This saberlike shape minimizes solidity. Depending upon the size and its construction, the blade may also be twisted to optimize performance along its entire length. (Lockheed, DOE)

Blade speed is a function of rotor diameter and rotor speed. Both are described by a single term, the ratio between the speed of the blade through the air at the tip and the wind speed: the *tip-speed ratio*. Tip speed increases as either rotor speed increases or the length of the blade increases. Large diameter rotors have higher tip-speed ratios than do smaller rotors spinning at the same speed. And a rotor spinning at a faster rate than another has a higher tip-speed ratio. Consequently, for optimum performance, solidity decreases as tip-speed ratio increases. Drag devices operate at tip-speed ratios of one or less compared to lift devices which operate at tip-speed ratios of five or more.

Still with me? To summarize, lift devices are capable of extracting more power from the wind with less material than drag devices. For them to perform optimally they must operate at low torque but at a high tip-speed ratio. To achieve high tip-speed ratios and low torque the rotor must have low solidity. This is why modern wind turbines operate at high speeds — the rotor is more efficient. It is not, as has been said by others, because electric generators operate at high speeds. If this were the case you could always use a transmission to increase the speed to the generator.

Only one blade is needed to do the job. Two are used for reasons of static balance. Many use three blades because they give greater dynamic stability. In large wind machines the cost of the blades is a major portion of the total cost. By eliminating a blade the cost of the wind machine can be reduced.

Two blades are commonly found on intermediate-size wind machines above 10 meters in diameter. Two blades cause a dynamic imbalance when the wind machine changes direction. When the blades are vertical little force is needed to yaw or to swing the turbine around the tower. But when the blades are horizontal much more force is required because of the rotor's inertia. (The blades like to stay just where they are.) When the rotor is spinning and the wind changes direction the wind machine and tower are subjected to these varying forces twice per revolution. The rotor yaws unevenly. This is not a serious flaw because the yaw of larger machines is dampened. Rotors with three blades avoid the problem and so they are more often used on small machines where dampening would be too costly.

Low solidity rotors do have one drawback. They may not be self-starting. Remember that the apparent wind flowing over the blade is partly due to the blade's motion. When the blade is still, the lift on the blade from the wind alone may not be enough to start it moving. When the blades are fixed in pitch the only solution is to spin the rotor up to a speed where it can drive itself. This is a common practice for Darrieus rotors. There are also conventional wind machines that require motoring the rotor up to speed. But many designers of small machines are willing to sacrifice some performance to gain a self-starting capability.

Wind machines that are self-starting begin turning in winds of 8 mph to 10 mph depending on the brand. Don't be misled by glib talk promoting a "new" wind turbine that runs in low winds. Anybody can design a wind machine to turn in winds of 5 mph to 7 mph. But why bother? There is no energy in winds at these speeds. The rotor may spin but it won't do much else.

In the past, researchers have stated simply that Darrieus turbines are not self-starting. It is now known that Darrieus turbines can be self-starting under the right wind conditions. These conditions—though infrequent—do occur.

The blades on a Darrieus turbine produce lift throughout their orbit around the rotor. Even when the blades are moving downwind (retreating) they see an apparent wind due to their own motion that creates lift (see fig. 6-14). But the speed of the blade must be high in relation to the wind (have a high tip-speed ratio) so that the apparent wind acts on the airfoil at all times.

When the rotor is at a standstill, only the wind across the ground acts on the blade. Because the pitch of the rotor is fixed the blades stall and nothing happens. Normally, the rotor must be motored up to speed. But on 6 July 1978 all that changed and a new corollary was added to Murphy's law, "Wind turbines that won't self-start, will."

The Canadians were testing a 230 kW experimental Darrieus rotor on Magdalen Island in the Gulf of St. Lawrence. While repairs were underway the brake was released. Because it was thought the turbine could not start itself it was left unattended overnight. By the next morning the rotor was spinning out of control. Eventually the rotor spun itself off the tower and drove itself into the ground.

Another misconception about Darrieus rotors is that they are less efficient than conventional wind machines. Research by Sandia Laboratories indicates that under ideal conditions they can extract over 40 percent of the power in the wind. In other words, their performance is similar to that of conventional rotors.

Developers of the Musgrove H-rotor have also found that a vertical axis rotor with fixed pitch blades can be made to start itself. They discovered that decreasing the blade's *aspect ratio*—its height to its width—by shortening and widening the blades, created more lift while the rotor was at rest. The stubbier blades could start the rotor without robbing from performance at operating speeds.

In an H-rotor with articulating blades, such as McDonnell Aircraft's giromill in figure 6-15, the pitch of each blade is set according to a predetermined schedule and the position of the blade relative to the wind. The blade's angle of attack is optimized at each position of its orbit around

Evaluating the Technology • 171

Fig. 6-14. Darrieus rotor. The chord—width—of blades on Darrieus turbines is constant. Note the air brakes at the equator of this Darrieus rotor built by DAF-Indal. They work in a manner similar to the tip brakes on some conventional wind machines.

172 • Wind Energy

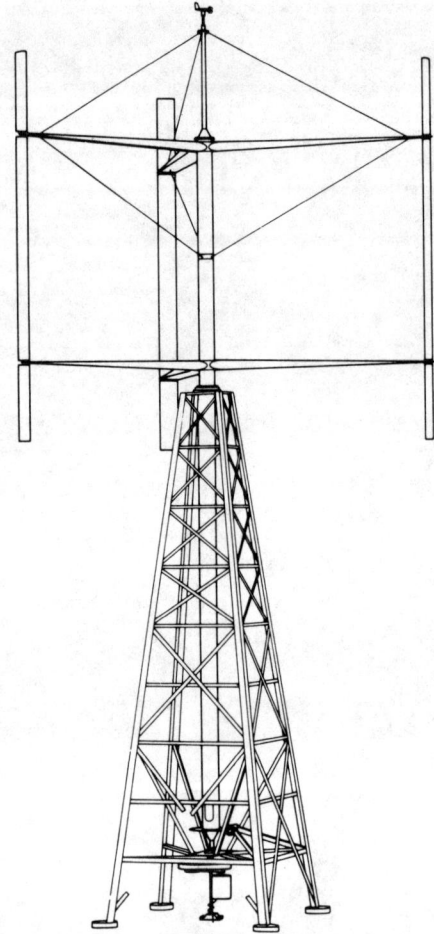

Fig. 6-15. Giromill. Also known among cognoscente as an articulating, straight-bladed vertical axis wind turbine. Note that like the Darrieus it uses blades that have a constant width. The rotor on this prototype is 54 feet in diameter and 42 feet in height. (McDonnell Aircraft, DOE)

the rotor. Controlling blade pitch with respect to the wind gives the rotor a reliable self-starting capability not found in the Darrieus rotor. It should also deliver better performance as well because lift can be maximized regardless of whether the blade is advancing into, across, or with the wind (see fig. 6-16). Giromills, however, have never lived up to expectations. Moreover, they are material-intensive because the cross-arm and torque tube must be large to support the loads on the rotor.

BLADES

Blades can be made from almost any material. Wood has always been popular. Early farm windmills used wooden slats (see fig. 6-4), and the windchargers of the thirties used wood almost exclusively. Wood is still the material of choice for many small wind machines. It is strong, readily available, easy to work with, relatively inexpensive, and has good fatigue characteristics.

For machines up to five meters in diameter single planks of Sitka spruce are carved by machine into the desired shape. They are then coated with a tough weather resistant finish to prevent warping, and a strip of aluminum or copper is glued to the leading edge. The leading edge cover protects the blade from wind erosion and hail damage.

You may find it hard to believe that the wind can erode anything, let alone the blade on a modern wind machine. It can, and it will. (If you need to be convinced of this, pay a visit to the High Plains during the windy season—but take your goggles.) I have seen blades where the leading edge had been chewed away after only two years of use.

Because wood planks are less than uniform it is difficult to control the blade's physical properties. Designers prefer laminated wood (like the construction of a butcher's block) in larger machines because they can control the blade's strength and stiffness better. Shrinkage and warpage

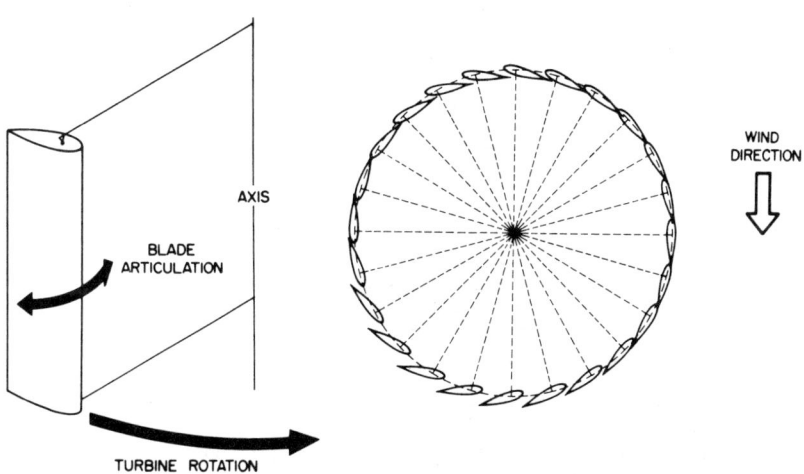

Fig. 6-16. Blade articulation. The mechanism for articulating the blades can be driven mechanically as in the Pinson Cycloturbine or electrically as in McDonnell Aircraft's giromill. (Courtesy Bergey Windpower Co.)

174 • *Wind Energy*

are also better controlled with laminations than with a single plank. In this process slabs of wood are bonded together with a resin forming a block that can be carved into a blade. By varying the types of wood, the direction of their grain, and the resin, a material can be produced that is stronger than any one part alone, and stronger than a single plank of the same size. Laminated wood blades are being used on rotors from 6 to 16 meters in diameter.

Thinner slices of wood are also used to produce a veneer. Layer upon layer of razor-thin slices are sandwiched together with a resin and molded into the airfoil shape. The process is widely used to produce the hulls of

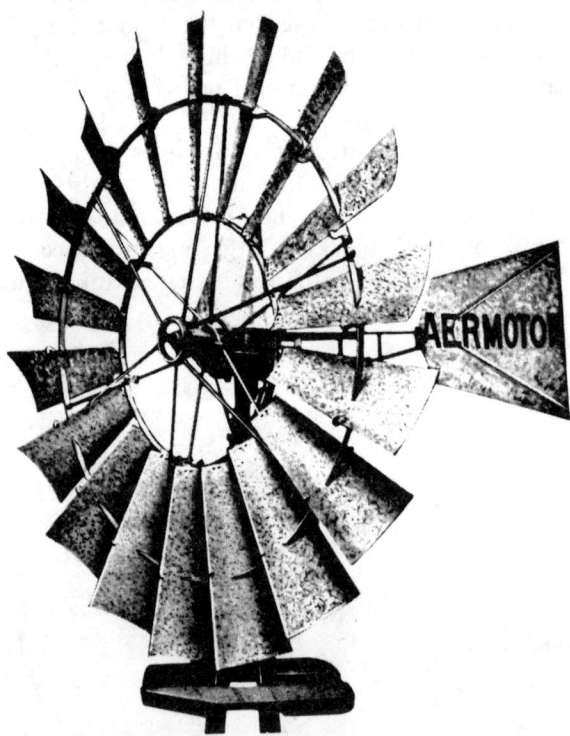

Fig. 6-17. Metal blades. In 1888, Aermotor introduced the first "mathematical" windmill. The curved metal blades resulted from the pioneering work of Thomas Perry who conducted 5,000 experiments on various "wheel" (rotor) designs. The Aermotor embodied all the principles Perry learned and was almost twice as efficient as the wood wheels then commonly in use. It also included "back gearing" which allowed the wheel to make several revolutions for each stroke. Since then Perry's design has been copied worldwide. (Courtesy Aermotor Division, Valley Industries)

sailboats and has been used successfully on blades for the 125-foot diameter Mod-OA wind machine in Hawaii (see fig. 6-4).

In the late nineteenth century galvanized steel began replacing the wooden blades on the farm windmill, and steel has been used ever since (see fig. 6-17). Steel is strong and we know a lot about it. That is why it was chosen by Boeing engineers for the blades on the 300-foot diameter Mod-2 (see fig. 6-35 later in this chapter). It is nothing fancy, just plain structural steel—the same steel used in bridges. Because steel is so heavy the hub, drive train, and tower must be stronger than on a wind machine with a lighter weight rotor.

Aluminum is lighter, and for its weight, stronger. It is used extensively in the aircraft industry for this reason. We can fabricate aluminum blades with the same techniques used to build the wings of airplanes: form a rib and then stretch the aluminum skin over it. The blades on the early Mod-OA were built this way. On smaller machines a simpler method is used. You stamp a curve into the leading edge, fold the sheet metal over the spars, then rivet it in place. One big plus for aluminum is that it can be extruded eliminating all the other fabrication steps. We can mass-produce extruded blades in the same way we manufacture drain spouts and window moldings by squeezing a hot piece of aluminum through a die (see fig. 6-18).

Aluminum, unfortunately, has two weaknesses. It is expensive and it is subject to metal fatigue. Ever take a piece of wire and break it by flexing it back and forth a few times? That is metal fatigue and it works the same way in the wing of an airplane or the blades of a wind turbine. Aluminum is a good material—when used within its limits.

Windcharger was successful in using extruded aluminum blades in their home light plants during the forties. In this era Alcoa and DAF have spent considerable sums developing extrusion technology for use in Darrieus turbines. It is thought that the Darrieus rotor is well suited for aluminum extrusions because the forces on the blades are in tension. The blades endure less stress in the Darrieus rotor than they would in a conventional wind machine. They can also best use a blade of a constant width such as produced by extrusion.

Windworks is one of the few manufacturers using an aluminum blade on a conventional wind machine (see fig. 6-19). Centrifugal loads are carried by a steel cable that runs the length of the blade, not by the blade itself. In this way the blade is spared stresses that have caused fatigue failure on other wind machines.

Fiberglass is increasing in importance. Like wood, fiberglass is strong, relatively inexpensive, and has good fatigue characteristics. It also lends itself to a variety of designs and manufacturing processes. Fiberglass can

176 • *Wind Energy*

Fig. 6-18. Extruded aluminum blades. Alcoa aluminum extrusions lie ready for assembly into a Darrieus turbine 17 meters (56 feet) in diameter. The blades shown are 2 feet wide.

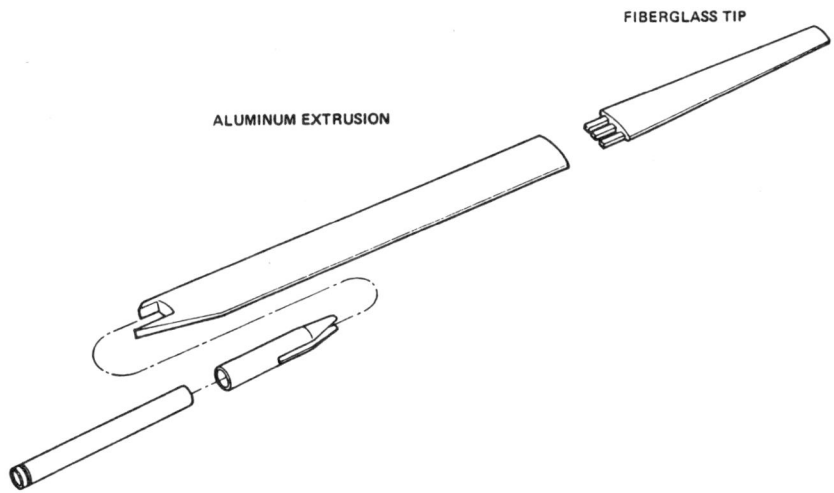

Fig. 6-19. Two-piece blade. Windworks uses a two-piece blade on their 10-meter (33-foot) turbine. Extruded aluminum forms the inboard section. Because extrusions are untapered and untwisted, the outer one-third of the blade is made from fiberglass. This allows Windworks to taper the blade towards the tip where most of the energy is captured. (Courtesy Windworks, Inc.)

be pultrusion, for example. Instead of pushing the material through a die, as in extrusion, fiberglass cloth (like the cloth used in fiberglass auto body kits) is pulled through a vat of resin and then through a die. This technique is used to produce the side rails for fiberglass ladders and is also suitable for mass production. Like extrusion the blades have a constant width and thickness.

Filament winding is another process for making blades from fiberglass (fiber-reinforced polyester or FRP to the Europeans). Originally developed for spinning missile cases, filament winding delivers high strength and flexibility. Often filament winding is used to produce the main spar which is then layered with fiberglass panels in a mold and covered with a smooth fiberglass shell (see fig. 6-20 and 6-21). This process is useful for small series production. It gives the designer the option of varying the blade's width, thickness, and twist, unlike pultrusion. Fiberglass blades laid up and molded with a filament-wound spar are usually found on wind machines larger than 10 meters in diameter, particularly European models.

Cloth is a blade material that should not be overlooked. Sailboats still use cloth successfully as do the Cretan windmills of the Mediterranean. Cloth is cheap. You have to replace it more often than other

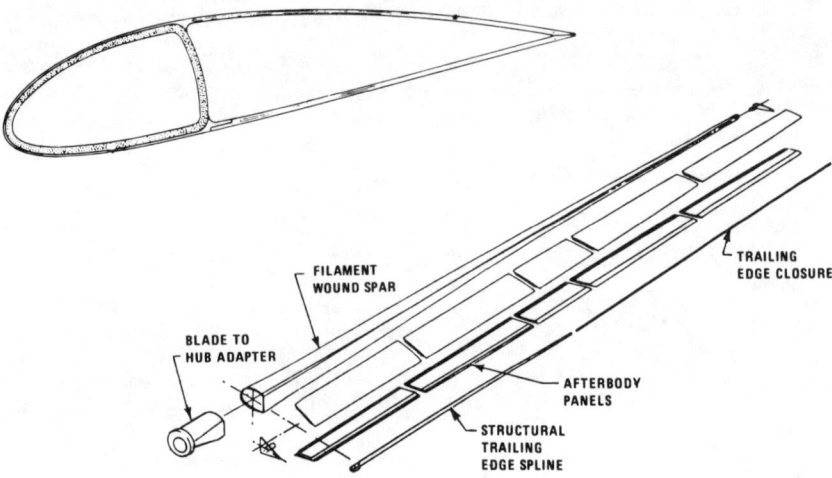

Fig. 6-20. Fiberglass blade construction. The nose or forebody on larger blades is wound from tape of fiberglass filaments. This tubular beam then becomes the main structural member of the blade. Plastic foam or balsa wood supports the aft part of the blade. The assembly is then covered with a smooth fiberglass shell.

Fig. 6-21. Fiberglass blade. Aero-Star 7.5-meter (25-foot) fiberglass blade is used on Vestas 15-meter (50-foot) wind machine as well as on other Danish wind turbines. The self-styled "Great Dane" of Danish blade manufacturers, Aero-Star produces blades in several sizes for stall-regulated wind machines. If the blade is subjected to rotor speeds 20 percent above normal, the blade tips (acting as air brakes) pivot 90 degrees to protect the rotor. (Courtesy Aero-Star)

materials but if it is cheap enough it may make the most economic sense. Cloth blades offer another advantage. They are self limiting. In high winds they simply tear apart leaving the rotor intact. If you are building your own wind machine, cloth is the ideal blade material.

HUBS

The *hub* is what holds everything together. It transmits the motion of the blades into torque. Two aspects of the hub are important: how the blades are attached, and whether or not the hub is flexible.

The blades can be cantilevered from the hub (supported at only one end) or attached with struts and stays (see fig. 6-22). Nearly all wind machines in this country use cantilevered blades. Rotors with struts and stays are more common in Europe following the pattern of the Danish wind machine at Gedser. Struts increase the drag on the rotor. This disadvantage is balanced by the reduction of bending loads on the root of the blade where it attaches to the hub. Consequently, the spar does not have to be as massive as on a cantilevered blade.

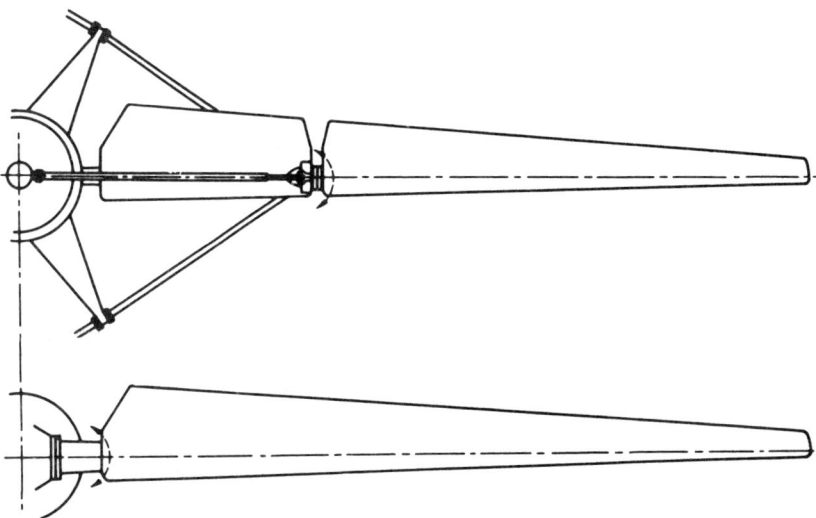

Fig. 6-22. Blade attachment. Two types of blade attachments are shown here. The upper blade is braced with stays. This reduces the loads on the root shaft where the blades join the hub. The lower blade is cantilevered (attached at only one end) from the hub. The root shaft must be considerably stronger than in the stayed blade because it bears all the load. (Danish Ministry of Energy)

180 • *Wind Energy*

Most hubs are rigid. The blades may change pitch but they don't change from the plane of rotation. This can be seen in figure 6-23 where the struts on the Wind-Matic rotor prevent the blades from flexing into the tower. In this example the blades are bolted directly to the hub as in many small wind machines. They do not change pitch. On the other hand, the blades of some stayed rotors are designed to change pitch. But the stays keep the blades in the plane of rotation.

On some large wind machines a rotor made up of two blades may "teeter" about the hub. The rotor, as a unit, swings in and out of the plane of rotation like a teeter totter. This is believed to relieve gust forces on the blade and the effects from tower shadow.

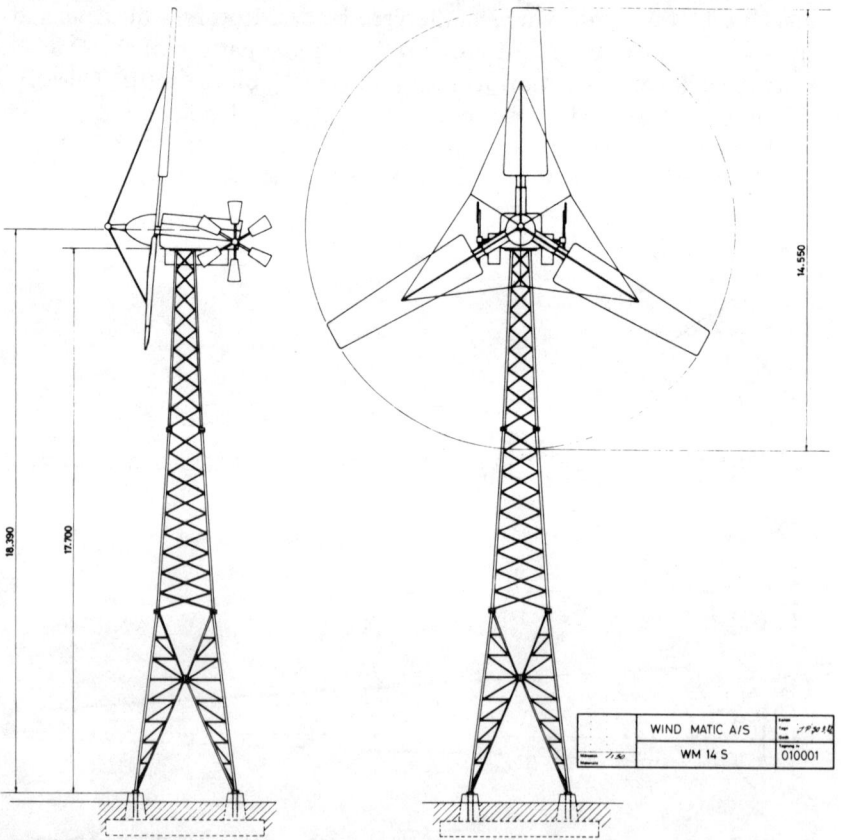

Fig. 6-23. Fixed-pitch rotor with stays. The laminated wood blades on this 15-meter (50-foot) Danish wind turbine are rigidly attached to the hub and braced with stays. The Wind-Matic typifies this feature as well as the use of fan tails to orient the rotor into the wind. (Courtesy Holec, USA)

Evaluating the Technology • *181*

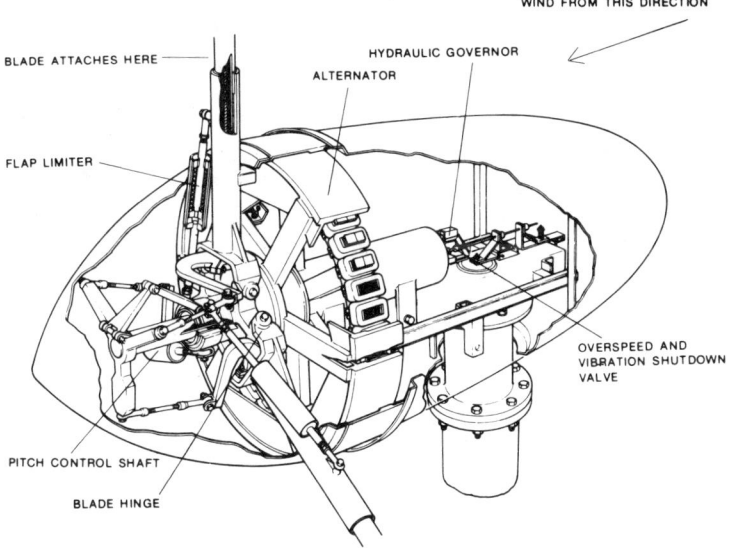

Fig. 6-24. Flapping rotor with direct drive. The hinged blades on this rotor are allowed to cone downwind several degrees in gusts. This sheds bending loads on the aluminum blades and smooths yawing of the wind turbine in gusty winds. The flapping is dampened by a shock absorber. The 10-meter (33-foot) rotor drives a 9 kW permanent-magnet alternator directly without a transmission. Overspeed is prevented by feathering the blades. (Courtesy Windworks, Inc.)

Only one small wind machine today uses a fully flexible hub. Shown in figure 6-24 is the complex hub used by Windworks on their 10-meter downwind turbine. Windworks opted for complexity to reduce stresses on the rotor. Each blade can individually flap 8 degrees downwind and 2 degrees upwind. In gusts this permits the blades to shed some bending loads by coning with the gust forces. The flapping motion is restrained by a shock absorber and the rotation of the blades. As the rotor spins, an equilibrium is reached between centrifugal forces trying to keep the blades in the plane of rotation and gusts pushing the blades downwind.

Following the hub the remainder of the drive train consists of the main shaft to which the rotor is attached, the transmission (where used), and the generator (see fig. 6-25).

TRANSMISSIONS

There are three ways to transfer power from the rotor to the generator. The simplest method is to drive the generator directly with the

182 • *Wind Energy*

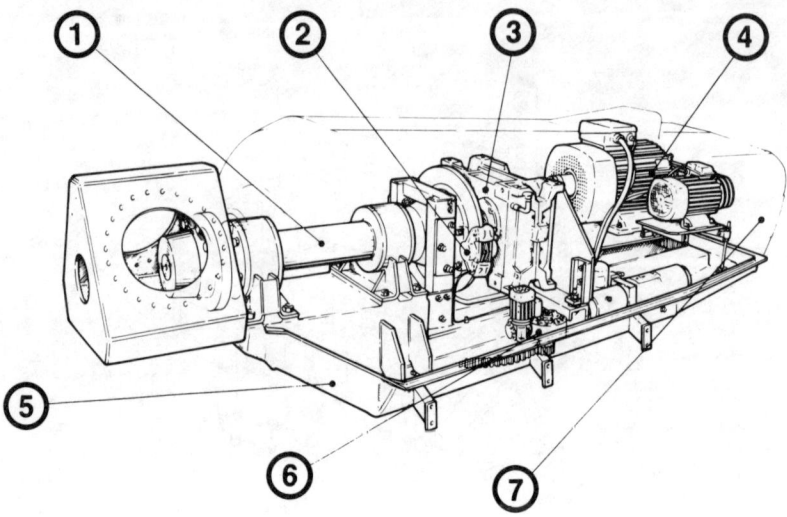

Fig. 6-25. Drive train for induction wind machine. Typical drive train on wind machines using transmissions and induction generators. (1) Main shaft. Note that the bearings supporting the main shaft are independent of the transmission. On some wind machines the bearings in the transmission housing support the rotor. (2) Disc brake. Note position on the main shaft. On some wind machines the brake is located on the output side of the transmission. (3) Transmission. (4) Induction Generator. European wind machines, such as this one, commonly use two generators. (5) Bedplate (frame or strongback). (6) Yaw or slewing drive. Many wind machines in this country are passively oriented into the wind. (7) Nacelle. (Courtesy Vestas a/s)

rotor. This eliminates the need for a transmission and reduces the complexity of the drive train. Another argument in favor of direct drive is higher conversion efficiencies because no power is lost in a transmission. Direct-drive though requires a slow-speed generator that may be larger and require greater amounts of expensive materials than a comparable generator driven at higher speeds via a transmission. The most successful of the pre–REA era windchargers, the Jacobs home light plant, used direct drive. The largest direct-drive wind machine today is a 10-meter turbine built by Windworks. Only a few other manufacturers, notably Northwind Power Company and Bergey Windpower Company, produce direct-drive wind machines (see table 6-1).

Most of the many products on the market use mechanical transmissions as speed increasers to step-up the speed of the main shaft to that required by the generator. The rotor speed on even the smallest wind

TABLE 6-1
Transmissions and Generator Characteristics

Wind Machine	Rotor Size m	Turbine rpm	Transmission	Generator rpm	Type
Bergey	3	800	direct-drive	800	permanent magnet alternator
Sencenbaugh	4	350	3:1	1100	alternator, 12 pole, 3 phase
Enertech	4	170	11:1, two stage	1800	induction generator
Northwind	5	250	direct-drive	250	alternator, 12 pole, 3 phase
Enertech	6	106	17:1, one stage	1800	induction generator
Windworks	10	160	direct-drive	160	permanent-magnet alternator
Vestas	10	84	21:1[a]	1800[a]	induction generator
	15	53	34:1[a]	1800[a]	induction generator

[a]Inferred

machine in table 6-1 does not exceed 800 revolutions per minute (rpm) yet most generators must be spun at speeds approaching 1,800 rpm.

The chief competitor of the old Jacobs generator, Wincharger, took the transmission approach. They used one large helical gear on the main shaft of the rotor to drive a small gear on the generator. Sencenbaugh Wind Electric's Model 1000 uses a similarly simple approach. The transmission increases the 350 rpm speed of the wind turbine to 1,100 rpm at the generator. Even with this 3:1 gear ratio a low-speed alternator must still be used relative to generators running at 1,800 rpm (see table 6-1).

On small machines such as the Sencenbaugh or old Wincharger it is also possible to use belts and chains instead of gearing. Most kits and homebuilts use belts and pulleys because they are cheap and readily available. Cogged belts have been used in commercial wind machines. They are presumably more efficient at transferring power than V-belts. Chains and sprockets, in contrast to belts, are noisy and more difficult to tension. No one uses either cogged belts or chains today on commercial wind machines.

As rotor size increases the speed of the main shaft decreases. Consequently as wind machines become larger the need for transmissions increases as does the gear ratio of the transmission (see table 6-1). A standard generator in a 6-meter turbine running at 100 rpm requires a transmission with a higher gear ratio (18:1) than a 4-meter turbine whose rotor spins at 200 rpm (9:1).

For small machines, transmissions with one or two stages of parallel shafts dominate. At the upper end of the size range, planetary or epycyclic transmissions are more cost-effective.

Hydraulic transmissions are another option. They can be more easily matched to the torque characteristics of a wind turbine rotor and over a wider range than a mechanical transmission. They should also be simpler and lower in cost. These advantages are offset, however, by greater transmission losses. The only large-scale test of hydraulic transmissions, the Bendix-Schachle turbine of Southern California Edison Company, ended in failure.

GENERATORS

The *generator* converts the mechanical power of the spinning wind turbine into electricity. In its simplest form a generator is nothing more than a coil of wire spinning in a magnetic field. Consequently, whether generating direct current (DC) or alternating current (AC), a generator must have

1. coils of wire in which the electricity is generated and through which it flows,
2. a magnetic field, and
3. relative motion between the coils of wire and the magnetic field.

By varying each of these conditions you can design a generator of any size for any application.

Power in an electrical circuit is the product of current and voltage. In a generator the coil of wire where output voltage is generated and through which current flows to the load is called the *armature* (see fig. 6-26). The portion of the generator where the magnetic field is produced is called the *field*. Relative motion between the two is obtained by either spinning the armature within the field as in figure 6-26 or the field within the armature. As you would expect, the stationary part of the generator is the *stator*; the spinning part is the *rotor*.

Generators are not perpetual motion machines. They transfer power, not create it. Power must be delivered to a generator before you can get power out of it. (In our case the prime mover, as it is called, is the wind turbine.) Nor are generators 100 percent efficient at transferring this power. The wind turbine will deliver more power to the generator than it produces as electricity. As a consequence the size of a generator only indicates how much power it is capable of generating if the wind turbine is big enough and there is enough wind to drive the generator at the right speed.

The power produced by a generator depends on the size and length

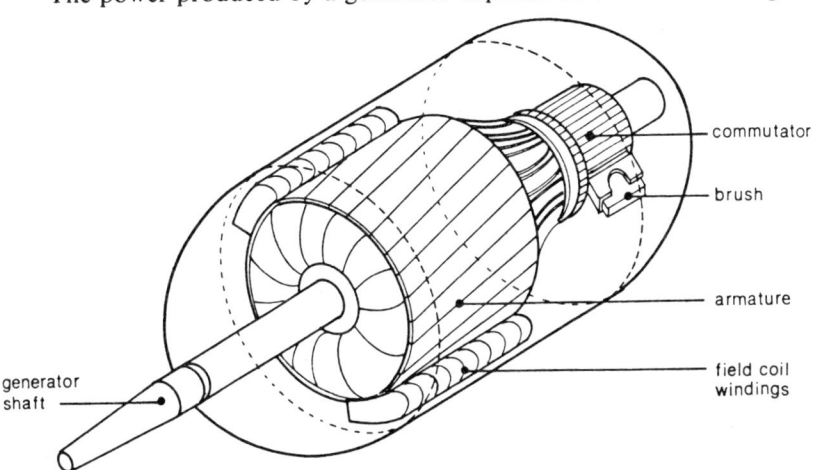

Fig. 6-26. **DC generator. Sketch of direct-drive generator on Jacobs windchargers. Power is drawn off spinning armature through brushes. (Energy Task Force, CSA)**

of the wires used in the armature, the strength of the magnetic field, and the rate of motion between them. Increase any one, and you increase the power the generator is capable of producing.

The size of the wire in the armature determines the maximum current that can be drawn from the generator before it overheats, melts its insulation, shorts-out, and otherwise destroys itself. The heavier the wire, the more current it can carry. As long as the wind turbine continues to provide greater and greater amounts of power as wind speed increases, the generator will continue to produce more current as the load demands it until the generator overheats. To prevent such occurrences generators have some mechanism for overcurrent protection that limits current to a safe maximum.

Generators are rated in terms of the maximum current they can supply at a specified voltage and (for AC generators) at a specific frequency. This rating is given on the name plate as a list of amps, volts, (and frequency where appropriate), as kilowatts and volts, or as kilovolt-amperes (kVA). The generator may be rated for the current it can supply continuously or the current it can supply for only a short period. If generator size is of concern to you, always check which rating is being used. Reputable manufacturers rate their generators for continuous rather than intermittent duty.

Let's turn to voltage, the other half of the power mix. Generated voltage depends on the rate at which magnetic lines of force are broken by the wire loops in the armature. Designers alter voltage by changing the magnetic field, by changing the rate of motion between them, or by both.

The generator's field is provided by magnets. With electromagnets, some power is used to "excite" or "energize" the field around the armature. The strength of this field is a function of the length of wire in the field windings and the current flowing through them. For example, double the length of wire in the windings and you double the strength of the field doubling generated voltage.

Many of the windchargers built during the thirties produced 32 volts. Resistance losses are high when transmitting low voltage power. Because of this, most reconditioned windchargers were rewound for 110 volts. The old wire was stripped off and replaced with more turns of thinner wire. Less current could be drawn through the smaller wire than before, but the increased length of wire produced a stronger field increasing the voltage. Generating capability was not affected, power from the generator remained the same, but the balance between the voltage and current changed: voltage increased and current decreased by an equal amount.

Permanent magnets can also provide the field. They do not require power for excitation. They are inherently—permanently—magnetic. The only way to increase field strength with permanent magnets is to use magnets with greater magnetic density.

The lines of force cut by the armature can also be increased by adding more field coils (or more permanent magnets) or by increasing the speed at which the armature windings pass through the field. This can be accomplished by spinning the rotor faster or by increasing the diameter of the generator.

Believe it or not all this has some bearing on the design of wind-driven generators. To get a feel for how, let's examine two popular wind machines of the thirties. Because both Jacobs and Wincharger used approximately the same size rotor (14 feet) the power available to the generator and the speed of the wind turbine were roughly equivalent. Yet Jacobs chose to use a direct-drive generator (see fig. 6-26) where Wincharger chose a transmission. To produce the same power as Wincharger at the desired voltage, Jacobs had to design a generator that would operate at lower shaft speeds than Wincharger's generator.

Jacobs increased both the diameter and the length of his generator over that of Wincharger's. This allowed the use of more field coils (six to Wincharger's four) with more wire. The greater diameter also increased the speed at the periphery of the armature where it passed the field coils. Doubling the diameter doubles the rate at which the armature cuts through the field with the same effect as a 2:1 transmission that doubles the speed of the rotor. All in all the Jacobs generator was considerably larger and used much more copper and iron than Wincharger, but it turned slower. Jacobs chose a slow-speed generator for long bearing life and simplicity. These advantages were offset by greater cost. Wincharger opted for a less expensive generator but one that would, presumably, have a shorter life.

"There's no such thing as a free lunch," puts it succinctly. Whether it is the design of generators or any other wind machine component there is always a trade-off. You gain by giving up something else. You hope that what you gain is more valuable than what you have given up. It is as true today as it was during the thirties. Designers who stress long life and low maintenance choose lower generator speeds at increased cost. For example, commonly available induction generators operate at 1,800 rpm or 3,600 rpm. But only the 1,800 rpm generators find application in wind machines using "off-the-shelf" components. (See table 6-1; note those with induction generators.) Designers seeking even greater reduction in wear and potential service calls, such as for those wind machines intended for remote sites in harsh environments, may opt as Jacobs did, for building a slow-speed generator tailored to their wind turbine. That's just what

Bergey, Sencenbaugh, and Northwind have done. All three build one-of-a-kind slow-speed generators (see table 6-1). Two, Bergey and Northwind, chose to use direct-drive.

Windchargers produced DC by spinning the armature within the field (see fig. 6-26). Power was drawn off the rotating armature through brushes. In the auto industry alternators began replacing DC generators during the sixties. Though alternators produce AC they offer several advantages over DC generators. For a given output, alternators are easier and cheaper to build than generators. They produce more power for a given rotor speed than a comparable generator. Brushes in an alternator also last much longer, because they do not carry the alternator's output.

In today's alternator the field revolves rather than the armature. Power is drawn off the stator from fixed terminals. There are no brushes or commutator to wear from the passage of high current. There is no arcing at the brushes. Excitation of the field is provided through slip rings on the rotor. But only enough power passes through the slip rings to excite the field (a small percentage of the alternator's output). Wear is negligible in comparison to brushes on a DC generator. There are no slip rings—no moving contacts—in a permanent-magnet alternator (the field is permanently excited).

As the name implies, *alternators* generate alternating current. As the rotor spins, current rises and falls like waves on the ocean (we believe that electrons in the armature are first jostled in one direction, then alter course and are jostled the other way). The rate at which current rises and falls or changes direction is called *frequency* and is given as cycles per second (Hertz). The speed of the rotor and the number of poles determines the alternating current's frequency. Most alternators use four poles so frequency is directly proportional to rotor speed. Drive the alternator faster and frequency increases; slow the rotor and frequency decreases.

In a simple alternator the four poles are wired together in series as a single circuit producing single-phase AC. When three groups of poles are arranged symmetrically around the stator, the alternator produces three-phase AC, each phase one-third out of sync with the next. Most alternators used in wind systems produce three-phase AC (see table 6-1). Output for residential use is converted to single phase.

Three-phase alternators do more with less. The designer can more efficiently pack poles within the generator. Power is determined by the rate at which lines of force are cut by the armature. Thus we can increase power by increasing the number of poles to take up all the available space within the generator.

If you have ever spun the shaft of a toy generator in your hand you remember how it felt when the rotor would "stick" slightly as the coils in the armature aligned with the magnets in the field. As the coils passed the magnets, the shaft would turn more easily. This same effect, cogging, occurs in large generators and motors. Cogging is of interest in wind machines because it can retard the start-up of the wind turbine in light winds when the poles are aligned. Increasing the number of poles by arranging them in three-phase reduces cogging.

There are, in a broad sense, two approaches to electrical generation with a wind machine:

1. *asynchronous*; where the speed of the wind turbine varies with the speed of the wind, and
2. *synchronous*; where the speed of the wind turbine remains constant as wind speed fluctuates.

When an asynchronous wind machine drives an alternator both the voltage and frequency vary with wind speed. The output is what I label "junk electricity." It is not compatible with the constant voltage, constant frequency AC produced by the utility. If you used it in your home, clocks would gain and lose time and lights would brighten and dim as wind speeds rose or slackened. Eventually you would burn up every motor in the house. But it can be used where high-grade electricity is not required such as in heating hot water with resistance coils.

The output from asynchronous wind machines can also be "conditioned" for charging batteries or for producing utility-compatible electricity. Batteries cannot use AC. As in your car, the AC output from the alternator is rectified to DC by *diodes*—electrical check valves that permit the current to flow in only one direction. The DC is then used to charge the battery.

Either the AC or rectified DC output can also be used with a synchronous inverter to produce constant voltage 60-cycle AC like that from the utility. These inverters are usually line commutated; that is, they must be interconnected with the utility to operate. The utility's alternating current provides a signal to fire a bridge of electronic switches (thyristors or transistors) within the inverter transferring the junk electricity at just the right time to produce 60-cycle AC at the same voltage as the utility. No utility power is consumed in the process; it is merely used as a signal to coordinate the switching.

Technically, the term synchronous inverter applies to the power conditioning required for DC. Most small machines such as the Bergey, in table 6-1, rectify the alternator's output at the wind machine and use a synchronous inverter on the DC output from the wind machine. On larger machines, such as the Windworks turbine, three-phase AC is brought down the tower where it is conditioned—cleaned up into a usable form—by a "frequency changer." Since it is already AC, the power does not have to be inverted as such.

Synchronous wind machines do not require power conditioning. They produce power synchronized directly with that from the utility. There are two ways in which this can be done. The first is to drive an alternator at a constant speed so voltage and frequency remain constant. Synchronous alternators are used primarily in the utility industry. Power to the generator is controlled precisely by limiting steam to the turbines. This keeps the generator spinning at the right speed to maintain synchronization with the utility's other generators. On a wind machine it is much more difficult to control the turbine's speed because its source of power, the wind, constantly varies. Complex hub mechanisms for changing the pitch of the blades or electrical controls for increasing or decreasing the load are required to maintain a constant generator speed. To date, only large wind machines, those 100 to 300 feet in diameter, have been able to use or justify the cost of synchronous alternators successfully.

Induction generators are used in the second method. They have two advantages over alternators: they are cheap, and they can supply synchronous power without sophisticated blade or load control. Induction generators are induction motors (like the motor in your refrigerator) in disguise. The difference is that an induction motor becomes a generator when driven above its synchronous speed. Plug an induction motor into an outlet and the motor will turn at 1,800 rpm consuming power. Leave it plugged in, but now drive the motor at 1,800 rpm. The motor will no longer consume power from the outlet. You are now supplying it. Spin the rotor just a little faster, say at 1,820 rpm, and it not only won't be consuming electricity it will be generating it. As you try to spin the motor faster it gets harder to turn. The utility consumes the additional power as you produce it without rotor speed appreciably increasing. The beauty of an induction generator is that the load automatically increases as more torque (power) from the wind turbine is applied when wind speed increases. This continues until the generator reaches its limit and slips—breaks away from the grip of the utility and runs wild.

Induction generators have proven popular. First, they are readily available in a range of sizes and are produced by a number of manufacturers. Secondly, no development is needed. Design your wind turbine, find the right transmission, and pick an induction generator. Development and lead time are minimized. Third, interconnection is straightforward and does not require any inverters. Literally, plug it in and go. Early Enertech promotions showed their wind machines being plugged into a wall socket. Interconnection is a little more sophisticated today (the machine is wired to a dedicated circuit in your service panel) but the principle remains the same. This is an advantage not so much from a cost standpoint but because it helps when dealing with the utilities. They understand induction generators. Synchronous inverters however, remain mysterious.

From the early to mid-seventies several firms began rebuilding the DC generators from the Windcharger era. As the supply diminished two of these companies (Dakota Wind and Sun and Independent Energy Systems) began to manufacture DC generators patterned after the Jacobs home light plants seemingly ignoring the superior performance and lower cost of alternators. Few improvements were made. These machines were intended originally for charging batteries at remote sites. As demand grew for wind machines capable of generating utility-grade power these DC generators were adapted for use with synchronous inverters. The DC generators never performed well in interconnected applications, and both firms had problems with brush wear and quality control that forced them into bankruptcy. Today only one manufacturer still produces a DC wind generator: Winco. The wind turbine is small, a mere toy, and is marketed to hobbyists.

During the fifties wind machines using alternators were developed. The output was rectified to DC for charging batteries at remote sites. Two of these machines, the Swiss Elektro and the Australian Dunlite, are still in production.

The industry today is divided among manufacturers who use alternators or induction generators. Yet there is still a great deal of diversity in design. Consider two wind machines in table 6-1 — Bergey and Windworks. Both use permanent-magnet alternators driven directly by the rotor. But where Windworks opted for high magnetic density, rare-earth magnets, Bergey chose more commonplace materials.

In a conventional alternator the field revolves inside the stator. But both Bergey and Windworks spin the permanent-magnet field outside the stator (see fig. 6-24). The blades of the rotor are attached to the magnet ring that spins around the armature. Power is drawn off from inside the generator. Once again, though, Windworks and Bergey differ in their approach. Windworks uses a large diameter magnet ring, but only one of them. This gives them a high peripheral speed and room for more magnets, thus maximizing the rate at which lines of force are cut by the armature. Bergey increased the number of magnet rings instead by stretching the alternator's length.

When looking at a wind machine's generator there is no need to be dazzled by the technology employed. Your primary concern is what kind of power it produces, the form that it takes. If you want utility-compatible power then you cannot use a wind machine with an alternator that does not also include a synchronous inverter. And this inverter should not be something slapped together just for the occasion. Experience with inverters has shown that they must be carefully tailored to the generator. That is why, elsewhere in this book, I suggest using rebuilt windchargers only for heating water or charging batteries. Where inverters have been used with these machines (as with more modern machines) for which they were not designed, they have performed poorly if they have worked at all. Today, manufacturers don't design their generator and then go looking for any old inverter to use with it. They design the two as a team.

The inverter not only produces 60-cycle AC at the correct voltage but it also performs an important function. It loads and unloads the generator as more or less wind is available. When the inverter and generator are improperly matched, the wind machine will not perform optimally. The turbine may take higher winds to start than necessary if the generator is highly loaded. It may never reach the tip-speed ratio where it performs most efficiently. At the other extreme the inverter may not load the generator sufficiently to extract as much power as it can.

Likewise, if you want a wind machine for charging batteries at a remote hunting cabin you will not be able to use an induction generator. They only work when interconnected with the utility.

You also have to look at what is available. Wind machines larger than 10 meters in diameter use induction generators almost exclusively. If you want a slow-speed, permanent-magnet alternator on a wind machine larger than 10 meters in diameter you are out of luck.

As mentioned in an earlier chapter, do not be swayed by the size of the generator alone. It is only an indication of how much power the generator is capable of producing, not how much it will generate. Ask a Danish manufacturer what size generator they have in their machine and they will look at you quizzically and ask, "Which one?"

Danish wind machines often use two induction generators, one for low winds and another much larger generator for higher winds. Generators operate inefficiently at partial loads. For wind machines with a generator designed to reach its rated output in a 25 mph wind the generator would be operating at partial load most of the time. Rather than use one generator they instead bring a smaller one on line first so that it operates at nearly full load in low to moderate winds. As wind speed increases the smaller generator drops out and the larger or main generator is energized. Thus, both generators operate more efficiently and overall performance of the wind machines is improved, they assert. The generators may be in tandem and driven by the same shaft or they can be side by side. Usually both generators are spun at the same time and are not brought on line mechanically but by energizing the field electrically.

ROTOR CONTROLS

The most basic and foolproof method for controlling rotor speed is to decrease the area of the rotor intercepting the wind. As frontal area decreases less wind acts on the blades reducing the rotor's torque, power, and speed. The thrust on the blades (the force trying to break the blades off the hub) and the thrust on the tower (the force trying to knock the tower over) is also reduced. This permits the use of lighter weight and less expensive towers than on wind machines where the rotor remains facing into the wind.

Halladay's wind machines exemplify the concept. They automatically opened their segmented rotor into a hollow cylinder in high winds (see fig. 6-4) letting the wind pass through unmolested. Each segment was composed of several blades mounted on a shaft allowing the segment to swing into and out of the wind. Halladay's umbrella mill looks like any other water-pumping windmill when the segments are closed. But in high winds, thrust on the segments would force them to flip open. This action was counterbalanced by weights so the farmer could adjust the speed at which the windmill would stop running.

Later developers, such as the Reverend Leonard Wheeler, chose to use the same technique (changing the area of the rotor intercepting the wind) but in a different manner. Rather than swing segments of the rotor out of the wind, Wheeler thought it simpler to swing the entire rotor—in one piece. He could not do this with the downwind rotor used by Halladay. Instead he used a rotor upwind of the tower with a tail vane that kept the rotor pointed into the wind. The tail vane and rotor (windwheel to old-timers) were hinged to permit the rotor to swing sideways towards the tail. As the rotor folded or furled towards the tail the rotor disc took the shape of a narrower and narrower ellipse gradually decreasing the area exposed to the wind. The mechanism for executing this was the pilot vane, the small vane below the tail vane in figure 6-27. The pilot vane extended just beyond and parallel to the rotor disc. Unlike the tail vane it was fixed in position relative to the rotor. Wind striking the pilot vane pushed the rotor towards the tail and out of the wind. In the folded position the rotor and pilot vane were parallel to the wind like the segments of the Halladay rotor. The thrust on the pilot vane was counterbalanced with weights. By adjusting the weights the farmer could determine the wind speed at which the rotor would begin to furl.

As the American farm windmill evolved, the pilot vane went the way of hand starters on automobiles. Offsetting the axis of the rotor slightly from the axis about which the wind machine yaws or pivots around the top of the tower produced the same results: self-furling in high winds.

194 • Wind Energy

Fig. 6-27. Pilot vane. The small vane below the tail on the "OK" windmill in the foreground is the "pilot vane." In high winds it pushes the rotor towards the tail as shown.

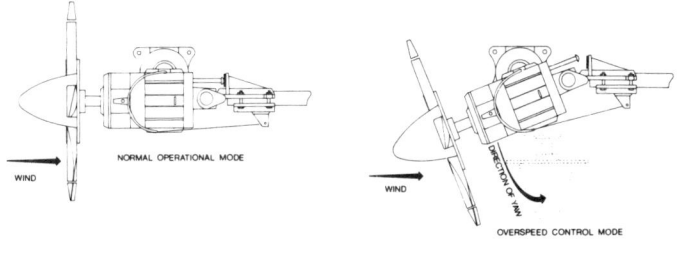

Fig. 6-28. Self-furling overspeed control. Again, note that the axis of the rotor is offset from the pivot or yaw axis. In high winds, thrust on the rotor forces it to fold towards the tail as shown. (Courtesy Sencenbaugh Wind Electric Co.)

When the rotor axis is offset from the yaw axis, the wind's thrust on the rotor creates a force acting on a small moment arm (lever) represented by the distance the two axes are offset. This thrust turns the rotor out of the wind (see fig. 6-28).

As before the tail vane is hinged to allow the rotor to furl. But there are no weights and levers to counteract the furling thrust. Instead modern water-pumping windmills use springs. By adjusting the tension in the spring the farmer controls the windspeed at which the rotor will furl. To see this for yourself find an operating farm windmill and watch it in high winds. It will constantly fold towards the tail and reopen without any intervention. Millions of machines using Wheeler's approach to overspeed control have been put into operation around the world. It is what you might call a proven concept. And if it worked reliably for all those machines for all those years it should still work today. It does.

The Sencenbaugh Model 1000 and the Bergey Model 1000 are two modern wind machines that use the same principle. In figure 6-28 the self-furling action of the Sencenbaugh 4-meter turbine is illustrated. Self-furling provides accurate and predictable control over rotor speed and power. Both the Sencenbaugh and the Bergey (see fig. 6-29) wind machines have operated unattended in wind speeds over 100 mph. Both were also designed for applications requiring high reliability with low maintenance. Their designs stress simplicity. In both wind machines gravity is used rather than springs to return the rotor to the running position. The hinge pin for the tail vane on these machines is skewed several degrees from the vertical. When the rotor furls in high winds it lifts the tail vane several inches. As the wind subsides the weight of the tail pulls itself down into position forcing the rotor to swing back into the wind.

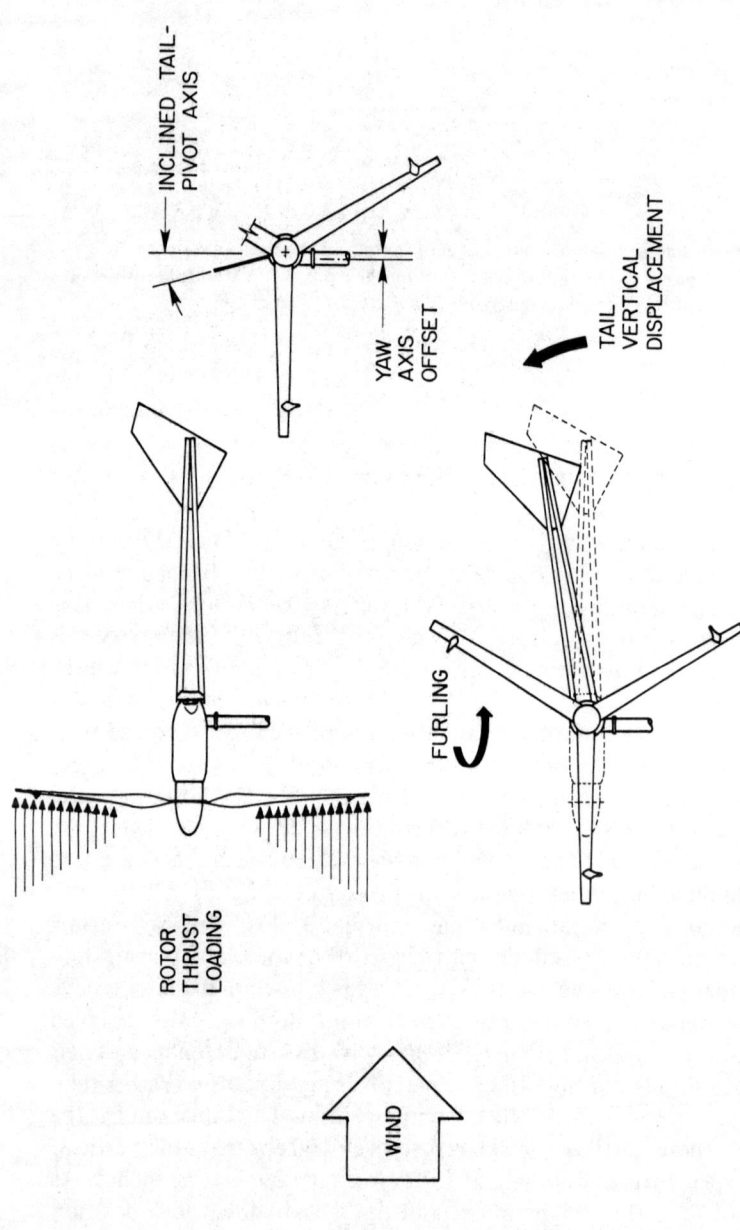

Fig. 6-29. Bergey Model 1000. Like the Sencenbaugh 1000 this wind machine is self-furling in high winds. Note that the yaw tube (the tube connecting the wind machine to the tower) pierces the nacelle off-center. As before, the rotor axis is offset from the yaw axis causing the rotor to fold towards the tail. The weights near the blade tips twist the flexible fiberglass blades to improve performance. (Courtesy Bergey Windpower Co.)

Wind machines using this approach, like the water-pumping windmills before them, can be controlled manually by furling the rotor with a winch and cable. The rotor does not come to a complete stop when furled. It continues to spin, but at low speeds and power.

During the thirties, Parris-Dunn built a windcharger that used a variation on the furling theme. Rather than turning the rotor parallel to the tail they chose to tip the rotor up out of the wind (see fig. 6-30). It would look something like a helicopter in high winds. As the winds decreased the rotor would rock back towards the horizontal. Northwind's HR 2 is a modern wind machine that has adopted the same technology. The only difference has been Northwind's addition of a large return spring and shock absorber to return the rotor smoothly to the running position. Gusty winds can cause the rotor to tip up then quickly rock forward dropping the rotor and generator onto the wind machine's frame severely jarring the blades and the rotor's main shaft. Parris-Dunn simply used a rubber pad to

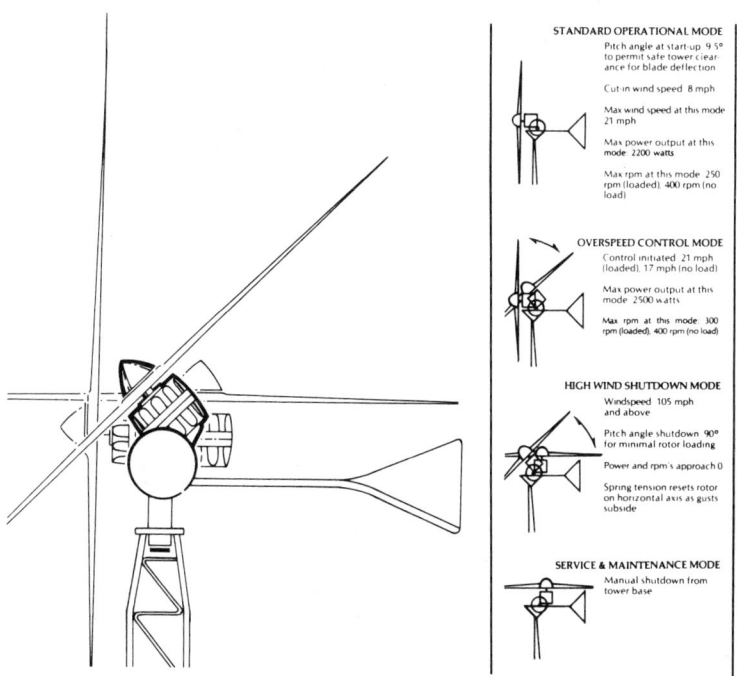

Fig. 6-30. Variable axis overspeed control. Rather than swinging the rotor to the side, Northwind Power Company's HR 2 tilts the rotor skyward following the example of a thirties era windcharger. (Courtesy Northwind Power Co.)

198 • Wind Energy

cushion the blow but this was not enough protection for the 5-meter Northwind machine. The wind speed at which the rotor begins to pitch back is governed by tension in the return spring.

Rotor speed can also be controlled on a vertical axis wind turbine by varying intercept area when the rotor uses straight blades (see fig. 6-31). As mentioned before, the H-rotor offers several advantages over a conventional Darrieus turbine. Its disadvantage is the tremendous forces trying to bend the blades at the juncture between them and the cross-arm. These bending forces can be reduced and the speed of the rotor controlled by hinging the blades.

In an English design, the blades are hinged to the cross-arm in such a manner that the portion of the blade above the cross-arm is not equal to the portion below. As the rotor spins, centrifugal force throws the heavier portion of the blade away from the vertical, varying the geometry of the rotor. The wind and rotor speed at which this occurs is determined by the weight of the blade and the tension in a spring restraining the blades in the upright position. At high wind speeds the blades approach the horizontal reducing the intercept area. We learned earlier that swept area in an H-rotor is the product of the height of the blades and the diameter. Though their length remains the same the height of the blades gradually decreases eventually reaching zero when the blades are horizontal. West Virginia University developed a variation of this concept by rocking the blades towards the axis of rotation.

Rotor speed can also be controlled by changing blade pitch. Like changing intercept area, changing blade pitch affects the power available to the rotor. By increasing or decreasing blade pitch we can control the amount of lift that the blade produces. This can be accomplished in two ways: by pitching the blade to stall, or to feather. Blade pitch is usually set a few degrees into the wind, for example 10 degrees from the plane of rotation. The blade is stalled by turning it towards the direction of

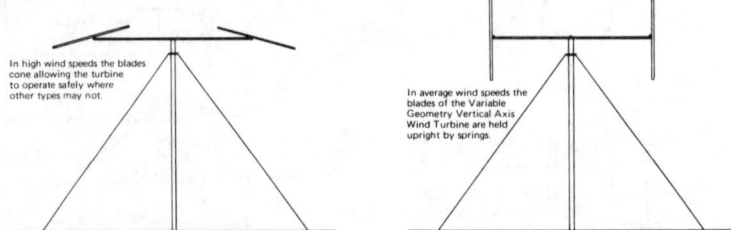

Fig. 6-31. Variable geometry H-rotor. In this ingenious design developed by Dr. Peter Musgrove of Reading University, the straight blades of the rotor are hinged so that they tilt toward the horizontal at high rotor speeds. (Courtesy P.I. Specialists, Ltd.)

travel and decreasing the pitch. When the blade is parallel to its direction of travel, pitch reaches zero. Thus, to stall the blade, it need be turned only a few degrees. The blade is feathered, on the other hand, when it is at right angles to its direction of travel (90 degrees pitch). To feather a blade it must be turned much further (in our example, a full 80 degrees) than when stalling the blade, causing the pitch mechanism to act through a greater distance.

Stall destroys the blades' lift limiting the power and speed of the rotor but it does nothing to reduce the thrust on the rotor or the tower. Though it is simpler to build a mechanism for stalling the blade than it is to build a feathering governor the technique is less reliable. On upwind machines, thrust on the blades bends them towards the tower. Brands dependent on blade stall as the sole means of overspeed protection have a poor survival record. The blades have a nasty habit of striking the tower. Downwind turbines using stall regulation have fewer problems because the blades are forced to cone farther downwind and away from the tower. Where changing blade pitch is the primary means of control on an upwind rotor the blades should rotate towards full feather. By doing so they reduce the drag on the blade to one-fifth of that on a blade flat to the wind.

Governors appear in a variety of forms. During the thirties, the Jacobs brothers popularized the flyball governor. Above normal rotor speeds the weights would feather all three blades simultaneously via the mechanical linkage (see fig. 6-32). (It is important that all blades change pitch at the same rate. If they do not, the rotor will become imbalanced causing severe vibrations.) This massive 70-pound governor protected the 14-foot rotor reliably when carefully adjusted. On later models, Jacobs also marketed a clever version called the "Allied" (after a windcharger on which it first appeared) or blade-actuated governor.

Why use weights when you don't have to? The blade-actuated governor uses the weight of the blades themselves to change pitch. Unlike the blades on the flyball governor, the blades not only turn on a shaft in the hub, they also slide along the shaft. Each blade is connected to the hub through a knuckle and springs. The knuckle in turn is attached to a triangular spider. As the rotor spins, the blades are thrown away from the hub causing them to slide along the blade shaft. When they do, the blades pull on the spider which rotates the blades about the shaft towards feather. The springs govern the rotor speed at which this occurs.

During the seventies several small firms began to produce a modern equivalent of the Allied governor to meet a demand for spare parts. (At the time many people were reconditioning used Jacobs windchargers.) Soon two firms began to build copies of the Jacobs generator complete with Allied governor. New wind machines were also developed that used the

200 • *Wind Energy*

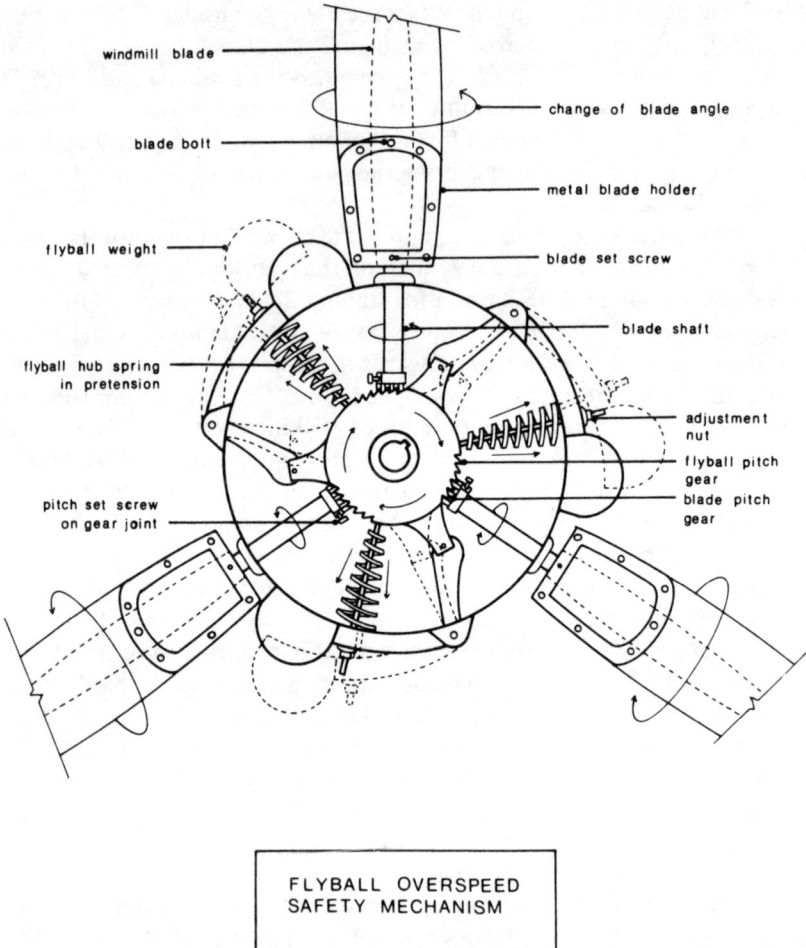

FLYBALL OVERSPEED
SAFETY MECHANISM

Fig. 6-32. Jacobs flyball governor. Centrifugal force throws weights — flyballs — away from the governor changing the pitch of the blades via a mechanical linkage. (Energy Task Force, CSA)

same governing concept. None of the companies, however, using the Allied governor as the sole means for controlling rotor speed has endured. Like their wind machines they did not survive. There are numerous reasons why they failed but one common factor has been problems with the governor. For example, the governor would stick occasionally allowing the rotor to reach excessive speeds that often destroyed the wind machine before the blades would feather. Poor materials and quality control were also to blame. Knuckles failed and the blades would go soaring into the heavens. But whenever these defects were corrected new governor problems would crop up to plague the wind machine again.

The new Jacobs Wind Electric Company is the only firm still in business that uses the Allied governor. But Jacobs' 7-meter wind machine does not depend solely on blade feathering to control rotor speed. The wind machine is also self-furling. The governor feathers the blades to limit power output to the alternator. Overspeed protection is provided by furling the rotor towards the tail. (I distinguish this company and its product from that which produced the Jacobs windcharger. The new firm was founded by Marcellus Jacobs, one of the original Jacobs brothers. He is also largely responsible for the design of the 7-meter turbine. However, this wind machine differs markedly from the much smaller Jacobs windcharger. The two companies also differ. There was a 20-year hiatus after the Jacobs Wind Electric Company ceased production in the mid-fifties and the new company was organized.)

It is also possible to change the pitch of the blades without using a governor. In the bearingless rotor concept the blades are attached to the hub with a torsionally flexible spar. At high speeds the blades twist the spar towards zero pitch stalling the rotor. Weights in or attached to the blade provide the necessary force. There are no moving parts in the hub; no bearings, knuckles, or sliding shafts.

Carter Enterprises' 10-meter turbine (see fig. 6-33) best represents this control strategy. The filament-wound, fiberglass spar permits the blade to twist torsionally reducing pitch. The flexible spar also permits the blade to cone progressively downwind of the tower in high winds like the fronds on a palm tree during a hurricane.

Almost all wind machines without governors use blade stall to some extent for limiting power from the rotor. This is particularly true of fixed-pitch rotors driving induction generators.

Stall regulation is more important and more helpful on synchronous wind machines because the rotor operates at constant speed. In high winds, those above the rated speed, the tip-speed ratio declines as the speed of the rotor remains constant. The angle of attack consequently changes lowering the performance of the blades. On an asynchronous wind machine (those driving alternators) the effect from stall is less significant because the rotor increases its speed in tune with the wind.

Blade stall is seldom if ever the sole means of overspeed protection. Stall is most effective on induction wind machines with fixed-pitch rotors. Induction generators, though, are dependent on the utility for controlling the load. During a power outage the generator immediately loses this

Fig. 6-33. Bearingless rotor. Blade pitch can be varied without a governor. By using a torsionally flexible spar, the blade can twist the few degrees needed to stall the airfoil. Shown here is a Carter Model 25, a 10-meter (33-foot) wind turbine driving an induction generator. Note the tapered and twisted blades. The large cuff near the hub aids starting the rotor in low winds. The Darrieus rotor in the background was built by Alcoa for Sandia Laboratories. The rotor is 25 meters (83 feet) high by 17 meters (56 feet) in diameter.

load. The rotor is no longer restrained to run at a constant speed and it begins to turn faster. Stall is now ineffectual for regulating power because the rotor spins at a more optimum tip-speed ratio. The rotor performs better when least desired.

On an upwind machine with a tail vane the rotor can be prevented from destroying itself by furling the rotor out of the wind. Those machines without a tail vane must use another approach. Brakes are the most popular.

Once brakes have been selected as the means to limit rotor speed during a loss-of-load emergency, they are also frequently used during normal operation. In a typical fixed-pitch wind machine the brake is applied at the cut-out speed to stop the rotor.

Consider the example of a small downwind turbine driving an induction generator which is braked to a halt at the cut-out speed. When the wind machine was first introduced the manufacturer stressed that the rotor was stall regulated and that it could operate safely above the cut-out speed. The rotor was braked, they asserted, only to minimize wear on the drive train at high wind speeds. The amount of energy in the wind at these higher speeds, they said, did not warrant the cost of capturing it. Mother Nature soon gave them ample opportunities to test their design. The brake failed on several occasions and the rotor ran away. In a few cases, the wind machine destroyed itself. Stall was not enough. This example not only illustrates why stall regulation should not be the only means of control but also that where brakes are used back-up controls are required.

Brakes can be placed on either the main (slow-speed) shaft from the rotor or on the high-speed shaft. Brakes on the high-speed shaft are the most common because the brakes can be smaller and less expensive for the needed braking torque than those on the main shaft. When on the high-speed shaft, the brakes can be found between the transmission output and the generator or on the tail end of the generator. In either arrangement braking torque places heavy loads on the transmission. There is also the limitation that the brake exerts no control over the rotor should the transmission fail.

Brakes on the main shaft give more positive control than brakes on the other side of the transmission. (There is a greater likelihood of a transmission failure than there is a failure of the main shaft.) But the lower shaft speeds require more braking pressure and greater braking area. The brakes are larger and cost more than brakes on the high-speed shaft. Note that in figure 6-25, illustrating a typical drive train, the brake is on the main shaft.

Disc brakes are preferred over drum brakes. Braking area is easily increased by increasing the disc's diameter and by adding more calipers. Drum brakes are used on home-builts because of their availability.

Brakes can be applied mechanically, electrically, or hydraulically and most operate in a fail-safe manner. In other words it takes power to release the brake. The brake automatically engages when the wind machine loses power. Springs provide the force in a mechanical brake, batteries, the electricity in an electrical brake, and a reservoir, the pressure in a hydraulic brake.

Dry brakes are the most common. (Like the brakes on your car, asbestos pads ride against a steel disc.) But on wind machines larger than 10 meters in diameter a type known as wet brakes are finding application. These brakes depend on the resistance to shear of a thin oil film sandwiched between metal discs. (Dry brakes use the shear resistance between the particles in the brake pad and the metal disc.) This oil film makes it possible to stop the rotor under more controlled conditions than with dry brakes, assert oil film proponents. According to them, dry brakes stick, then slip when applied causing shocks to be transmitted throughout the drive train. On a rotor 15 meters in diameter weighing several hundred pounds even slight shocks can cause severe stresses on major components. (To get a feel for this, imagine trying to stop such a large rotor instantly. You may be able to stop the rotor but the blades would keep right on going—to the next county.)

The problem with brakes, any kind, is that they fail. Not often, it's true, but once is enough. Dry brakes require replacement or adjustment of the pads periodically, for example. After extensive use the calipers (or shoes on drum brakes) have to travel farther to reach the disc where mechanically applied. Pressure from the springs decreases and so does braking torque. In one 12-meter wind machine this led to numerous runaways in high winds. (The rotor was supposedly "stall regulated.") The brake just did not have enough power to stop the rotor from spinning, and there were no back-up devices to slow the rotor. Several machines ran to destruction but some were brought under control when the designer climbed the tower and wedged a lever into the brake and manually forced the pads against the disc. This dangerous practice eventually cost him his life (see chapter 11).

Besides brakes there are three common means for overspeed protection on wind machines without tail vanes and blade governors: spoilers, tip brakes, and movable blade tips. In practice these are used as back-up control measures for fixed-pitch rotors driving induction generators.

The Wind-Matic 15-meter turbine (see fig. 6-23) is typical of those using spoilers. This Danish wind machine uses a brake under normal conditions to stop the rotor, but in case there is a loss of load or the brake fails, centrifugal force activates spoilers along the length of each blade. The spoilers change the shape of the airfoil destroying its effectiveness reducing power and rotor speed.

Tip brakes are plates attached to the end of each blade. Like spoilers they are usually activated by centrifugal force. When deployed (see fig. 6-34) they slap or drag at the wind. They are simple, effective, and they have saved many a fixed-pitch rotor from destruction. Tip brakes, however, have been likened to keeping your foot on the accelerator at the same time you are stepping down on the brake. They keep the rotor from reaching destructive speeds but do nothing to reduce the lift of the blade or the thrust on the wind turbine and tower. Tip brakes also reduce the performance of the rotor slightly by increasing drag at the tip where blade speed is greatest. In their line of wind machines, Enertech, represented here by the 4-meter model in the foreground of figure 6-35, was the first to adopt tip brakes in this country.

Most of the power in the wind is captured by the outer third of the rotor. Consequently, it is not necessary to change the pitch of the entire blade to limit the rotor's power and speed. By changing pitch only at the tip, the blade can be attached to the hub without a conventional governor. Vestas typifies manufacturers producing fixed-pitch rotors that use movable blade tips. At above normal rotor speeds the tips are thrown away from the rotor by centrifugal force causing them to slide along a grooved shaft that feathers the tips.

The Boeing Mod-2 shown in the background of figure 6-35 uses the same approach. It differs by hydraulically driving the tips towards the feathered position. In contrast to the tip brakes and movable blade tips previously described, the Boeing blade tips are actively controlled. The others use passive overspeed controls. They function from forces inherent in the spinning rotor. They are not dependent upon an external power supply for their operation as is the Mod-2.

CONTROL GUIDELINES

As part of its standards development for the wind industry AWEA offers general guidelines on wind machine design. These guidelines are grouped by the operating condition of the wind machine.

Designing for normal operation is the first task of any wind turbine developer. How well they have succeeded determines the wind machine's performance and life expectancy. Have they, for example, minimized wear and fatigue on major components such as the rotor, transmission, and generator. Equally important, though, is how well the wind machine performs under abnormal or emergency conditions. A wind machine that runs

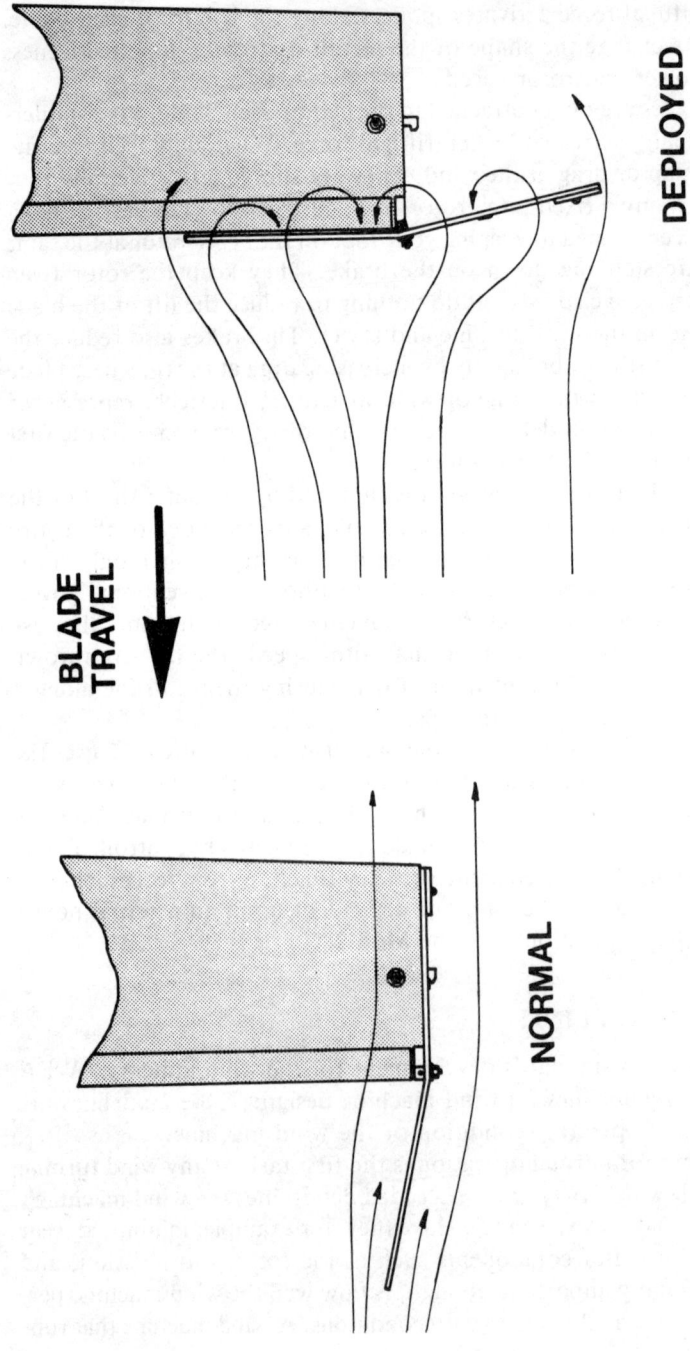

Fig. 6-34. Tip brakes. Enertech popularized the use of tip brakes in this country for overspeed protection on induction wind generators. (Courtesy Enertech Corp.)

Fig. 6-35. Tip brakes and movable blade tips. The Enertech Model 1800 in the foreground uses tip brakes for overspeed protection. The Boeing Mod-2 in the background controls its massive 300-foot diameter rotor with movable blade tips. These two machines and two other Mod-2's are located in the Columbia River Gorge near Goldendale, Washington.

well in moderate winds is of little utility and creates a hazard if it flys apart in high winds.

Abnormal conditions may lead to a catastrophic failure destroying the wind machine. These include high winds, ice, hail, and lightning. Emergencies can arise from abnormal conditions or due to the failure of a component during normal operation. For example, an electrical fault may cause the generator to lose its load. This fault frees the rotor creating the danger that it will reach excessive speeds. Or there might be a mechanical failure in the transmission between the rotor and the generator. An out-of-balance rotor is another serious fault that could threaten the life of the wind machine.

Schemes used to protect the wind system during abnormal or emergency conditions must be designed for maximum reliability, and AWEA suggests, in addition, that there be a redundant or back-up mechanism for controlling the rotor should the primary system fail. The back-up control will then prevent the wind machine's destruction. Whenever rotor controls have been applied due to an emergency, automatic operation should not resume unless the controls have been reset manually. In this way someone has to go out and physically inspect the wind machine, find the fault, and correct it before it is put back into operation.

Brakes are used for two purposes: first, to control rotor speed in high winds and during emergency conditions by bringing the rotor to a halt; second, to park the rotor for installation and service. Where used to stop an operating turbine, the brake must be more powerful than the rotor when it is running at 150 percent of normal speed, and the brake must be able to dissipate the energy without damage to the wind machine. Parking brakes must be strong enough to prevent the rotor from turning in winds up to survival speeds (120 mph).

Let's look at three types of wind machines each of increasing complexity and the different means they use to control rotor speed. The Bergey 1000 and the Sencenbaugh 1000 both use the same control strategy. In winds above the rated speed the blades stall and performance of the rotor decreases. Above 30 mph the rotor begins to fold towards the tail. Blade stall and furling are the only means of controlling the rotor during normal or emergency conditions. There are no brakes. Nor are there brakes to park the rotor for servicing or to prevent the wind machine from yawing about the top of the tower. The turbine designers stressed simplicity and ruggedness over greater rotor control.

The Enertech 1800 is more complex. Like the two preceding wind machines the blades are fixed in pitch and bolted rigidly to the hub. Above the rated speed the blades begin to stall dumping excess power. Where the Enertech differs is its use of a downwind rotor to drive an induction generator via a transmission. The downwind configuration precludes the use of furling to control rotor speed. There is no simple way for the rotor to turn out of the wind. Instead Enertech uses a fail-safe hydraulic brake on the output side of the transmission. The brake is applied in winds over 40 mph and during emergency conditions where the rotor exceeds its normal speed. For example, if there is a power outage the loss of line power unloads the generator allowing the rotor to speed up. The loss of line power or the abnormal rotor speed signals application of the brake. The brake can also be applied manually for servicing the wind machine.

The brake is normally "On" or engaged. It is powered by pressure within a hydraulic reservoir. The brake can be released only when there is electrical power to the wind machine. Thus, if there is a loss of power the brake automatically returns to its normal position stopping the rotor.

If the brake is unable to prevent the rotor from unsafe speeds (should hydraulic pressure, for example, be lost or the shaft between the rotor and the brake fail) Enertech uses tip brakes as an emergency control. These deploy slowing the rotor to a safe speed. The tip brakes must then be reset manually once deployed. Like the previous wind machines, the Enertech does not use a yaw brake or protect against an imbalanced rotor.

The Windworks 10-meter turbine is the most complex. It uses a hydraulic cylinder that changes blade pitch to limit rotor speed. Like the brake on the Enertech the hydraulic pitch control is fail-safe. It operates against a spring in the hub of the rotor. Should the hydraulic system lose pressure, the spring drives the blades to full feather. Rotor speed under normal and emergency conditions is controlled by taking advantage of this. In high winds, during excessive rotor speeds, or when there is an imbalance in the rotor, pressure is bled from the hydraulic cylinder allowing the spring to feather the blades. (Unlike the springs on blade-actuated governors the pitch control spring is protected from the elements.) The wind turbine can be shut down manually in the same way by releasing pressure on the hydraulic system. Once the blades have been feathered the rotor can be brought to a stop with a parking brake. A yaw brake is also included for servicing the turbine. Whenever the blades have been feathered due to an emergency condition the controls must be reset manually.

7

Towers

Towers are as integral to the performance of the wind system as the wind turbine itself. Without the proper tower your wind machine is not much more than an expensive lawn ornament and could even become a hazard to all in the vicinity.

Towers, as a rule, are one of the few wind system components where you have some choice. Unlike the selection of blades, transmission, and generator, which have been determined by the manufacturer, you have a variety of towers to choose from — at least for wind machines up to 7 meters in diameter. With small wind machines you can select the kind of tower as well as the height that best suits your site, temperament, and budget. The many options can be confusing and, for those installing the wind machine themselves, this could lead to a mismatch between the wind machine and the tower.

Fortunately, most manufacturers today carefully specify the towers that are compatible with their wind machines and insist that only these towers be used. And, for many of the wind machines over 7 meters in diameter, only one kind of tower is offered. The days of slapping a wind machine on any old tower that just happens to be lying around and sticking it in the air are over. We'll discuss the reasons why as we look at the different kinds of towers available.

Towers can be divided into two categories: free-standing and guyed. *Free-standing towers,* or self-supporting towers as they are also known, are just that—free standing. They depend upon a massive foundation to prevent the tower from toppling over in high winds, and must be strong enough internally to withstand the forces trying to bend the tower to the ground. *Guyed towers,* in contrast, employ several far-flung anchors and connecting cables to achieve the same ends. Free-standing towers are more expensive than guyed towers but take up much less space. Within each of these categories there are several subdivisions (see table 7-1).

TABLE 7-1
Types of Towers

Free-standing		Guyed	
Truss	*Pole*	*Lattice*	*Pole*
	Tapered tube		Tapered tube
	Pipe		Pipe
	Wooden pole		Wooden pole
	Concrete pole		

FREE-STANDING TOWERS

There are two types of free-standing towers. The most common is the *lattice* or *truss tower,* so called because it resembles the lattice work of an arbor or because it uses a triangular framework (see fig. 7-1). *Pole towers* are another form of free-standing tower but one which encompasses several different varieties. In general, truss towers are more rigid and stable than pole towers but also less appealing visually.

Towers can be designed to withstand any load. But as the size of the wind machine increases, so does the weight and cost of the tower supporting it. The same is true as the tower increases in height. The components become heavier, harder to move, and more costly to ship.

Truss towers are assembled from a series of 20-foot sections. For small wind machines, the sections may be pre-assembled and welded together prior to delivery. For larger machines, the tower is shipped "knocked-down" or in parts and must be completely assembled on the site. Unarco-Rohn's largest welded assembly weighs 230 pounds. (Unarco-Rohn is a leading manufacturer of both truss and guyed towers.) To say the least, it is more than a one-man job to move this 20-foot section around. Subsequent sections must be bolted together, but the legs alone on a large truss tower weigh more than most would like to handle. The lower section on

Fig. 7-1. Truss tower. Enertech E-44 mounted atop an 80-foot self-supporting tower by Unarco-Rohn, this country's largest supplier of both truss and guyed lattice towers to the wind industry. Steel pipe is used for the legs and angle-iron for the cross girts. Most truss towers in this country stand on three legs, whereas truss towers in northern Europe use four legs. Note the tip brakes on the 13-meter (44-foot) rotor.

an 80-foot tower can weigh well over 500 pounds. Pi-rod, another manufacturer of truss towers, assembles and welds all sections before they are shipped. A forklift or crane would be needed just to unload the sections from the tractor trailer.

Installation of truss towers usually requires a crane. The tower is assembled on the ground, then hoisted into place and bolted to the foundation. Some enterprising individuals assemble the tower piece by piece in the upright position with the aid of a gin pole mounted on the tower. This is time consuming and risky. It is not an approach for amateurs. Another method is to hinge the tower at its base. The tower is then bolted together on the ground, the wind turbine attached, and the whole assembly tipped into place with a gin pole and winch, or a small crane.

POLE TOWERS

To many observers free-standing pole towers are the most aesthetically pleasing. The pole can be made from tapered steel tube (fig. 7-2), or from steel pipe, wood, concrete, or even fiberglass. Like light standards or utility poles, these towers are what the engineers call cantilevered—they are supported at only one end.

Pole towers are not as rigid as truss towers. They sway visibly in the wind (truss towers do as well, but not as much). Nor are pole towers available in all heights and strengths. For example, the selection of wood and concrete poles suitable for wind machines is limited by the length of pole that can be shipped conveniently.

Like truss towers, pole towers, at least some of them, can be difficult to handle without a forklift. Because wood and concrete poles are shipped in one piece, they are the most difficult to handle. Tube towers, shipped in sections thirty to fifty feet long, are more manageable. Each section can weigh several hundred pounds. Pipe towers are the easiest to handle.

Installation, once again, usually requires a crane. A pole tower, though, can be hinged at the base and tipped into place with a gin pole. When upright, the tower is bolted to the foundation. One French wind machine manufacturer erects a hinged, tube tower with a wind turbine attached by a powerful hydraulic jack in the foundation (see fig. 7-2).

One tower supplier in this country ships the tower in nested sections. The telescopic tower is then extended at the site with either a crane or a tower-mounted gin pole. The supplier advertises that this telescopic tower can be erected in one day by one person without the need for a crane.

Pole towers of pre-stressed concrete have also been used. Several wind

214 • *Wind Energy*

Fig. 7-2. Free-standing tube tower. This tower of tapered tubular steel rises above the gorse and heather on Cap Frehel, a windy promontory jutting into the English Channel from the French province of Brittany. It supports a prototype 8-meter (26-foot) wind machine manufactured by Aeroturbine. The hinged tower can be lowered with a hydraulic jack that is embedded in the foundation.

turbines built by Grumman Aerospace have been installed on these concrete towers with no reported failures.

Wooden poles are another option. They are strong, rigid, and can be inexpensive when bought in quantity. But they are not much to look at, and the heights available are also limited to the length that can be shipped conveniently.

Free-standing wooden poles are installed by crane or utility truck with a boom for setting poles. For a sixty-foot tower a seventy-foot pole must be used. (Ten feet of the pole is embedded in a concrete foundation.) Longer poles are available but the cost rises rapidly with lengths beyond the standard sizes used by utilities.

GUYED TOWERS

Guyed towers are, by far, the most common tower choice. They offer a good compromise between strength, cost, ease of installation, and appearance. But they take up more space than free-standing towers and should a guy cable fail, the tower and turbine are lost.

Guyed towers are comprised of a mast, cables, and earth anchors. Most towers for wind machines less than 10 meters in diameter use masts of welded lattice (see fig. 7-3). This lattice is made from steel tube and rod. Because these masts are used throughout the telecommunications industry, they are available in a range of sizes from lightweight sections designed for CB antennas to heavy sections for mountaintop microwave dishes. These masts are shipped in easily handled 10-foot long sections weighing from 40 to 70 pounds. Tower height is practically unlimited. A guyed tower using a lattice mast can be assembled by bolting sections together in the upright position a section at a time with a tower-mounted gin pole or the entire mast can be assembled on the ground and set in place with a crane.

Masts of steel pipe and tube are gaining in popularity for machines 7 meters in diameter and larger (see fig. 7-4). Masts of the desired height are assembled from several sections bolted together (on the pipe masts) or slipped together (on the tube masts). Well casing is a frequent choice for pipe towers. It is readily available, inexpensive, and light enough that it can be moved by hand. Do-it-yourselfers often choose well casing for these reasons. Tapered tubes differ from well casing by their much larger diameter and their taper. Guyed towers using masts of steel pipe and tube are assembled on the ground and tipped into place with a crane or gin pole.

The second element in a guyed tower is the guys themselves. The guys are extra high strength (EHS) stranded steel cable from ¼ to ¾ inch in

Fig. 7-3. Guyed lattice tower. Bergey 1000 (3-meter diameter) mounted on a guyed tower using a Unarco-Rohn, 25-G lattice mast. The modular mast is produced in 10-foot sections and can be easily assembled by hand. This tower is capable of handling wind machines with rotors up to 10 feet in diameter or 300 pounds of maximum thrust.

Fig. 7-4. Guyed pipe tower. Jydsk Vindkraft's 8-meter (26-foot) wind machine on a guyed tower using steel pipe at the Danish Small Windmill Test Station near Roskilde, Denmark. Note the use of four guy cables. This allows the hinged tower to be tipped into place with a simple gin pole. Also note the tapered and twisted blades.

diameter (up to 1 inch on some larger machines). There can be from three to five guys at each level and two or more levels. Most use three guys at each level because they do the job at the least cost. The old Winchargers, though, were installed with five guys equally spaced about the tower at the topmost level. If one guy was lost, the tower would remain standing and the installation protected. Four guys are sometimes used where the site or method of erection requires them.

Guy radius, the distance from the tower to the anchors, is dependent upon the site, the loads imposed on the tower by the wind turbine, and the stiffness of the mast. Some wind turbine manufacturers specify the guy radius precisely. Others provide a range of minimum and maximum distance from the tower to allow the installer some flexibility. Guy radius is limited by the compressive loads on the mast and what it can withstand before it buckles and also on the anchor construction and its resistance to being pulled out of the ground. You can bring the anchors in so close to the tower, for example, that the mast will buckle in high winds or the anchors will pull out. But the guy radius can be as great as you want. The tension in the guy cables and the compression on the tower continues to decrease as distance from the tower increases. Usually there is no reason for going beyond three-fourths of the tower's height. As a rule of thumb, the guy radius should not be less than half the height of the tower.

There are two or more guy levels on most towers using a lattice mast. One anchor is used to guy all levels. The topmost guy prevents the tower from overturning and the lower guys prevent the tower from buckling. When the tower is stiff, like the tapered tube towers used on some larger machines, only one guy level is needed. For small machines installed on lattice masts, this is not the case. The mast is too soft and additional guy levels are used to keep it from bowing. On the modular masts made by Unarco-Rohn, guys are called for every 25 to 30 feet. For an 80-foot tower supporting a 3-meter turbine (maximum rotor diameter of 10 feet), we are looking at three guy levels 25 feet apart.

NOVEL TOWERS

If you can think of it, it has been tried. Trees are a good example. Yes, trees. Wind machines have been mounted atop trees—but not successfully. First, for a tree to be of long-term use it must remain alive. After you finish hacking away at it, though, its days are numbered. Unless the tree has been cut and treated with a wood preservative (then it ceases to be a tree and becomes, instead, a wood pole) it is a poor idea for a tower. Second, trees seldom occur right where you would like your tower. Nor is there usually one lone tree standing out in the middle of a large clearing that towers above all surrounding objects. Usually it is just one of many in a forest of trees equal in height. Third, the turbine will be difficult to install and service. Fourth, the manufacturer will not issue a warranty for such an installation. They know what will happen.

Another equally troublesome idea is to mount wind machines on buildings. Forget it! Seems like a fine idea at first glance: the building gets the turbine above the ground and eliminates the need for a tall tower. The building, though, creates its own turbulence. The wind machine must be installed above this turbulent zone to perform well and to insure long life. This often negates any potential savings on the tower. And, can the building take it? That is, can it support the loads created by the tower and the wind turbine. This would exclude the wood roofs of single-family residences. They can't. But a reinforced concrete roof in a commercial or industrial building might be strong enough. Can the roof, then, handle the dynamic loads—the vibrations? The tower will transmit vibrations from the wind machine to the roof and on to the rest of the building unless precautions are taken. If the building is an unoccupied warehouse, the vibrations will not bother anyone, but if it is an office building they may prove annoying.

Much has been made of the small Jacobs windcharger installed on a tenement in the Bronx. True, it was done once, and it can be done again. But what were the objectives? The installation was intended as a demonstration of alternative energy in the city and as a challenge to Consolidated Edison Company, the local utility. The project was highly successful in meeting these goals. As a practical option, however, it leaves a lot to be desired.

Farm silos also seem ideal for a low-cost tower. They are already in place, usually stand tens of feet above surrounding farm structures, and are relatively close to where the power will be used. Alcoa (Aluminum Company of America) tried it (see fig. 7-5). The results were inconclusive.

Alcoa mounted their Darrieus turbine on a short tower section and strapped the tower to the silo. One was installed at Clarkson College and another at Agway's research center (a large Mid-Atlantic farm supplier) in New York State. The silo experiments were abandoned when Alcoa withdrew their turbine from the market. The option has not been pursued any further.

TOWER SELECTION

Choosing a tower for a small wind machine requires juggling your various needs against the site, what you can afford, and what is avail-

220 · *Wind Energy*

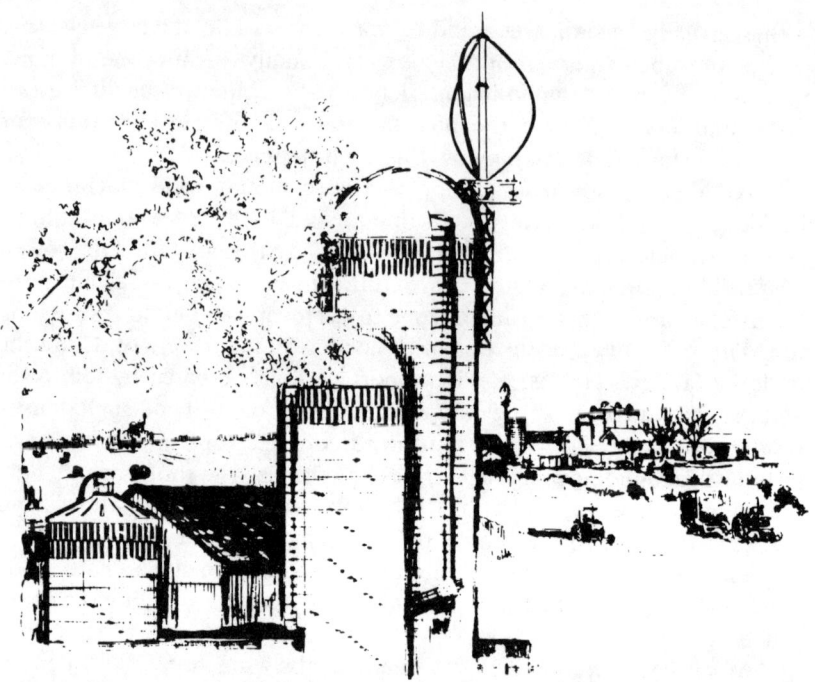

Fig. 7-5. Silo mounting of Darrieus. Alcoa experimented with mounting their small 10 kW rotor on farm silos. Not all silos are suited for such use. Only those built from concrete staves (castellated blocks) are strong enough. The concept was abandoned when Alcoa withdrew from the market. (Courtesy Alcoa)

able. Several factors enter into the choice of a tower. First and foremost is whether the tower is available in the height desired.

As the wind industry has matured—and that of wind system users as well—selecting a tower of the proper height has become increasingly important. In the early seventies anything that would get the wind machine off the ground was acceptable—a vestige of the windcharger era.

Towers for wind machines used on the Great Plains during the thirties were never very tall. The flat terrain and the few obstructions present did not call for towers taller than 60 feet. Even so, by the late forties Wincharger (one of the era's major brands) was installing guyed towers 85 and 105 feet in height. Parris-Dunn was instructing its customers that "the higher the tower the greater the power."

As we have become more sophisticated, we have learned that economic power generation and good performance are only obtained on a tall tower. We have known for some time that wind speed and power increase with height, but it did not begin to sink in until wind systems started to be installed in numbers across the country. We gained more experience with power-robbing turbulence and what it can do to a wind machine's performance. As a result, recommended tower heights have gradually increased. Manufacturers want their products to perform well and want to minimize turbulence-induced service and warranty claims. Dealers are going to taller towers for the same reasons but also because greater height allows more flexibility in siting. If buildings and trees are present — and they usually are — a tall tower can redeem an otherwise unusable site. A minimum by today's standards is 60 feet. Most dealers are installing towers 80 to 100 feet in height or taller. And when trees are nearby, a 100-foot tower is the norm.

Strength, or the ability to withstand the forces acting on the tower in high winds, is probably the next most important factor. Towers are rated by the thrust load they can endure at the top of the tower without buckling. The lightest weight guyed tower used with wind systems, the Rohn 25-G, is rated for a maximum of 300 pounds thrust at the top of the tower. The next heavier size, the Rohn 45-G, is rated for 900 pounds thrust, and so on.

Industry standards call on wind system manufacturers to design their wind systems to withstand 120 mph winds without damage. The thrust on the tower at this wind speed depends on the size of the wind turbine and its mode of operation under such conditions. Two wind turbines of the same size may require entirely different towers because of differing approaches to protecting the rotor in high winds.

All towers flex to some degree. Truss towers are the most rigid but they do sway in the wind. Free-standing pole towers deflect visibly even more so than truss towers. Guyed towers fall somewhere in between, depending upon the stiffness of the mast and the tension in the guy cables.

An acquaintance of mine installed a turbine 7 meters in diameter on a 100-foot, heavy-duty truss tower. He was very eager to see his new — and costly — wind turbine in operation (an affliction Jack Park diagnoses as "fire-em-up-itis"). The wind was strong, blowing near the rated speed of the self-furling turbine. To insure that all was well and to get a bird's-eye view of his investment, he climbed the tower to just below the rotor (this is strongly discouraged as you'll see why). He decided to check the operation of the feathering governor by unloading the generator and letting the rotor speed increase. When the disconnect switch was thrown he suddenly found himself hanging on for dear life as the blades feathered and the tower sprang several feet back into the wind like a giant whip.

Deflection is not a problem unless the turbine and tower are mismatched. If the tower or the blades flex too much and at the wrong time, the blades might strike the tower. Because of this, dynamic interaction between the wind machine and the tower is a major concern.

As the rotor and tower deflect in the wind they begin to oscillate to and fro like the swaying spans of a rickety suspension bridge. Should the turbine and tower begin to sway in harmony, the oscillations will build upon themselves and could gradually increase in magnitude until they eventually destroy the wind machine. It has happened too often to ignore.

Though no records exist, the reported accounts of tower failures in wind system applications include about an equal number of truss towers and the apparently less secure guyed towers. In one widely discussed case, a truss tower failed when the bolts holding two 20-foot sections together sheared. The tower manufacturer asserted that the tower was overloaded by the dynamic interaction of the turbine and the tower. Witnesses noted that the tower was vibrating wildly prior to the accident. The turbine manufacturer countered that the failure was due to "bad steel" in the bolts. Whether bad steel or not, a wind system vibrating in resonance can exert tremendous force on the tower creating loads well beyond design limits. Another lesson learned from this and similar episodes is that truss towers can fail even though they look strong enough to take whatever Mother Nature can throw at them.

Dynamic interaction between the wind machine and the tower is the reason that manufacturers are picky about the tower their wind machines are mounted atop. Each tower type and height must be specifically "tuned" for their wind machine. This is particularly true for wind machines larger than 6 meters in diameter because of the large forces involved.

Reputable manufacturers today honor their warranties only when the wind machine has been installed according to their specifications including the type of tower used. They like to know how the wind turbine and tower will interact, how much deflection will occur, and whether dangerous harmonics will be encountered. They want a tower that is predictable. Consequently, not all towers in all heights will be approved by the manufacturer. Your choice may be limited severely.

If you plan to use a tower not recommended by the turbine's manufacturer, here are a few questions to consider. Has this type of tower been used for wind machines by anyone else? If so, how has it performed? What, if any, were the problems encountered? What does the manufacturer think? Will they still honor their warranty? If not, why not?

The site or local ordinances may further limit your tower options. For example, the size of the lot may preclude a guyed tower. A free-standing tower is your only option. Your choice is now between a truss or pole tower. The decision will be affected by an evaluation of the relative cost, appearance, and ease of installation between the two.

Guyed towers can be adapted to small lots by pulling the anchors in closer to the tower. If you must do so, check with the manufacturer. They may have no objections to your plans. Or, contact a structural engineer to run through the numbers for you. He may find that the tower has an ample safety margin with the shorter guy radius. On the other hand, if you are approaching the limits of the mast, you may be forced to use a free-standing tower after all.

Sites with a lot of traffic, either vehicular or pedestrian, will also limit your selection to free-standing towers. Several times while scavenging for used windchargers in Montana, I came across the remains of a guyed tower felled by a rancher on his tractor or by a heifer seeking a back rub. Not only is there the danger of losing your wind system, but the guy cables may present a hazard to unwary passersby. Snowmobilers are the worst. They are always getting strung up on someone's fence; a guy cable would become a prime target. Kids on bikes or running about the yard are as hapless. To avoid tragedy, guy cables should be kept out of traveled ways.

Another factor to consider in tower selection is maintenance. Will the tower have to be painted? The steel used in most towers is galvanized. Individual members or welded subassemblies are dipped into a molten batch of zinc which completely encloses all exposed metal, including the welds. This coating offers protection against corrosion because the zinc is more reactive than steel, and the corrosive elements in the atmosphere prefer to react with the zinc than with the steel. When shipped from the plant, galvanizing presents a shiny silver finish. It soon oxidizes outside, however, and takes on a dull grey luster. Galvanized surfaces do not require painting or other treatments.

Some truss towers and pole towers are now being assembled from Cor-Ten steel. Cor-Ten was developed for outdoor applications requiring low maintenance. Cor-Ten oxidizes to an attractive dark brown finish to protect the underlying steel from further corrosion. Essentially it develops a skin of rust and never needs to be painted or treated in any other way. Cor-Ten's only disadvantage is that rain running off the steel leaves a trail of rust for the first few years. This is not a problem unless you plant the pole in the middle of a concrete lot. The concrete will soon look like the best Georgia red clay.

You may be able to save a few dollars by buying a tower that is untreated. It may be primed but will need to be painted. Do not even consider it for a truss tower. You will go crazy trying to paint the nooks and crannies around the cross-girts. But painting a pole or tube tower is not unrealistic. If trouble-free operation is what you seek and not the least-cost installation, galvanizing or using Cor-Ten steel are your best options.

Access to the turbine should not be overlooked. The turbine will require at least occasional inspection if not maintenance. Some provision must be made for getting to it, without the need to hire a crane every time you want to get up there. Truss towers can be ordered with climbing lugs or rungs. Guyed towers using lattice masts can be climbed using the lattice work (cross-girts). Tube towers, whether free-standing or guyed, require the addition of climbing lugs if they are not part of the package. From a service perspective, hinged towers are superior to those fixed in place. Instead of climbing up to the turbine you bring it down to your level. This is much safer and less demanding than being forced to scale the tower.

The tower should be easy to install. You may have found an inexpensive tower that will do the job you want, but if it takes a lot of effort to install, the savings may be offset by greater installation costs. And if the tower sections are so large you need a crane or other heavy equipment just to get them off the truck, you will need an experienced crew with equipment. You will pay more for specialized handling and shipping than when the sections can be moved around by hand.

When considering how the tower is installed, also look at how the wind machine is mounted on top of the tower. A *tower adapter,* sometimes called a stub tower, is needed between the tower and the wind machine. One popular approach is to use a steel plate on the topmost section. The wind machine is mounted onto this plate with a series of bolts and leveling nuts. (This plate is predrilled to accept the turbine's mounting hardware before it is shipped from the manufacturer). The leveling nuts permit adjustment to insure that the machine is level. If not level, it will tend to stick in one direction and yaw unevenly. Another method is to use concentric tubes. For example, in one design, a tube extends from the bottom of the wind machine. This yaw tube is then slipped inside a split pipe at the top of the tower. The turbine is held in place by clamps compressing the sides of the split pipe against the turbine's yaw tube. Wind turbines using this mounting method are easier to install than those using a bolted yaw plate, but there is no way to level the turbine if the tower is not vertical.

The tower should also have a pleasing appearance. Wind systems need not look as though they have grown out of an adult-size Erector set. Nor do they need to look like a transmission tower designed in the thirties. There will be fewer objections from your neighbors if the tower offers a pleasing appearance. Besides, you will feel a thrill every time you look at it. The installation should be something to be proud of and not something to hide. It is not good enough to look around and point an accusing finger at obtrusive objects (such as utility poles and transmission towers) on the horizon that have found acceptance.

Last, but not least, is cost. Because the strength of the tower has been set by the turbine manufacturer to meet the expected static and dynamic loads, tower selection is more often based on cost than any other factor, site permitting. Let's see how the various towers stack up in relation to cost.

We will use an example requiring a 60-foot tower (a site on an open knoll) because that is the minimum tower height recommended and also because a 70-foot wood pole is about as long as we can conveniently work with. (Ten feet of the wood pole will be embedded in a concrete foundation.) The wood pole, a truss tower, and a free-standing pole tower will all require an equivalent amount of foundation work. We will also use a guyed tower. Its foundation costs are considerably less but we will assume that the cost of concrete anchors makes up the difference. Thus, the installation costs of each tower should be roughly comparable if they are all installed with a crane.

First, consider a small wind machine that produces a maximum thrust at the top of the tower of 300 pounds or the equivalent of a wind turbine with a rotor 3 meters in diameter. The steel towers to be used are rated for this load by the manufacturer. Wood poles are less commonly used as towers for wind machines so we will take a look at how they are rated.

Wood poles are classified according to their circumference measured 6 feet from the butt end. Poles of a given class and length are rated to carry approximately the same load regardless of the tree they came from. Each class is capable of withstanding a thrust acting on the top of the pole as given in table 7-2. They can handle even greater loads when guyed. Wood poles are stiff and more like truss towers in their rigidity than they are like guyed towers or steel poles. Lengths suitable for wind systems are found only in Class 4 or better. A Class 4 pole is more than strong enough for our application.

In table 7-3, shipping is extra for the truss, steel pole, and guyed tower. The wood pole was quoted as delivered. Transportation costs can vary

TABLE 7-2
Wood Pole Strength

Class	Thrust in Pounds
4	2,400
3	3,000
2	3,700
1	4,500

TABLE 7-3
60-foot Tower Costs
(1983 $)

Truss	$1,700
Steel pole	2,500
Wood pole	1,000 (or $600 each for five or more)
Guyed lattice mast	1,100

significantly. It is wise to take a close look at freight charges; they can alter the balance. In this example it does not pay to use a wood pole unless you plan to install five or more in a mini-wind farm.

Wood poles are unsightly to most people. It is hard to make an attractive match between the wind turbine and the top of the pole. Then there is the problem of the power cable snaking down the pole, climbing lugs jutting out at all angles, and anemometer supports. These galvanized parts contrast sharply with the black or brown preservative treatment of the poles. Even if you do not mind the looks others might—especially the zoning officer or your neighbors.

The nature of the site, the space available, or the type of use in the area of the tower (snowmobilers careening around at high speed, for example) may not permit the use of the less costly guyed tower. Some may simply fear the loss of the turbine should a guy cable fail or be vandalized. Thus, a free-standing tower may be the best choice.

Let's assume that you do not like the idea of guy wires cluttering up the yard and the wood pole is more than you can take visually. Your options are either of the free-standing towers or to modify the guyed tower so it is less obtrusive.

The pole tower is the most attractive but it is also the most costly (about twice the cost of the guyed tower). The truss tower is only 50 percent more than the guyed tower and meets our goals—a generally pleasing appearance without the space required for guy cables. The truss tower makes a good compromise between cost, aesthetics, and space requirements.

But don't drop the guyed tower yet. If the property bordering your yard is not occupied (say it's a field of corn) and your neighbors do not object, you can position the tower so that two of the guys are anchored along the fence line. Now, there is only one anchor in the usable area to contend with. If this arrangement is tolerable, a guyed tower is your best choice.

The numbers change when we consider a larger wind machine or when a taller tower is needed. As the size of the wind machine increases, the cost differential between each type of tower changes. The cost advantage of the wood pole becomes better, and the difference in cost between the truss tower and the guyed tower decreases. This trend becomes more marked with wind machines 6 to 7 meters in diameter. Several manufacturers of wind machines in this size range insist that their turbines be installed only on truss towers. They feel that the additional cost of the tower is justified by the elimination of the guy cables.

Towers also become a smaller part of the total cost of a wind system as the wind machine's size increases. For a small machine, the tower may cost 30 to 50 percent of the total installed cost, depending on whether it uses a guyed or truss tower. On the larger machines, the cost of the tower in proportion to the total cost is 15 to 25 percent or about one-half as much. The cost difference between the two types of towers is less significant as the size of the machine increases.

Other manufacturers, particularly those of wind machines over 10 meters in diameter, specify only one tower (sometimes guyed, sometimes free standing) because the turbine and tower are carefully matched. You don't have a choice.

8

Cutting Costs

If you are handy with tools, don't mind heights or hard physical labor, and have plenty of time, then this chapter offers several ways you can cut the cost of installing a wind system. You gain by replacing the skill, time, and energy of the dealer-installer with that of your own.

Working with a wind generator, tower, or (it seems) any of its subsystems requires muscling around some hefty components in awkward places. But it can be good exercise — a few trips up an 80-foot tower will tone muscles you never knew you had — and a rewarding experience. You will develop a sense of accomplishment in doing it yourself, and you will learn more about your wind system than in any other way. You will know its strong points and also what can go wrong, where, and how much effort it will take to fix it. The whole process will also give you an appreciation for the amount of effort and the different skills required to produce your own electricity reliably.

First, what are the costs of today's wind systems, and where can we cut these costs? The wind machine itself comprises from 40 percent of the total installed cost of small wind systems, to over 60 percent in wind machines above 7 meters in diameter. Towers become a less significant percentage of the total cost as wind machines increase in size. Installation costs, as a percentage of the total cost, also decline as size increases (see table 8-1). But for small wind machines, those from 3 to 6 meters in diameter, installation accounts for 25 percent of the total cost.

TABLE 8-1
Approximate Cost

Rotor Size (meters)	Installed Cost (1983 $)	Installation (percentage)
3	6 to 8,000	25 to 35
4	10 to 12,000	25
5	12 to 15,000	25
6	15 to 20,000	25
7	20 to 30,000	25
10	30 to 40,000	15
13	40 to 50,000	15

BUILDING A WIND MACHINE YOURSELF

Because the bulk of the cost is in the wind machine, people frequently try to save money on this component by building their own. If this is your choice, ask yourself some hard questions—first, why? Because you like to tinker, work with your hands, or create some new and wonderful device, or is it to save money? Don't be embarrassed; tinkering is a valid reason. Then go ahead and try your hand. You will learn a lot about natural forces, mechanics, and Murphy's law (if anything can go wrong it will). But if you want to build your own wind machine as a shortcut to buying a commercially available wind system, forget it. You will produce nothing but headaches.

Very few home-builts work reliably, and those few that do, don't produce much usable power. They are more expensive than you might first imagine, and they can be dangerous. Manufacturers with teams of competent engineers have a hard enough time keeping their turbines in one piece. You must be exceptionally talented to do better with fewer facilities and no technical support.

This goes for towers, as well as the wind turbine itself. It is not too difficult a task to calculate the static loads operating on the wind turbine and tower. Anyone with a little background in math and physics can master the equations, but figuring out how to deal with the dynamic loads is something altogether more involved.

There are several rotating or moving subsystems comprising a wind generator and tower: rotor, transmission (where one is used), generator, yaw mechanism, and the tower—yes, the tower moves too. The interaction of all these moving components in varying winds is almost unpredictable. The point is that dangerous harmonics can develop between components, causing dynamic loads to exceed the static loads for which they were sized. Design teams try to predict when these harmonics will occur and how to prevent them from doing damage. You will be doing it by the seat of your pants.

If you must build your own wind turbine, choose a design that is small and simple, where the forces involved are manageable. A prime candidate would be any design featuring sail wings. Blades using sails are inexpensive, can be made in any size, and are easy to work with. They are also unlikely to fly apart and cause damage or injury. The beauty of sail wings is that under severe loads (read high winds), the sail cloth simply tears away leaving the rotor in one piece.

Plans for a Cretan sail mill can be obtained from the National Centre for Alternative Technology in Wales (see appendix D). The horizontal axis rotor is patterned after those on the island of Crete and is used to drive an automobile alternator. The design calls for readily available materials and will cost less than $1,000 to construct. A similar design for a vertical axis sail windmill has been developed by Low Energy Systems of Ireland (see fig. 8-1).

Plans for more complex wind turbines are on the market but are not recommended because of the more costly materials used, the higher level of skill required, and the greater hazards due to their use. Should one of these plans be your choice, however, then carefully review Jack Park's *The Wind Power Book*. He provides much needed information for do-it-yourselfers on materials, design approaches, and how to make the necesary calculations.

Don't skimp on the tower either. Numerous homeowners try to reduce costs by building their own towers or using whatever just happens to be lying around (a CB tower in one case). By building your own tower you may not only be shortening the life of your wind machine, but you may also be endangering yourself and your neighborhood. One day you may find the whole wind machine crashing down on your head. As Murphy himself would say, there's never enough to do it right, but there's always enough to do it over.

I believe that building a wind machine and tower that will work reliably and safely together is beyond the skills and knowledge of most homeowners. I have seen too many worthless home-built contraptions for me to think otherwise. There is, however, another option: buying a used wind machine. This approach is also hazardous, but it gives you a better starting place. You begin with a (presumably) workable wind machine. Whether it is a used modern wind machine, or a used windcharger from

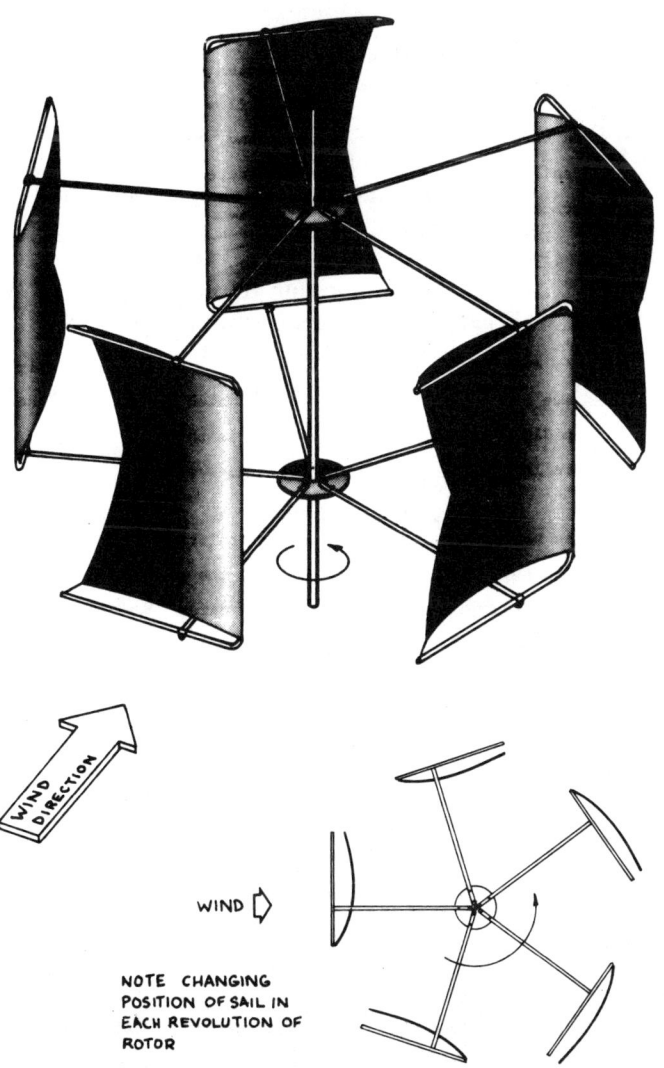

Fig. 8-1. Low-cost sail wing rotor. If you want to build a wind machine yourself use a sail wing rotor. They're inexpensive, easy to construct, and much safer than wooden or metal rotors. (Courtesy Low Energy Systems, Ltd.)

the thirties, you can find out how well it worked in the past and what the problems are, rather than starting from scratch with a home-built.

BUYING A USED WIND MACHINE

When buying a used wind machine, whether or not it is from a dealer, the rule is *caveat emptor* (let the buyer beware).

After years of shopping in supermarkets and department stores, few of us have well-developed bargaining skills. Our trust in products offered for sale has grown through extensive advertising and standardization. But not long ago, everyone was haggling with the vendors at the local market. There, you examined the goods carefully, decided how much they were worth to you, and began the exchange. It is still that way with used cars. Much remains hidden beneath the hood. So a great deal of faith is placed on the truthfulness of the seller with regard to its inner workings. The same is true with used wind machines. Not only must you make a careful examination of the goods, but also a careful examination of the seller. Are his claims reasonable and verifiable?

Another important question is how you plan to use the wind system. Will it heat water, produce line-quality power for supplementing existing utilities, or power a remote homestead? The answer will determine the degree of reliability required. If, for example, the wind system is to be used as a sole source of power for a remote homestead, then it better work reliably. If, on the other hand, you plan to use it for supplemental power, then reliability is less crucial.

The simplest use of a wind system is to supplement existing power from the local utility. To do so requires a wind system that generates line-quality power. Most modern wind systems are designed to do just that. By use of either an induction generator, or a synchronous inverter, these systems produce 60-cycle AC at the proper voltage. So if you are buying a used modern machine, use it as it was intended, in an interconnected application.

Older wind generators, those from the thirties and many of those from the early seventies, though, produce variable-frequency, variable-voltage electricity. Don't try to find a synchronous inverter that will interconnect these generators with the grid. Leave well enough alone. It is better to use them for charging batteries (for which most of them were designed) or to heat domestic hot water through resistance heaters. The installation of electric resistance heaters is straightforward, and you will avoid a confrontation with the local utility over power quality.

Related to how you plan to use the wind system is why you want it in the first place. Look deep down inside. If you want a wind machine principally to tinker with (sure, you'd like it to be a paying proposition as well), then you are freer to take more risks than if your primary concern is a wind turbine that will consistently generate usable energy.

Used Modern Machine

Now that the wind industry is beginning to mature, used turbines are becoming more plentiful. This is due in part to design improvements, such as the introduction of larger more economical units, that result in customers trading in their earlier models. It is also due to the failure—bankruptcy—of some firms as the industry goes through periodic shakeouts. These failures leave a number of units in circulation as their former owners tire of a constant hunt for spare parts.

The first question to ask the seller, then, is why is the wind machine up for sale? Is it because the previous owner traded up, died, moved, or simply got fed up? If the latter, what's wrong with the turbine? What caused the problems? What were the headaches? Did the previous owner, for example, tire of climbing the tower every month to change the brushes or tighten some bolt? If so, how do you plan to deal with this problem?

Are parts readily available? Where can parts easily be found? How much do they cost? If parts are unavailable because the company went bankrupt, how difficult will it be to make them? Someone out there may be making parts for his own unit and could easily make a few more if need be. How will you find them?

The seller should be familiar with these questions and have answers for you or at least tell you where to look for the answers. The classifieds in *Alternative Sources of Energy* magazine or other energy publications are a good place to check for spare parts (see Bibliography). If you cannot find them there, you will have a hard time finding spare parts—period.

Does the seller offer a warranty? If so, what kind: the manufacturer's, or the seller's own? How long does it last? In some cases the manufacturer's original warranty has not expired and may still be honored if the wind system is installed according to the company's specifications. When in doubt, check with the company.

A used wind machine may be bought directly from the previous owner or through a dealer. By buying the machine directly, you save by cutting out the middleman's markup. At the same time, buying direct forces you to install the turbine and tower yourself, as the dealer probably will not want to install a turbine on which he has not made a commission.

By buying through a dealer you may pay more, but you gain some assurance that the turbine will work and will be installed properly. The dealer can also be held accountable more easily than an individual owner, because the dealer has a reputation to protect. In a direct sale, the previous owner may want to dump the wind machine and wash his hands of the whole affair as quickly as possible.

Take a hard look at the economics. Are you saving enough by buying a used machine to justify the risk that it will not work, or work as well as it should? Check with local dealers. You may find that it does not cost much more to buy a new wind system instead.

Buying an old windcharger is another possibility, if you don't mind a little sweat and toil.

Used Windcharger

"You'll never get there from here," said the rancher in disbelief, "not without a four by four."

"We'll give it a try anyway," said the Easterners.

But after an hour of searching the banks of the Powder River, they were about to give up in frustration. There were no street signs to give the dudes directions. With nothing left to lose, they decided to follow a hunch.

If you were a homesteader, and you were settling this land in southeastern Montana, where would you plant yourself? they asked. They pointed their small truck towards the horizon beyond the breaks. Heading across the dry rangeland they bounced over pungent sagebrush and crashed down steep arroyos, wheels churning in the sand. Soon they could see an old water-pumper in the distance. Then the sod house came into view, and finally the object of their search, an old windcharger. There it sat on a rusting tower, a home light plant with a shed full of glass batteries at its base. A shoulder-high pile of antlers was stacked nearby. It was absolutely still, and then a slight breeze caused the old mill to creak. A coyote howled in the distance.

Tracking down an old windcharger can be a rewarding experience for the whole family. It can also be a monotonous and tiresome task, or it can lead to tragedy. The outcome depends on thorough preparation, persistence, and a good measure of luck. First, a little background.

Change was racking the country during the thirties. Familiar institutions were failing and new ones being created overnight. Ranchers and farmers on the Great Plains and in other remote areas of the nation wanted to keep abreast of far-off events and to learn how they would be affected. The radio was their link with the outside world — it was also a major source of family entertainment. Radios then, like they do today, required electricity.

Prior to World War II the only source of electrical power on remote homesteads was that which could be generated on-site, or that delivered by batteries. On-site generation was impractical for all except the wealthy, because of high gasoline costs. Batteries were the only choice for most.

Batteries would eventually run down, and then have to be hauled into town for recharging. That is, until some genius bolted a two-by-four onto a Model-T generator and the first windcharger was born. Windchargers became even more popular when Zenith used them as a promotional gimmick to market their radios to rural America. For a time, they offered a windcharger with each radio purchased.

The radio not only brought news of the day's events, but also news of another way of life, one with electrical appliances and conveniences unknown to the homesteader. Before long, the rural family's need for electricity grew, as the man of the house realized how electric tools would be time savers. Women, meanwhile, saw the benefits of electric lighting and refrigeration that until that time were provided by kerosene or propane.

As their needs grew, so did the size of their windchargers and the companies providing them. From models with rotors four to six feet in diameter with outputs of only a few hundred watts, windchargers by the end of the era had reached 14 feet in diameter with ratings up to three kW. Jacobs, Wincharger, Parris-Dunn, Aero-Electric, Wind King and others became the center of a new and indigenous Midwestern industry providing independent power for lighting homes and running appliances. As recently as 1950, there were an estimated 50,000 of these home light plants scattered across the nation's heartland.

Remote areas of the country were not served by utilities until President Roosevelt's Rural Electrification Administration (REA) brought federally subsidized power to the hinterlands through the Rural Electric Cooperatives. But, many areas did not receive power before the late forties or early fifties.

Most windcharger manufacturers saw that the end was coming and closed up shop in the late forties. The Jacobs Wind Electric Company, one of the more successful, held out until the mid-fifties. Interestingly enough, Wincharger has continued limited production of its small battery charger to this day.

As the rural co-ops extended service to more and more remote locations, the home light plants were no longer needed. Many were taken down and sold for scrap. Some were sold to neighbors who had not yet been "electrified." Others fell to the ground as rust, disrepair, and violent storms took their toll. A few stand just as they did then, albeit more weather-beaten than before.

Where to Look

You find windchargers where you find them. It's trite but true. One can be anywhere: packed away in the backroom of an old store, buried beneath a farmer's junkpile, hanging in the barn, lying at the base of its tower, or setting up there on top. Some have been found still in their original packing crate.

Many of the locales where these relics were once abundant have been picked clean by professionals during the early and mid-seventies. At one time, a small industry locating and rebuilding these pre-REA wind machines prospered. They sent out teams of scouts who would comb the countryside. The upshot is that nearly all of the easily accessible generators, those close to paved highways, have been bought, and if they have not, their owners do not intend to sell, and no amount of persuasion will convince them otherwise.

There still are machines out there. To find a salvageable generator now, though, you must select a promising area and painstakingly search it. Regions likely to have wind generators are those that are still remote or received centrally generated power very late in the REA program.

To find such a place, talk to the people who should know—the rural electric cooperatives. Contact the state rural co-op association or bulk power supplier. From there, move on down the line to the local rural co-op office. Or talk to the people at farm machinery dealers, at service stations, at hardware stores, at lumber yards, and at appliance outlets. Keep in mind that electricity came first to towns along major river courses and along the transcontinental railroad lines. Consequently, windchargers in these areas were never needed and the chances of you finding one are pretty slim. Look for remote areas. The more remote the town the better. Take Broadus, Montana, as an example.

Broadus sits on the Powder River ("too thin to plow, too thick to drink") in the very southeastern corner of the state. It is eighty miles from the nearest railroad or bus line. The town has a population of approximately 500, yet Broadus is the county seat. Utility service provided by the local rural co-op did not reach much of this area until the early fifties. One area resident was using his Jacobs home light plant as late as 1975 when power was finally extended to his ranch.

When the REA entered an area, ranchers were forced to disconnect their windchargers. Naturally, if one rancher no longer needed his windcharger, a rancher from a neighboring area not served by the rural co-op would buy it at a very reasonable price. Since the buyers could be choosy, they usually bought the best: the Jacobs. Jacobs wind plants thus migrated away from the rural co-op towards the more remote homesteads. As a result, there are not only more windchargers available in remote areas where utility service came late, but there is also a higher concentration of Jacobs generators than found elsewhere.

Once you have located a promising region, follow up any lead you get. A good way to do this is to check the names and phone numbers of prospects and give them a call first—let your fingers do the driving. It could save you a lot of trouble. If you can't generate any leads, try an ad in the local paper, or try posting a flyer at the most popular restaurant in town, or the local bowling alley. If all else fails, hit the road with a pair of binoculars and a good map. This is definitely the hard way. You will need to drive miles of dusty back roads and dirt tracks that do not even qualify as roads, and you will see more nondescript landscape than you ever thought possible. But you will also see sights that are beyond the range of most tourists.

After you become familiar with the appearance of distant water-pumping windmills—they are a common sight in much of the country—you won't need the binoculars quite so often. But in the beginning, it is hard to tell the difference between a broken down water-pumper and a windcharger.

Good maps are essential to determine the best way to get from where you are to the next promising prospect and to keep from getting lost. They are also helpful to mark the location of a find should the owner not be there when you are. And, there is always an owner. Just because it looks abandoned, doesn't mean that it is. As a crime, taking a windcharger without permission doesn't rank with stealing horses, but it is a crime nonetheless.

What To Look For

Not all windchargers are created equal, and it saves time to know what you want and what it looks like. The most desirable of the old windchargers is the Jacobs home light plant (see fig. 8-2). It was the most solid, the most reliable, and one of the largest generators built during this period. The most common brand, on the other hand, is the Wincharger. It produced less power, was less reliable, and more prone to failure. While the Jacobs may have been known as the "Cadillac of home light plants," the Wincharger has been called by aficionados as the "Chevrolet." Several other brands are lower on the list because of poorer quality or fewer numbers.

There are plenty of Winchargers available. During the heyday of the salvage operations in the mid-seventies, this brand was often passed over in deference to the more valuable Jacobs generator. Wincharger produced

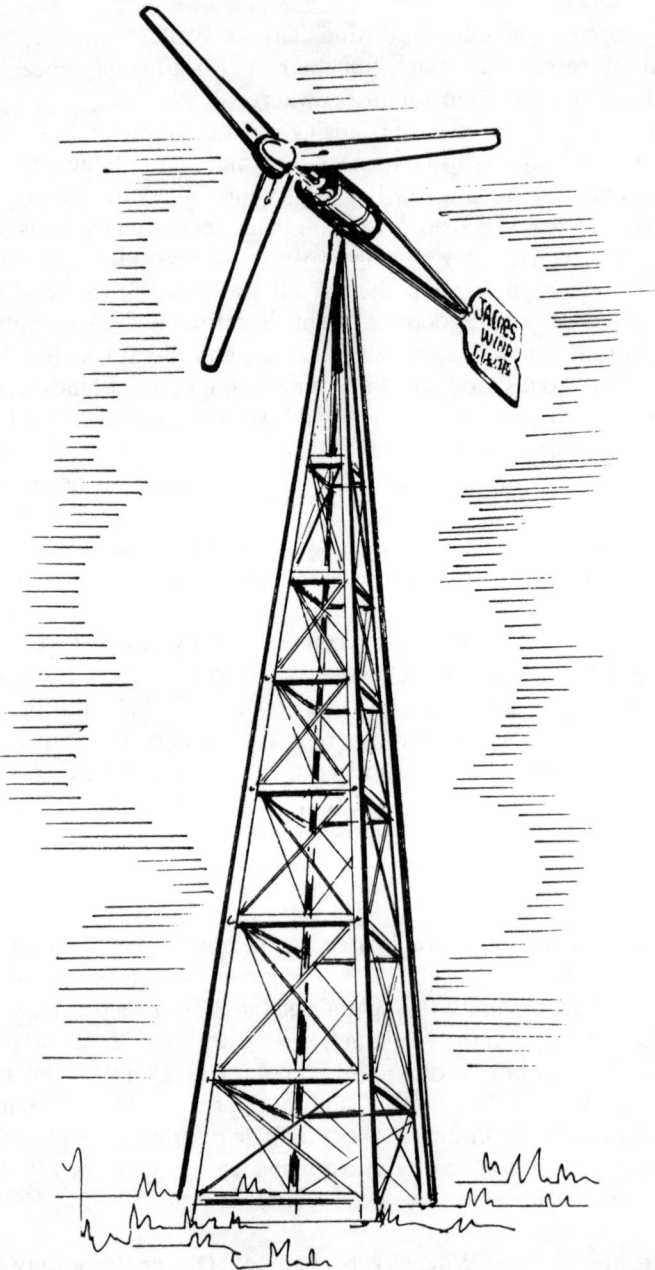

Fig. 8-2. Jacobs windcharger. The Jacobs home light plant has a distinctive tail vane and overall appearance that differentiates it at a distance from other windchargers. Jacobs wind plants, as they were called, were always mounted atop truss towers.

a wide variety of models. The early ones used a two-blade wooden rotor counterbalanced by dual air brakes or buckets. These models ranged from six feet to twelve feet in diameter, with power ratings from 200 to 1,000 watts. The smaller ones were direct-drive: the blade was bolted directly to the shaft of the generator. Later models became more sophisticated and sported four extruded aluminum blades 14 feet in diameter (see fig. 8-3). Rather than air brakes, this model varied the pitch of two blades (not all four) to control rotor speed via two heavy weights in the governor. The rotor drove a 1,000- to 1,500-watt DC generator through a single stage transmission. Most models were painted with the same yellow color scheme and used a triangular tail vane.

Jacobs took a different approach to building a home light plant, and the appearance of their turbines reflect this. Most obvious are the three-bladed wooden rotors that span 12 to 14 feet in diameter. Rotor speed in high winds was controlled by varying the pitch of all three blades simultaneously. Early models used what Jacobs labeled the flyball governor. This distinctive—and heavy—governor is spotted easily from the road. Later versions used a blade-actuated governor that took advantage of the centrifugal force acting on the blades themselves, rather than a group of weights, to change the blade pitch. The tail vane is also easily identified.

Unlike the Wincharger, Jacobs did not use a transmission. The rotor directly drives the DC generator. As a consequence, the generator is a massive affair of copper and iron. The turbine with blades, governor, generator, and tail vane weighs between 500 to 600 pounds depending upon the model.

Jacobs produced several versions. Most were 32-volt, but a late model introduced to compete with the REA generated 110 volts DC. The models were rated at 1,500, 1,800, 2,800, and 3,000 watts. Both the 2,800- and the 3,000-watt models were available in 110-volt versions for the modern homestead. The 3,000-watt model was the top of the line.

The Jacobs generator is the most sought after because it was built to survive the elements—and has. (Some Jacobs generators have been running for over thirty years.) Its reputation is well earned, and it is probably the best windcharger of the pre-REA era you can find. But the Cadillac slogan was the result of skillful marketing by Marcellus Jacobs, one of the two brothers who founded the firm.

Another windcharger worth mentioning is the Parris-Dunn. It used a two-blade wooden rotor from 6 to 12 feet in diameter. Very few of the larger models still exist, but you may stumble across one of the smaller ones. The Parris-Dunn was unique because of its approach to limiting rotor speed in high winds. Like the Jacobs, it too was direct-drive. But the blades were bolted rigidly to the shaft of the generator. They did not change pitch.

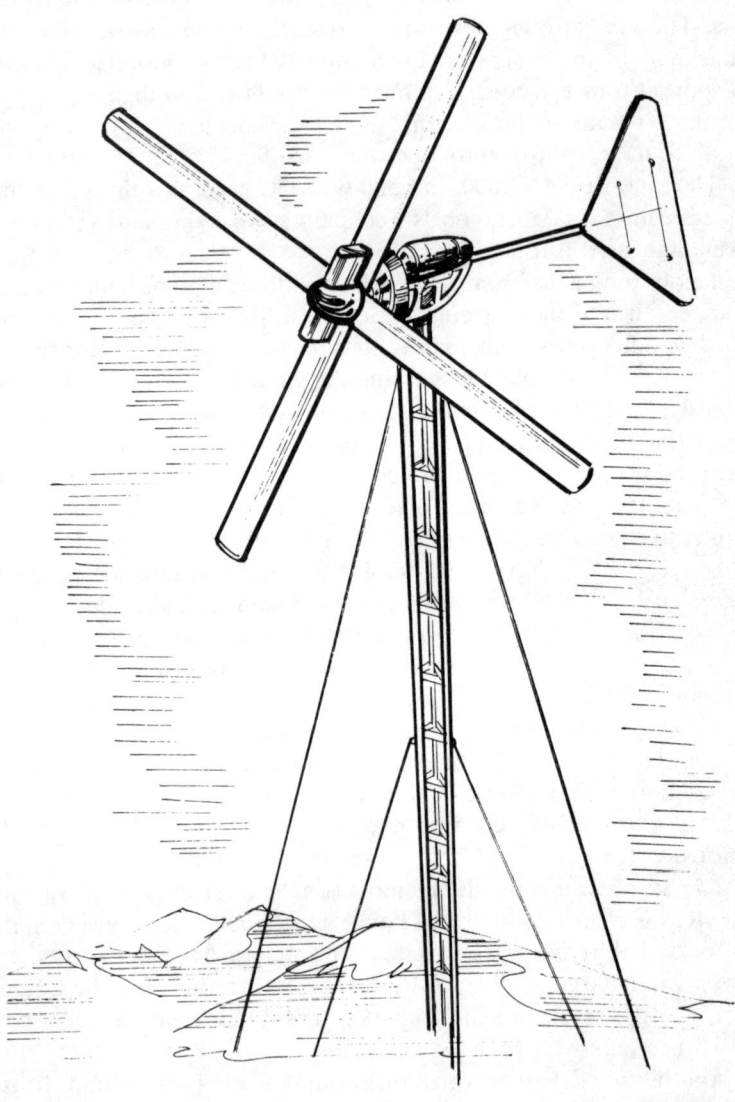

Fig. 8-3. Windcharger is the most abundant of the pre-REA wind machines remaining on the Great Plains. The unique rotor with four aluminum blades and triangular tail helps to identify the later Windcharger models. Earlier models often used a one-piece wood blade. Many Windchargers were mounted on guyed towers.

Instead, in high winds the hinged generator and attached rotor gradually tilted up out of the wind. When the winds subsided, the rotor-generator combination fell back towards the horizontal running position. Though extremely simple, this system worked reliably. Because of this simplicity, they are easy to rebuild—a good reason to pick one up if you have the chance.

Towers tell a lot about the windcharger that may no longer be there. First, windcharger towers are taller than those used for water-pumping windmills. They are noticeable from a distance because they rise up above the shelter belts commonly found around farmhouses on the windswept plains. Winchargers were installed on either free-standing or guyed towers. Jacobs were only installed on free-standing towers. If you find a guyed tower, then it's a good bet that the generator is a Wincharger. If it is a free-standing tower, closer inspection is necessary. Wincharger used a lattice tower with three or four legs. Each leg was made from a unique three-sided, lightweight angle-iron. The Jacobs, though, nearly always used four legs made from heavy $5/32$-inch thick angle-iron $2\frac{1}{2}$ inches wide.

Suppose you find a Jacobs or a Wincharger, but the nameplate vanished years ago. You have no idea what model you are looking at (not an uncommon situation). If it is a 110-volt model, don't worry. The rancher will surely remember—it was a sign of status in its day. But you can also check the voltmeter on the control panel, assuming it is not buried in the remains of the battery shed. The control box should have an ampmeter as well. If the ampmeter registers 70 amps, for example, it is a 2,800-watt generator (40 volts $\times$ 70 amps = 2,800 watts).

Making An Offer

If you have found a generator and the owner is interested in selling it, you need to take stock of the situation. Is the turbine still on the tower? If so, you've got to take it down (no small feat). If it is on the ground, are all the parts present? What size is it? Is it 32 or 110 volts? Were the blades stored away or just left to rot? These are some of the questions you need to answer before bargaining with the owner.

As with everything else, inflation has taken its toll on the going price for used wind machines. Renewed interest in windchargers has also forced farmers and ranchers to be a little more cautious in their dealings with salvagers. For starters, figure $100 as a minimum offer if much of the machine is present and still in one piece. If it is on the ground and in good shape expect to pay more. Larger generators are worth more than smaller ones.

A 2,800-watt Jacobs is worth more than a 1,500- or 1,800-watt model. Not only is the larger generator capable of producing more power, but it can also be rewound to generate 110 volts instead of 32—an important consideration. Should you come across the extremely rare 110-volt species, tread carefully. You've hit the jackpot; it does not need to be rewound.

Winchargers and other brands are worth less than the Jacobs. Some you may be able to pick up for nothing at all. In the case of the Wincharger, you may need two or three fairly complete machines to obtain enough salvageable parts for one wind turbine.

While you are striking a bargain with the rancher, you might ask if the tower is part of the package. Most will not sell the tower. They make excellent lightning rods and are ideal places for TV antennas as well. The angle-iron making up the tower is also in demand elsewhere on the ranch for various repairs. In addition, taking down a rusted tower can be more hassle than it's worth, but if you can buy a good 60-foot tower for a few hundred dollars, you have found a deal.

Ranchers, like everyone else, do not like to be rushed. Resist the urge to stuff a wad of bills into the owner's hand and haul away "your" windcharger. That machine you covet has probably been on the family homestead for thirty years or more, and he just might like to mull the whole deal over. His father may have installed that windcharger and he might want to "leave it right there where Dad put it."

The Plains are a great place to roam, and the people living there will make a trip in search of a windcharger a memorable one. Bargaining is half the fun of buying, and good bargaining requires consideration of more than self.

Getting It Home

Even after buying the generator, you are still a long way from erecting it in your backyard. First, you have to get it off the tower. To do that you will need pulleys, rope and/or cables, gin pole, safety harness, hard hats, utility belts, and assorted tools. Add to that list an adventurous spirit (tempered with caution) to work on a narrow platform (should there happen to be one) 60 feet in the air. For more details on equipment and safety practices read chapter 11.

Lowering a wind generator from a tower is dangerous work. There is no chance to test the equipment under full-load until the generator swings free. Everything must work then, or else. An excellent source of information on both how to remove a windcharger, as well as refurbish it is Michael Hackelman's *The Homebuilt Wind Generated Electricity Handbook*.

Hackelman devotes a whole chapter to the mechanics of raising and lowering windchargers. This book is required reading.

Safety around old windchargers cannot be overemphasized as the following news item testifies.

> CONRAD, MT (AP) — Strong winds may have blown a 29-year-old Choteau man, Timothy McCartney, to his death, the Pandera County coroner said.
>
> McCartney fell 24 feet to his death from the tower of a wind-powered generator. The coroner said McCartney apparently lost his balance and fell, possibly because of the strong winds that buffeted the area . . .

Once your windcharger is off the tower, you've got to get it back home. The next best thing to transporting it in your own truck is to ship it home via commercial freight. This is about the only way to ship more than one generator economically.

Crate all major components securely, box loose parts, fix shipping labels or tags to all parcels, and take the lot of them to the nearest trucking company that serves your hometown. Crate the windcharger yourself to insure that the job gets done right. Do not ask the freight company to build a crate. You'll regret it.

Restoration

Restoring a windcharger is not difficult for anyone familiar with electric motors. But because of their size, windchargers can be a little unwieldy. If you plan to use 32-volt DC and the generator's wiring is in good shape, you can do most of the work yourself. The job consists mainly of replacing the bearings and brushes, then cleaning, painting, and replacing worn or broken parts. Again, Hackelman's book is the best source at this stage.

If you want to convert the generator to 110 volts, then you will need to take it to a motor repair shop and have them rewind the generator. This will cost several hundred dollars, but it is better left to a professional than to try and do it yourself by hand.

While the generator is being rewound, repair the blades, governor, and tail assembly. If you were lucky enough to find the original blades in good shape, simply refinish them. Severely cracked aluminum blades or severely weathered wood blades should be discarded. Several firms carve wooden blades for the Jacobs windcharger, and spare blades for the Wincharger are available. Consult the classified ads in the alternative energy publications for sources (see Bibliography).

244 • Wind Energy

If you are pressed for time, or not inclined towards a long drive to Montana for that windcharger you have been dreaming about, you can find rebuilt units from the comfort of your armchair. As with components, classified ads will tell you who has old windchargers. Contact the advertisers and determine exactly what they are offering: rebuilts, or "as found." If rebuilt, what has been done to the turbine? A new paint job is not sufficient. What kind of warranty, if any, is offered?

If your rebuilt windcharger costs you $2,000, and you bought and installed a guyed tower for $3,000, the total cost for the system would amount to $5,000. A comparable small wind system from a dealer of new wind machines would cost from $6,000 to $10,000. But remember, the two systems are not directly comparable. The windcharger will not be as reliable, on the whole, as the new wind turbine, and parts are more difficult to come by. More importantly, the power produced may not be of the same value. Your windcharger will be producing power that can be used only for charging batteries or heating domestic hot water; a modern wind system may be able to perform all these tasks and also deliver line-quality power for supplementing that from the utility.

ASSEMBLING A KIT

Kits are another way many homeowners try to cut costs by avoiding a commercial wind machine from a dealer. Most, if not all kits currently available comprise a few cheap, shoddy, and poorly designed components intended for the hobbyist and priced accordingly. Set aside any notion of using such contraptions for supplemental heating or for producing utility-grade power. They are too small to make a noticeable dent in your consumption and too unreliable to be in service any practical length of time. These so-called kits are best suited to tinkerers and high school science fairs.

From these kits, you can assemble a crude windcharger with two to four blades generally less than 10 feet in diameter. I say windcharger because that is about all they can be used for — charging a small set of batteries. You can get all this for under $1,000. They are no bargain though. If you are a moderately skilled tinkerer, you can do as well on your own — and learn a lot more besides — by consulting Jack Park's *The Wind Power Book*.

Kits for the serious wind enthusiast or do-it-yourselfer should contain — at a minimum — a professionally designed turbine. Because the design and construction of the wind turbine is the most demanding of all the components in a wind system, it is best left to those who know what they are doing. All that you should be required to do is bolt on the blades and hang the tail (if one is used). A complete kit should also contain a tower designed for the wind machine (not a string of TV antenna masts as I've seen in some ads), all tower hardware, wiring, conduit, and electrical connectors needed, and a detailed assembly and installation manual. That is a tall order. No manufacturer or distributor presently provides such complete kits. (You are essentially asking for the same service and packaging a wind machine dealer receives from the manufacturer, but you need it if you want to do the job right.)

Many of the smaller items, such as the wiring and the conduit, can be purchased locally. But unless you are familiar with the ins and outs of wiring and your local electrical code, you could run into problems. This is where a good installation manual becomes important. It not only tells you exactly what you need to do the job right, but warns you where problems may develop and how to deal with them.

Buying a wind system kit is similar to buying a minicomputer through the mail. You can save a significant amount of money, but you do not have ready access to someone who can help you correct a problem. When you buy a wind machine from a dealer, you always have someone to turn to for repairs.

With a professional kit, you are not so much building a wind turbine as you are providing final assembly and installation. For example, Bergey Windpower offers a kit of their 3-meter turbine and a Rohn 25-guyed tower. The components can be handled easily and installed with a minimum of risk by following the instructions in their installation manual. The wind machine comes assembled, except for the addition of the blades and the tail vane. Because installation accounts for 25 percent to 35 percent of the total cost for this size wind machine, assembling and installing it yourself can produce considerable savings.

Kits such as these are limited to wind machines less than 4 meters in diameter. The larger wind machines are more difficult and dangerous to install and, consequently, the manufacturers do not offer them for owner installation. But even on the larger wind machines, there may be aspects of the installation that you can do yourself.

PREPARING THE SITE YOURSELF

Depending upon the dealer and the wind system you are planning to install, you may be able to do much of the site preparation. You surely can clear the site and lay out the tower and anchor locations. But you may also be able to excavate the foundation, pour the concrete, and set anchors where used.

The wind industry is changing rapidly and becoming more sophisticated. One result has been an increasing emphasis on proper installation due to the high cost of warranty claims. Some manufacturers insist that only their authorized dealers install the wind system. A few bypass dealers altogether. They deliver and install the wind system directly from the factory. In either of these cases, the amount of site preparation you can perform and its value is limited. Then there are those that operate at the other end of the spectrum and permit you to install the entire wind system yourself.

Remember though, that you assume liability for any work you perform. For example, say you prepared the guy anchor foundations, and one windy day an anchor pulls free and your nice new wind machine topples to the ground. The warranty provided by the dealer will not cover such an accident because you were responsible for properly installing the foundation and the anchors—even if you followed directions to the letter. The consequences are a pile of scrap metal and costly litigation.

Most of the cost in site preparation is in excavation, trenching, and pouring the concrete. You can cut costs here by using your own equipment if you have it, by borrowing it if you can, or by finding a lower cost contractor than that used by the dealer. Supplement the procedures outlined in chapter 10 with detailed drawings provided by the dealer or manufacturer. Do not skimp. If the plans call for re-bar, do not leave it out. Follow the plans religiously; your investment's in the balance.

Specifically, here is what you can do. Lay out and stake the position of the tower base pad, anchor locations, and cable runs. This may require accurately surveying the site where precise anchor location is required. Next, excavate for the tower pad and anchors. Build the forms for the concrete and lay in the reinforcing bar as specified by the manufacturer. Pour the amount and type of concrete called for. While the concrete is curing, you can dig a trench for a buried wire run from the tower to your house. (Aerial cable runs are less expensive and easier to install, but most dealers are moving towards underground runs because of the improved appearance. There is also less likelihood that the cable will be damaged by falling limbs and high winds.) The dealer may go so far as to permit you to lay the cables yourself. Most, however, will want to do this job themselves.

Depending upon the size of the wind turbine, the type (variable speed or induction), and the location, you can save anywhere from 5 percent to over 10 percent by preparing the site yourself.

Don't be fooled by the easy sound of it. Preparing the site or installing the entire wind system yourself can be backbreaking work. Halfway through you may want to call in the cavalry, but this is one avenue for cutting costs that is realistic and attainable, and one where the savings are substantial.

QUANTITY DISCOUNTS

There is one remaining method for reducing the cost of a wind system—buy more than one. This doesn't necessarily mean that you, yourself, buy and install more than one, though that is a possibility. You can buy more than one yourself, or get your neighbors to band together and buy several wind systems from the same dealer—or directly from the manufacturer—at the same time. By forming a buying cooperative, or club, you have more leverage when negotiating price with the dealer. Because the dealer gains by selling more than one machine, he is willing to listen and may be able to offer a discount for multiple orders within a certain geographical area within a limited time frame. The potential for discounts is greatest for installing multiple machines in one location, such as on a wind farm. You may be able to save from 5 percent to 10 percent, possibly even more, by buying more than one unit.

SUMMARY

If you want to cut the cost of installing a usable wind system, first and foremost find out how much of the installation you can do yourself. If possible, buy a kit from a reputable manufacturer and install the entire wind system. Join with your neighbors and buy more than one if you can. If you are less interested in reliable performance, consider buying a used wind machine. But build the wind turbine yourself only if you want to experiment.

9

Buying a Wind System

He that will not be counseled cannot be helped.
— John Clarke, *Paroemiologia*

Selecting and then buying a wind system entails gathering information (as you're doing now) sorting through it (weeding out the hype), and then determining which combination of product, manufacturer, and dealer best meets your needs with the least risk and the best chance of success.

CHOOSING A PRODUCT

First determine the size range that meets your energy needs while remaining within your budget as explained in chapter 2. For example, a homeowner without electric heat will be looking at wind machines from 3 to 6 meters in diameter. Those with electric heat will be considering those 6 meters in diameter and larger.

Next, find the product that offers the most for your money. Do not limit your evaluation only to energy output. Reliability, maintenance, and the soundness of the firm manufacturing it are equally important.

To gauge reliability "ask the man who owns one," advises Jon Traudt of Windcatcher Company. Track down owners of the same wind machine or those owning models by the same manufacturer. Ask them how well it has performed. What kind of problems, if any, developed? Are they satisfied? If they had to do it over, what would they do differently?

Call the local utility for information, but consider the source when assessing their response. I have found them dependable for general information on where specific brands are located, for example, and whether there have been any unusual problems of the sort, "Yeah, the tower fell over last year."

The state energy office and the American Wind Energy Association (AWEA) will also alert you to any obvious cases of fraud. Several scams involving pyramid schemes for distributorships of fictitious wind machines have been hatched. Though they did not last long, they bilked a few unsuspecting people out of their savings. If it sounds too good to be true, it probably is, warns the Postal Service.

Locate any test reports on the product you can find. Some universities have tested specific wind machines (West Texas State University is a good example), but most testing, of the little that has been conducted, has been by government agencies. In this country, DOE's Rocky Flats Small Wind Systems Test Center is the main source of objective reports on machine performance and reliability. Unfortunately, most of the wind machines they have tested are no longer on the market. The Canadians operate a similar but smaller facility on Prince Edward Island. They test both Canadian and American products. An excellent source of information on Danish wind machines is the Test Plant for Small Windmills near Roskilde, Denmark. The addresses for these test stations can be found in appendix D.

Where test reports are not available, which is often the case, talk to the manufacturer or their dealer. Ask them what kind of tests have been run on the wind system, for how long, and the highest winds experienced. Not all wind machines are built to the same standard. For example, I once dealt with a manufacturer who designed their wind machine to withstand a maximum wind speed of 90 mph. They asserted that no winds above that speed had ever been measured in their region. I personally had recorded wind speeds greater than 90 mph in the area they were referring to. The design of this product was highly suspect after this revelation and after considering that all other manufacturers were designing their products for a maximum speed of 120 mph.

Knowing how long the tests were run, or the length of time that a particular model has been in service, is especially important. Several times I have seen a wind machine being promoted as "extensively tested" when it had not been. In one case, the new product had only been in operation during the summer in an area of moderate winds. The first time this extensively tested product was operated in a good wind regime it flew apart. In another case, the extensive tests were conducted on a bench-scale model. Not even a working prototype had actually been tested. Reputable, well-tested products, in contrast, have been in unattended operation for months if not years at numerous sites and have endured winds observed to exceed 100 mph.

When evaluating performance and reliability, do not be alarmed by occasional reports of defects. You are looking for trends. If every one produced so far has thrown a blade and they are still throwing them, then there is a good chance the one you are looking at will, too. But wind turbines should not be held to any higher standards than we hold other machines. After more than eighty years of development, automobiles are still being recalled by the thousands for manufacturing and design defects. Yet we continue to buy and use them. We try to minimize the risk of buying a lemon by trying to select a model with the least potential for problems. Reputable manufacturers of wind machines make mistakes like everyone else. Your goal is to find one that makes fewer mistakes than the rest.

Design defects and maintenance problems show themselves either immediately or after several months of operation. Like automobiles, new products must undergo a period of debugging. Unexpected problems show up and must be corrected. These problems are greatest when the product is first introduced and decline thereafter. Ideally, you would like a wind machine that has been on the market for several years and one that is operating successfully in a range of environments. Because the wind industry is so young and because it is changing rapidly, this is not always possible.

Take a good close look at the wind turbine itself. How is it put together? Does it look like it was welded together in a backyard shop? Do the parts fit snugly or have they been made to "press fit" with a hammer? Now step back and look at the wind machine as a whole. How do you feel about it? Is it something you would be proud of or will you have to put it "out back" so no one will see it? Wind machines are highly visible objects on the landscape. Their appearance is important for community acceptance.

Examine the promotional literature describing the wind machine. Are the estimates of energy output reasonable? Do they stress generator size while ignoring energy output altogether? Most manufacturers will present, with their product literature, a list of parameters suggested by AWEA

that describes succinctly how the wind machine performs, how it functions, and what it can be used for. The minimum required by AWEA is:
1. annual energy output,
2. power curve, and
3. power form.

AWEA intends that the annual energy output replace the rated power at rated wind speed method prevalent in this country to describe the size of a wind machine. They would like the AEO to be presented for three average wind speeds: 10.1 mph, 12.3 mph, and 14.6 mph. (The standard is in meters per second, not miles per hour.) Where only one rating number is used, AWEA suggests that it should be the AEO at an average wind speed of approximately 12 mph. The AEO may also be presented in a graph (see fig. 5-3 in chapter 5).

Power form is a description of the form in which the power is delivered to the load. For wind systems interconnected with the utility, this will be given as the nominal voltage and frequency as well as the phase of the electricity generated.

AWEA also recommends that the following be included.
1. average noise level in dBA
2. cut-in, cut-out wind speeds
3. maximum power
4. overspeed control
5. maximum wind speeds: design and tested
6. rated rotor speed

Overspeed control is a concise description of the method used to protect the wind turbine in high winds or during a loss of load. The *design wind speed* is the maximum speed for which the turbine was designed to operate unattended without suffering damage. The *tested wind speed* is the maximum observed or recorded speed the wind machine has survived without major damage.

For synchronous wind machines the *rated rotor speed* is the average revolutions per minute from cut-in to maximum power. In an asynchronous wind machine, this is given as a range of rotor speeds from cut-in to maximum power.

The following is typical of information you will find on promotional brochures using the AWEA format.

AWEA estimated, Annual Energy Output
8,000 kWh @ 10 mph
12,000 kWh @ 12 mph
16,000 kWh @ 14 mph

Power form: 3 phase, 220 volts AC, 60 hertz (cycles/sec)
Average noise level: 50 dBA @ tower height of 80 ft.
Cut-in speed: 10 mph
Cut-out speed: 45 mph
Maximum wind speed: 120 mph design, 112 tested
Overspeed control: Blade stall, brake, tip flaps
Rated rotor speed: 200 rpm

Ask to see a copy of the owner's manual. Is it well written and sufficiently detailed to tell you what is needed to maintain the wind system and how to operate it safely? Are there instructions for starting and stopping the wind machine, for example? Does it provide a parts list?

Engineers, when ordering expensive equipment, often ask for a copy of the owner's-operator's manual from each company competing for the sale and compare them. They not only gain a better understanding of the equipment by doing so, but they also develop a feel for the manufacturer's approach to problems that may be encountered by the user and their attention to detail. I know people who have bought computers in large part due to the thoroughness of their owner's manuals. Don't expect the dealer to give you a manual free, but they should offer it to you at a nominal charge.

Now examine maintenance. What maintenance is required, for example? How often must it be performed? How difficult is it to perform? Must someone climb the tower or is a cursory examination from ground level adequate? If parts must be replaced, are they readily available or must they be specially ordered? These are a few more questions to answer when evaluating the product.

EVALUATING THE MANUFACTURER

After you have dissected the technological aspects, you must evaluate less tangible factors such as the durability of the manufacturer and the reputation of the dealer. Businesses are like children. Some grow up to become healthy adults. Others never grow up, instead stagnating in extended adolescence. Others become sick, wither, and die. Business is a rough and tumble world in the best of times but it is worse when the field is new, the market just emerging, and money tight. Your task is not only to find the right wind machine, but also to find one that is built and distributed by a company with good management and finances. You want a firm that has put together a good sound business that is going to last twenty years. You want them to be around as long as your wind machine will be.

How long has the company been in the wind business? What is their track record? Do they have sufficient financial resources to honor their warranty commitments? There is no easy way to find answers to these questions. In most cases you will be dependent upon the dealer for information. Even when you do get the answers, it is hard to determine what is important and what is not. For example, a well-established company that has been in business for several years is a better risk than one just starting out. Likewise, a company that is partially or wholly owned by a major national corporation usually indicates that there are ample financial reserves for it to survive a major warranty recall. Nevertheless, a corporate executive who has no personal stake in the company can much more quickly make a decision to cut its losses during hard times and cease production than the owner-entrepreneur who has put his own sweat and blood into the business. Sole proprietors tend to hang on until the very end. Only you can decide on which type of business you should place your bets.

One minimum test if you need reassurance about a manufacturer is to ask for their certificate of insurance. This is simply a statement by an insurance company describing the type of coverage the firm carries. Backyard garages and fly-by-night operations will not be able to provide this statement (they may not even know what it is) and demonstrate that they carry product liability insurance.

EVALUATING THE DEALER

The dealer you choose is determined primarily by the wind machine you want and where you live. Most dealers represent more than one company to round out their product line. Even so, within a certain locale there will be only one dealer for each brand. (Manufacturers want to insure a healthy dealer network so they limit the number of dealers selling their product.) Proximity is important. If repairs or service are needed, particularly during an emergency, you do not want a dealer who lives on the other side of the continent.

Determine if the dealer is reputable by checking with their previous clients. Have they been prompt in making repairs, for example, or have they taken their time while hustling new sales? The dealer should have references available for such an inquiry. If not, are they willing to provide them? The dealer should also provide professional and character references if you ask for them. Call the state energy office and the local solar (or wind) association if there is one. Ask them what they know about the dealer.

Also call the manufacturer and check whether they are an authorized dealer. In one instance, a "dealer" was selling a popular brand without authority to do so from the manufacturer. This dealer had declared bankruptcy previously leaving a number of clients high and dry without spare parts or service for their ailing wind machines. In this case, he was selling a used wind machine as new. The customer bought his whole line — and the wind machine — that is, until the authorized dealer blew the whistle. The whole sad affair could have been avoided by a single phone call to the manufacturer.

Don't be misled by membership in various organizations as a claim of legitimacy. I know dealers (and some manufacturers for that matter) who use membership in AWEA and a host of solar associations as a promotional tool. It is one of the oldest marketing tools in the book and most often used when no other credentials exist. Anyone can join an organization. All you need is money.

WHAT TO EXPECT FROM A DEALER

You have a right to expect that all work will be performed according to standard practices, local building codes, and local electrical codes. The work should also be performed in a timely manner and the dealer should clean up the site before leaving.

Don't expect overnight miracles. If delivery of a component has been delayed due to circumstances beyond the dealer's control (a trucker's strike for example), you should not hold them accountable. The dealer should make a reasonable effort to expedite the installation of your wind system or its repair, but do not expect them to jump at your every request. Keep in mind that dealers operate a business and that they have other commitments, and that they must be paid for work that is not under warranty or a maintenance contract. At the same time, the dealer should fulfill those obligations stated in the contract or implied during negotiations.

SIGNING ON THE DOTTED LINE

To insure that you get what you pay for, put it in writing. Demand a written contract and warranty and have your attorney look them over. Installing a wind system is a major investment akin to buying a car (or a house). It is worth the added cost of getting good legal advice. You may also need advice from a tax consultant about the method and timing of payment to keep the IRS happy.

First, what is specifically included in the price you have been quoted? If the dealer is to install the entire wind system they should do everything from preparing the site to cleaning up afterwards. If you plan to do any of the work yourself, the contract is necessary to spell out exactly where the dealer's responsibilities end and yours begin. For example, if you are going to install the wind machine, will the dealer provide the gin pole (if you use one)? If so, is there an additional charge?

When performing some of the work yourself, the contract should also describe exactly what you must do to meet the terms of the warranty. Who has the final say, for example, as to how the work should be done? (Usually, the dealer does.) How will disputes be resolved should they arise?

What is covered by the warranty? What is not? How long does it last? There is a one-year warranty on all components for most small wind machines. The one-year comprehensive warranty is also offered on larger machines, but an extended warranty limited to major components is also included. These extended warranties cover repair or replacement of the rotor and drive train, but not such items as fuses and relays. Some manufacturers are offering extended warranties up to five years long. Because of the difference in warranties between manufacturers, it is wise to read the fine print. Is the warranty transferable or assumable by the manufacturer if the dealer goes out of business, for example?

In multiple machine purchases such as for a wind farm, the buyer has more leverage with the manufacturer and can obtain written assurances of performance that the wind machine will generate power as advertised and that it will be available to generate power a minimum percentage of the time. Unfortunately such assurances are not offered to purchasers of small wind machines in single units—at least not yet.

Another aspect is the terms of sale. The contract should state the amount, how, and when payments should be made. In general, you will pay for the wind machine and installation in advance. (You don't drive off the lot with a new car until you have handed over your check. Similarly, you should not expect the dealer to install the wind system without your first paying for it.)

Terms vary from one dealer to the next. Usually a down payment is made to secure your order. Then payment in full for the turbine and tower is required when they are ready to be shipped. Some dealers require payment for the turbine, tower, and installation in advance. In most cases 5 percent to 10 percent of the total contract is held by the buyer until the wind machine has been installed and is operating.

EXAMPLES

In a classic example of how not to go about it, a rural cooperative bought a wind system from a local manufacturer. The co-op did not contact anyone about the company or its product. Nor did they investigate the company's claims. If they had, they would have found that the wind machine was a prototype (not a well-tested wind machine ready for commercial sale), that the manufacturer could not possibly build and install the wind machine for the contracted price (one of those "it's too good to be true" deals), and that the wind machine could not do what the manufacturer said it would. In short, neither the rural co-op nor the manufacturer knew what they were doing. The machine was installed and—of course—it never worked. It still stands as a testament to ignorance. Then there are those who take a more studied approach.

Capitola Reece, 74, is a retired school teacher. She knows what she wants and she wanted a wind turbine. It had to pay for itself and it had to work reliably. One thing was certain; she was not about to climb the tower and fix it herself. She also realized she didn't know the first thing about wind machines or even where to buy them.

In her search for information she ran across an article about a fellow who was promoting them and wrote to him. After thoroughly reviewing a package of literature he sent her, Cappy, as she's called, arranged for him to visit her site. She was ready. If it was going to cost her fifty dollars to get him there, she was going to get her money's worth.

She had the site picked out, copies of all her utility bills, and a notebook full of questions. How much wind do you think I have? How much does it cost for an anemometer? For a wind turbine? What tax credits are available? Is it noisy? What maintenance is required? How many have been installed; how many have you installed? How well have they worked? How much energy could one produce here? What does the utility think about all this? How will it affect my taxes and insurance rates? What happens when the utility lines go dead; will it still work?

After the inquisition, she took him to her proposed site. Bad news. He would be glad to install an anemometer, but he would just be taking her money. Too many trees nearby. Though not tall, they were tall enough to block the flow at the anemometer. The results would be less than the speed at the nearest airport. Would that data do? he wondered. He gave her his estimates based on the airport, and left.

"Well," Cappy thought, "we'll just check this guy out." She called the state energy office, the manufacturer, AWEA, and a previous client with a similar wind turbine. He seemed all right. In fact, he was highly regarded by those references he had given her. Still, she wanted her attorney to look over his contract and offer of warranty. She also wanted to talk to the township supervisor about the need for a zoning variance and to the utility about the interconnection. No variance was required for the rural site. The utility did not know a thing about the particular wind machine or the dealer, but did warn her that the few wind machines installed in their area had not worked well. That didn't deter her. She thought they would be less than thrilled with the idea.

The attorney had some objections; so did her bank. The contract called for a sizable amount of money for a rather novel purchase and it called for most of it up front. "It won't do," said the attorney. He demanded changes in the contract and the terms of payment. He wanted to pay after installation. The dealer balked—too great a risk for his small business. But a compromise was reached. A portion of the payment would be held in escrow by the attorney until the wind machine was installed and operating. The dealer agreed. His needs were met by knowing that the money was earmarked for him and was in safe keeping.

The three parties met and signed the contract, and the equipment was ordered. The contract stipulated that the dealer had ninety days to install the wind machine and get it running. Within two months it was operating, but just before the final payment was made a problem developed. Because the escrow account had not been released, the dealer hustled to make the needed repairs. The wind machine has run unattended ever since.

10

Installation and Maintenance

Small wind machines in the 3- to 5-meter size class may be installed by the homeowner or hobbyist who has basic construction skills. The work can be dangerous as the following account warns, but no more so than other self-help projects around the home or farm. With proper respect for the hazards involved and attention to detail a small wind machine can be safely installed.

Larger machines (those 6 meters in diameter and larger) entail greater risks. The components are heavier and may require special equipment and techniques unfamiliar to most do-it-yourselfers. It is usually best to leave the installation of these machines to experienced dealers.

The following sections provide detailed information required by any installer. If you plan to do the work yourself this information will help you to select the tower, anchors, and erection methods that best suit your talents and the conditions at the site. After reading this chapter you may choose not to do the work yourself. The information gained, however, will enable you to track the progress and evaluate the performance of the contracted installer. For instance, the following dialogue would have greater meaning.

He loosened the last bolt. The generator was now ready to swing free.

"All ready," he yelled.

"Yeah, let 'er rip," replied the ground crew.

"You sure that pulley's secure?" he asked, his voice less certain now.

"Yeah, it's not going anywhere. Let's get this one down and go home."

The old generator rocked on its saddle. Slowly it rolled off towards the gin pole. Suddenly there was a loud twang and the squeal of steel cable over pulleys as the 400-pound mass of copper and iron whizzed by his head to crash through the platform next to him.

He looked about in dazed silence.

"Are you all right?" they asked from below.

He glanced at his feet. Yep, all still there. Then to his hands. They were too, as were all his fingers. Lucky this time he thought.

"I'm okay, what the hell happened anyway?"

"That pulley broke loose from the tower, like you were afraid it might."

This incident actually took place. It happened to an experienced crew working professionally. Though it occurred removing rather than installing a wind machine it illustrates what can happen without thorough planning, preparation, and—equally as important in this case—execution.

Planning is necessary to determine what is needed to do the job, when, what is the best method, and what skills are required. You must anticipate what will be needed at each step during assembly, the problems you may encounter, and how you expect to deal with them. You must coordinate the schedules of your subcontractors, suppliers, and erection crew to keep the project moving smoothly. You must also choose an erection procedure that best suits you, your site, and the crew you will be working with. And if you lack any of the required skills you must find someone who has them.

Figure it will take you twice as long as you expect. A skilled two-man crew can install a small wind machine in one day. It may take a novice a week if he is cautious. Climbing an 80-foot tower is tiring, particularly if you have never done it before. Do it several times in one day and you may have had enough. As you tire you begin to make mistakes. Don't take chances on you or your crew making mistakes because you tried to do too much in one day. Give yourself plenty of time.

Prepare for the installation by collecting the parts, fittings, and tools for the job. Learn how they are to be assembled, in what way, and with what tools. Make sure you have met all legal requirements and that you are insured for any accidents that may occur. If you are installing the wind system yourself it is a good idea to check whether your insurance will cover hospitalization and liability for your friends who lend you a hand.

Without proper execution all your planning and preparation may be for naught. You may know the right way to do a task and have the right tools to do so but if you do not follow through under the press of time and conditions (you may be tired and a trip down the tower to check a pulley in the previous incident may seem like more trouble than it's worth) the results may be disastrous.

UNPACKING AND INVENTORY CONTROL

Installation can be hindered and operation of the wind turbine prevented by components damaged in shipment. Before accepting delivery from the freight carrier examine the invoice or billing form to determine the number of crates shipped. Make sure all are present, and then carefully examine the crates for external damage. If damage is found, open the crates and look at the contents. Hold the shipper until you have determined the extent of the damage, if any, to the contents. The crates are designed to take some abuse while still protecting the product. Note any damage as precisely as possible and immediately contact the dispatcher at the freight company. Instant photographs can be helpful in verifying claims.

Tower sections and sensitive electronic components are the most easily damaged during shipment. The welded lattice tower sections commonly used in this country for guyed towers can be crushed or bent from other goods on the truck. Check for proper alignment and any bent cross-girts. (It's hard to install a straight and true tower when a section is twisted or bent.) Damage to control boxes and synchronous inverters is much harder to determine. The best you can do is look for loose parts rolling around inside.

Catalog the serial numbers on the generator, blades, control panel, synchronous inverter, and tower. If you have to make a warranty claim, the numbers are much easier to find in your files than at the top of the tower. Serial numbers will aid trouble shooting if problems develop.

Make an inventory of all parts received as soon as possible. Many manufacturers provide a parts checklist for this purpose. Use it. The time to realize that an important bolt is missing is prior to installation—not while you are hanging 80 feet in the air.

For those wind systems where the manufacturer does not also build the tower (on small wind machines), the tower and wind turbine will be shipped separately and will be delivered by separate carriers. Unless you have a special reason for removing the contents from the crates (for an inventory possibly) leave them as delivered until you are ready for the installation.

FOUNDATION PREPARATION

Clearing the Site

The tower and guy cables (where used) must be kept clear of vines, trees, and shrubs. It may be necessary to clear the site of any trees or shrubs that could eventually interfere with the tower or guy cables. The site does not have to be level so there is no need for grading.

Selecting the Foundation

To prevent the tower from overturning, anchors are placed in the ground. Anchors resist uplift. Piers, on the other hand, resist loads in compression. Consider a guyed tower, for example. The anchors hold the tower upright and resist the forces trying to knock the tower over. The pier beneath the central mast supports the weight of the tower and wind turbine and resists the reactive forces trying to drive the mast into the ground. On free-standing towers, the legs act alternately as piers and as anchors depending on the direction of the wind.

The type of anchor or pier used is contingent upon the tower and the site. If you plan to install a guyed tower there are several anchoring options to choose from: concrete, screw, expanding, and rock anchors. The best choice for your site is determined by the engineering properties of the soil, the depth to bedrock, and the power equipment available in your area. For a free-standing tower, the choice is limited to concrete.

Anchors

Anchors must withstand the static and dynamic loads acting on the wind system, under all weather conditions, for the life of the system. They must do so without appreciable creep towards the surface or settling. The holding power of anchors depends on the area of the anchor, its depth, the soil in which it is embedded, and the soil's moisture content. Weight is a factor as well, but it is not as important as most think.

Soils vary tremendously in their ability to resist creep. Resistance to creep and to some extent settling is controlled by the soil's shear strength — the resistance of soil particles to sliding over one another. Shear strength is a function of soil type and whether the soils are wet or dry. Shear strength ranges from a maximum in solid rock to a minimum in muck or swamp soils.

In an engineering sense soils can be divided into two broad categories according to one anchor manufacturer: cohesive and noncohesive. Cohesive soils stick together; the particles cling to each other. Soils with a high clay content are considered cohesive. Sandy soils, those which crumble in your hand, are noncohesive; the particles slide right by each other. Wind system manufacturers specify that their standard anchor designs are intended only for normally cohesive soil. Such a soil has a high clay content.

Anchor holding capacity decreases as moisture content increases. Creep can be troublesome in saturated soils as the particles become fluid and tend to flow around the anchor and also because the water is buoyant decreasing the weight of the anchor. Holding capacity can be reduced by 50 percent in wet soils. Wherever possible anchors should be placed below the level of periodic saturation from heavy rains — but above the water table.

Frost heave causes similar problems. When the soil freezes it expands slightly (ice occupies a greater volume than water). If the anchor is not below the frost line the cycle of freezing and thawing will heave or jack the anchor towards the surface. This is more of a problem for anchors than piers or footings because of the existing loads acting to pull the anchor out of the ground. The forces on piers act counter to frost heave. The frost line varies from year to year and depends on the severity of the winter and the soil cover. Bare soil freezes quicker and to a greater depth than a soil with a grass cover. (The grass and the organic soil it grows in act as an insulator slowing the soil's winter heat loss.)

Engineers have created several classification systems to describe the properties of soils. A. B. Chance, an anchor manufacturer, has devised a soil-grading system for the application of their prefabricated anchors by the utility industry in the construction of power lines. Chance's soil groups range from a Class 0 for solid rock to a Class 8 for peat and muck soils that form in swamps. Solid rock provides the greatest holding capacity, peat and muck the least. In the Chance system, a normally cohesive soil falls into Classes 3 to 5.

To determine the soil-holding capacity at your site you can test the soil with a probe, examine nearby road cuts or — better yet — talk to the people who work with the soil: the Soil Conservation Service (SCS), the county extension agent, or the office of the conservation district. Explain your plans to them. Describe what it is you want to know and why it is important (you don't want the anchor pulling out of the ground). They will not only be able to tell you what kind of soil you will be working with but also the depth to the water table and the average frost penetration.

The SCS has prepared soil maps of much of the country. It will be helpful if you can find your site on these maps (they are on public display at the local office). The SCS will interpret the information on the map for you. Local excavation companies are another good source. They have a feel for subsurface conditions since they work with them daily. They are in business to make money, though, not give out free information. If you want their help, you should hire them.

The requirements for piers are less stringent than those for anchors. Most soils are strong in compression. With an adequate bearing surface, concrete piers of standard dimensions are used throughout the country.

The most common method for anchoring a tower or constructing a pier is to excavate a hole and partially (sometimes completely) fill it with reinforced concrete.

Working with Concrete

Concrete is literally man-made rock, conglomerate to be specific. It is strong in compression (as when squeezing an accordion), weak in tension (as when expanding an accordion). Thus, it works well as a pier or foundation but poorly as a beam. Tensile strength is improved by reinforcing the concrete with steel bars or *re-bars* (reinforcing bars).

Concretes are rated by their compressive strengths. Most construction uses concrete that obtains a minimum, ultimate strength of 3,000 psi after curing for twenty-eight days. Strength is a function of the water-cement ratio and the degree to which curing has taken place. The lower the water-cement ratio (the more cement in the mixture) the stronger the concrete. Strength also increases with curing time.

Curing is rapid in the first few days. (Concrete sets or becomes rigid within an hour of adding water.) Hydration does not go forward if too much water evaporates in hot weather, or if the concrete becomes too cool in cold weather. Curing should take place for a minimum of seven days before any load is placed on the concrete. The concrete will continue to gain strength if moisture and temperature conditions remain favorable for complete hydration. Heeding the above precautions, concrete can be poured year round.

Concrete can be bought in bags and mixed by hand or bought directly from a ready-mix plant. The ready-mix plant will deliver, or in some communities you can pick up the concrete with a small trailer with a rotating drum. Concrete from a ready-mix plant is sold by the cubic yard (27 cubic feet) or "yard," in lots of less than two yards, two to five yards, and so on. The cost for delivery is included in the price for delivery within a limited distance of the plant. If they have to haul it further, it costs more. For any installation requiring one yard or more, ready-mix is the way to go. It is far easier than mixing it by hand, it is quicker, and it offers a better way to control the quality of the concrete if you are inexperienced.

Installation drawings invariably show nice neat anchors and piers that look like they came out of a cooky cutter. Except where an anchor or pier is exposed at the surface this is not necessary. Where the soils are stiff and will not collapse into the hole, the concrete can be poured in place. For anchor blocks below grade this is superior because the concrete acts directly on undisturbed soil. Forms are necessary where the hole is larger than the anchor or pier desired and where the concrete will extend above the surface.

Installation instructions usually call for the concrete to be poured over a grid work or cage of re-bar to give the necessary tensile strength. Re-bar is designated by its diameter in eighths of an inch. Thus, specifications calling for a rectangular cage of #5 vertical and #4 lateral will use ⅝- and ½-inch re-bar respectively. The re-bar is tied together with wire.

Re-bar must be covered by at least three to four inches of concrete. When closer to the surface, acid-laden water enters the concrete and corrodes the steel. The re-bar expands slightly causing the concrete to spall or chip. For long life, the concrete must seal the re-bar from corrosion.

To insure that the re-bar stays where you want it when the concrete is poured, it should be staked down in the excavation or should be tied to the forms so it will not "swim" around. Pieces of rock or brick can be used to keep the re-bar cage off the ground. This keeps the re-bar from being too close to the concrete surface on the bottom.

Forms can be built from heavy plywood and a 2-by-4 frame. When the 2-by-4s are staked to the side of the excavation they will hold the plywood form in place while the concrete is hardening. Where the soil is stiff and the excavation is no larger than the pier desired, a short form can be made from 2-by-12 planks set in the excavation to a depth that gives the finished pier its correct height above the surface. Cylindrical forms can be purchased for pouring concrete columns.

Before pouring concrete, moisten the forms or the excavation to prevent them from absorbing water from the surface of the concrete mix and from reducing its strength. Avoid pouring the concrete from a height greater than four feet or the aggregates will begin to separate. Once in the form, work the concrete to eliminate air pockets by poking a board or shovel into the concrete. Work it around the re-bar and along the sides of the forms. Don't overwork, or the aggregates begin to settle.

You can simplify the whole process by hiring a contractor to excavate the hole and pour the concrete. Get firm quotes before you do and make sure they understand what the concrete will be used for. The fact that a lot is riding on their work may discourage them from cutting any corners.

Guyed Towers

Let us first examine the requirements of guyed towers because of their popularity. To support a guyed tower you will need a pier to set the mast on and at least three anchors to guy the mast.

The pier is simple. For most small wind machines, those from 3 to 5 meters in diameter, a pier 2½ by 2½ feet and 4 feet deep is more than adequate for towers up to 100 feet tall in normally cohesive soils. In weak soils, with an extremely tall tower, or with bigger wind turbines, a larger pier may be necessary. (As with all other aspects of installing a wind machine, follow the manufacturer's suggestions. They know what works best for their particular product.) It may also be necessary in some areas to extend the pier deeper to get below the frost line. Strengthen the pier with a crib of re-bar.

Guyed towers also need some way to keep the mast in place—to keep it from scooting off the pier. Before the concrete sets, insert a pin or threaded rod into the center of the pier. Most masts have a base plate that then fits over this pin and rests on the pier.

Choosing a guy anchor, though, is a little more complex. The type of anchor used depends on several factors: soil strength, the depth to bedrock, and your access to power equipment. In a normally cohesive soil you have a choice among concrete, expanding, and screw anchors.

Concrete anchors are used most often—first, because building inspectors and contractors are more familiar with them than with the other anchors; second, because the equipment needed—a backhoe—to excavate the hole is abundant even in the smallest community. Third, because concrete anchors can be easily adapted to weak or soft soils by expanding the bearing surface (lengthening and widening the anchor).

Backhoes work best at digging trenches, so they are well suited for making the anchor excavation. The concrete can usually be cast in place. Backhoes can also excavate the pit for the pier but the excavation will not be square. When using a backhoe for the pier, forms will have to be used.

Table 10-1 gives the dimensions of concrete anchors for wind turbines mounted atop an 80-foot tower in normally cohesive soil (see fig. 10-1).

An easier and quicker method is to auger the holes for the pier and anchors. Where the soils are not rocky and bedrock is well below the sur-

266 • Wind Energy

TABLE 10-1
Concrete Anchors for 80-foot Guyed Tower

Turbine Size	Anchor Dimension	Depth
3-meter	3 x 3 x 1	3
4-, 5-meter	3 x 4 x 1	4
6-meter	3 x 6 x 1½	4

NOTE: #4 re-bar in top and front face on one-foot centers, laid each way.

face, a truck-mounted power auger can be used. Linemen use these power augers to set wood utility poles in minutes.

For the pier, a 36-inch auger is best. It excavates a hole large enough to form the pier. Square the hole to the proper dimensions by chipping away at the sides allowing the dirt to slough off into the hole. Set a short form into the top of the hole so that it extends above the surface 4 to 6 inches, place the re-bar, and cast the concrete in place. If the auger is smaller, it will take more work to widen the hole, but it is still simpler to cast in place than building a form for the entire pier.

The auger is then used to drill holes for expanding anchors (see fig. 10-2). Expanding anchors work much like toggle bolts and similar fasteners used in plaster walls. The hole is drilled just large enough to admit the anchor. A hand-operated power auger can also be used for drilling anchor holes.

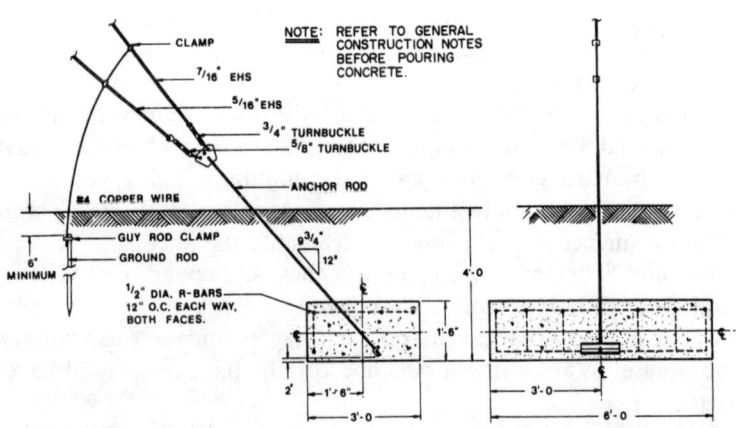

Fig. 10-1. Concrete anchor detail. Note position of reinforcing bar and attachment of ground wire. (Courtesy Enertech Corp.)

Installation and Maintenance • 267

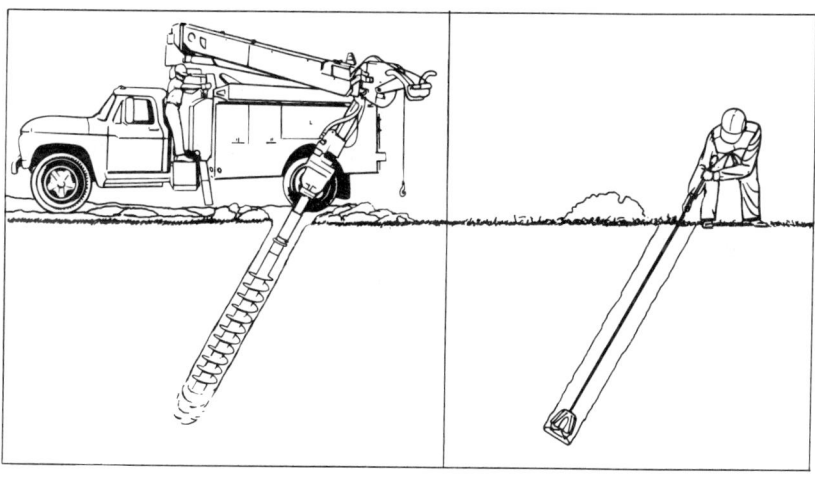

Fig. 10-2. Expanding anchor installation. *Top left,* augering hole for anchor; *top right,* inserting anchor into hole; *lower left,* expanding the blades of the anchor with tamping bar; *lower right,* backfilling hole and tamping. (Courtesy A. B. Chance Co.)

Anchor strength is controlled by soil type, the size of the expanded anchor, and the firmness of the backfill. Expanding anchors hold more in heavy, stiff soils, less so in sandy or swampy soils. The larger the diameter of the anchor, the greater its holding power. But the anchor must be fully expanded into undisturbed soil to reach its rated strength. Compaction of the backfill is also critical. Anchor manufacturers note that

268 • Wind Energy

a compact backfill can actually improve the soil's holding capacity over that of undisturbed soil.

The anchor is set at the bottom of the hole and then expanded by a tamping bar that forces the blades to fan out into undisturbed soil. (You can check whether the blades are expanded by shining a flashlight down the hole.) Once fully expanded, the hole is backfilled in increments of no more than six inches at a time while being compacted with the tamping bar. And that's it.

Table 10-2 was derived for 80-foot towers and anchors in soil ranging from a Class 7 (sandy or swampy soils) on the low end to Class 3 (representing hardpan, dense clay, and hard silts) on the upper end. The anchors are sized to give a margin of safety in the weaker soils. They are more than strong enough in the heavier soils.

TABLE 10-2
Expanding Anchors for 80-foot Guyed Tower[a]

Turbine size	Anchor size (diameter) (closed)
3-meter	8-inch
4-, 5-meter	12-inch
6-meter[b]	—

[a]For taller towers anchor size depends upon specific soil condition.
[b]Consult manufacturer for anchor size. Expanding anchors can be used only in certain soils.

The construction of high-voltage transmission lines using guyed towers has led to the development of screw anchors. These anchors are literally screwed into the ground with the anchor rod attached. Three anchors can be installed in less than thirty minutes.

Many truck-mounted augers have been adapted to drive screw anchors by replacing the auger bit with a special tubular wrench. The hydraulic boom controls both the angle at which the anchor enters the soil and the rate. The driving rate determines whether the soil is churned by the anchor acting like an auger. The screw anchor achieves its rated strength only when acting on undisturbed soil (see fig. 10-3). But, screw anchors can't be driven in all soils. Rocks can thwart the anchor from advancing.

Screw anchors are sized by the number and diameter of the screw or helix. Their holding strength is, once again, based on the cohesiveness of the soil. See table 10-3 for screw anchors that could be used on small machines under a range of soil conditions (from Class 7 to Class 3) with an adequate margin of safety.

TABLE 10-3
Screw Anchors for 80-foot Guyed Tower[a]

Turbine size	Anchor size (diameter)
3-meter	single 8-inch helix
4-, 5-meter	single 12-inch helix
	dual 10-inch helix
6-meter[b]	—

[a]For taller towers, anchor size depends upon specific soil conditions.
[b]Consult manufacturer. Screw anchors can be used only in certain soils.

Fig. 10-3. Screw anchor installation. Truck-mounted power augers can be adapted for driving screw anchors. Utilities frequently use screw anchors because they can be installed in the least time and with the least effort of any anchor while providing equivalent holding power. (Courtesy A. B. Chance Co.)

To give you an idea of how soil strength affects the anchor's holding capacity consider the anchors for a 3-meter turbine in a soil of Class 3; the 8-inch screw anchor has a holding capacity over three times greater than the maximum load on the anchor. The 8-inch expanding anchor has a safety factor of five. But in the weak Class 7 soils the safety factor drops to 1.8. In heavy soils these anchors are more than sufficient. In weaker soils, though, you should check with the turbine manufacturer.

If you are unfortunate enough to encounter solid rock at or near the surface, none of the preceding anchoring methods can be used. You will have to drill a hole at the proper angle with an air drill and compressor. A rock anchor and rod are then inserted down the hole and wedged in place. These anchors have a high holding capacity but installation is time consuming (see fig. 10-4).

In most installations one anchor rod is sufficient for all guy levels. The anchor rod must depart from the ground at an angle that coincides with the resulting angle of tension in the guy cables to minimize bending of the rod. Bending weakens the rod over time. The angle of departure is from 45 to 70 degrees depending upon the height of the tower and the guy radius. In the field it is easiest to use a 45-degree angle of departure by measuring a rise of one foot over a run of one foot, but it is better to follow the manufacturer's recommendations on guy spacing and the appropriate guy angle. The size anchor rods most likely to be used are given in table 10-4.

TABLE 10-4
Anchor Rods

Turbine Size	Rod Diameter	Length
3-meter	5/8-inch	5 feet
4-, 5-meter	3/4-inch	7 to 10 feet
6-meter	1-inch	10 feet

Fig. 10-4. Rock anchor installation. Drill hole by hand or with an air drill, insert rock anchor, and then expand by turning the anchor rod. (Courtesy A. B. Chance Co.)

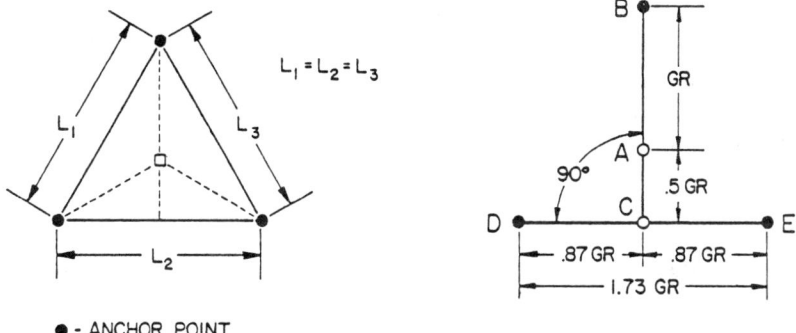

Fig. 10-5. Guy anchor layout. (Courtesy Bergey Windpower Co.)

Anchor Positioning for Guyed Towers

Most guyed towers use three anchors in the form of an equilateral triangle; that is, the distance between each anchor is the same and the angle between them is 120 degrees. The mast is centered within this triangle. Some towers require four anchors.

If the manufacturer specified that the anchors are to be installed with precision — and some do — a surveyor's transit will be needed. The transit is used to position accurately where the anchor rods depart the ground. Drive a stake to mark the spot. You're not finished yet, though. During excavation for the anchor the stake will be moved. You can mark the correct location with four stakes driven at the perimeter of the hole, aligned such that a line between them crosses where you want the anchor. Now you can dig your hole and set the anchors accurately.

Where precision is unnecessary, simpler methods are used. Position the anchors on a tower with four guys by measuring the distance from the tower base to each anchor as directed. This is the guy radius, *GR*. Eyeball a right angle between them. The distance between each anchor should be the same and should be equal to 1.41 times *GR*.

For a tower using three anchors the procedure is a little more involved. This is what the people at Bergey Windpower Company use. The relationship between the anchors is illustrated in figure 10-5.

1. Start by locating one anchor (*B*) the specified distance (*GR*) from the base (*A*).
2. Next carry a line from the anchor (*B*) through the base (*A*) to point *C*, a distance of 0.5 times *GR*.

3. Project a line at right angles to that in Step 2 from point *C*. Measure 0.87 times *GR* along this line on both sides of *C* to the remaining anchor positions.

Check anchor alignment by measuring the distance between each. They should all be 1.73 times *GR* apart.

Free-standing Towers

Free-standing towers, whether they are truss or a tapered tube, require an excavation and the placement of concrete. The easiest method, where the depth to bedrock permits, is to auger a hole for a central pier to support the entire tower or for piers at each leg of a truss tower.

For a small machine (3 to 4 meters in diameter) this entails augering a hole 3 feet in diameter up to 10 feet deep. The uppermost section is squared up where specified by the manufacturer, the re-bar placed in the hole along with a short tower section, and the concrete poured. This applies whether a truss tower or a tapered tube tower is used. On larger machines that are installed on truss towers, the piers for each leg may be up to 3 feet in diameter and up to 10 feet deep. In these cases the holes are left unfinished; that is, in their circular form, the re-bar and J-bars for anchoring the tower leg added, and the concrete poured. Of course, a backhoe can be used to excavate a rectangular hole for footers when augering is not practical. Formwork may be necessary and a greater amount of concrete will be used than in a cylindrical pier.

It is important that the stub tower sections or J-bars do not swim around when the concrete is placed and that they accurately fit the foundation template provided with the tower. Otherwise you may have a rude awakening when you go to set the tower on the base—it may not fit.

Novel Foundations

If power-installed screws work well as anchors why couldn't they also be used for the pier supporting the mast of a guyed tower? In theory at least they can. The foundations for light standards and transformers at substations are now being installed in this way. No one, however, has adapted this technology to wind systems. When they do so, an installer will be able to drive the anchors and pier in just minutes instead of waiting days for the concrete in the pier to cure. The whole wind system could then be erected in one day even in areas with frost heave.

Using power-installed screw anchors to secure the legs of a free-standing truss tower is along similar lines. There may be engineering limitations particularly in weaker soils, but the advantage of quick and easy installation justifies a look into whether it is possible. Fast-setting plastics, such as high-density foam, also offer promise if their curing time is less than that of concrete for an equivalent strength. Again, this technology has not been adapted to wind systems.

GUYED TOWERS – ASSEMBLY AND ERECTION

Guyed lattice masts commonly used for small wind machines are most often assembled a section at a time with a tower-mounted gin pole. The entire tower can also be assembled on the ground and hoisted into place with a crane. The method you use depends on whether a crane is available and can get to your site and whether you want to pay for it. In either case the guy cables first need to be cut to length and attached to the guy brackets that fit around the mast.

On masts from 80 to 100 feet tall, three guy levels are necessary. For an 80-foot tower we are looking at three levels roughly 25 feet apart. This would give sufficient clearance on the topmost section for a 3-meter wind turbine (a blade length of 5 feet). Larger wind turbines will require lowering the topmost guy to compensate for greater blade clearance. Follow the manufacturer's recommendations. They are based not only on the buckling strength of towers but also on where stiffness is required to dampen tower vibrations (see table 10-5).

TABLE 10-5
Suggested Guy Levels
Height Above the Ground
(in feet)

	Tower						
	Rohn 25-G				Rohn 45-G		
	Maximum Size Wind Turbine						
	3-meter				5-meter		
	Tower Height (in feet)						
Guy Levels	60	80	100		60	80	100
1st	30	25	25	(31)[a]	25	25	25
2nd	55	50	50	(63)[a]	53	50	50
3rd		75	75	(95)[a]		72	75
4th			95				93

[a]For self-furling rotors.

Only extra high strength (EHS) cable is used for guying the tower. It is three times stronger than common grade steel cable and 40 percent stronger than high strength cable. If you are building your own tower do not skimp on the size and quality of the guy cable. For turbines 3 to 4 meters in diameter ¼-inch EHS cable will do the job. The topmost guy on a 5-meter turbine will require 5/16-inch EHS, while a 6- to 7-meter turbine will demand 7/16-inch EHS cable. The lower guys, because of the lower loads, need only 5/16-inch.

Steel cable is shipped in coils or on reels. Kinks damage the cable beyond repair. Avoid them by rolling the coil along the ground to lay out the cable. If the cable is on a spool use a spool stand to unreel the cable.

Calculate cable length by using the Pythagorean theorem:

$$\text{Guy Length} = \sqrt{GR^2 + GL^2}$$

where GR is the guy radius and GL is the guy level or height above the ground. Give yourself plenty of extra cable to allow for sag and for slight errors in the position of the anchors. For example, find the guy length needed for the topmost guy on an 80-foot tower with a 60-foot guy radius. We will use a 3-meter turbine and place the guy bracket in the middle of the topmost tower section, 75 feet above the ground.

$$\text{Guy Length} = \sqrt{60^2 + 75^2}$$
$$= \sqrt{3600 + 5625}$$
$$= 96 \text{ feet}$$

This tower will need three lengths of cable at least 100 feet long for the topmost guy level. If the lower guys are 25 feet above the ground, we will need three guys each 70 feet long and for guys at the 50 feet level, three lengths of cable 85 feet long. For an 80-foot tower with a 60-foot guy radius a total of 800 feet of cable (765 feet) will be used.

Mark this distance on the ground. Unreel the cable and cut it to length with bolt cutters (18-inch bolt cutters will handle up to 5/16-inch cable). Next, bolt the guy brackets to the tower sections and then attach the guy cables to the guy brackets.

Cable attachments is an area, like anchors, where installation is being simplified. Developments in the utility industry are being adapted for use with guyed wind systems. Guy cables can be attached with U-shaped cable clamps (known as Crosby clamps), strand vises, and with preformed cable grips.

Past practice dictated using the cable clamps. The guy cable would be wrapped around a metal thimble and passed through two or more clamps. The clamps prevented the cable from slipping off the thimble which distributed tension uniformly over a large radius. (If the cable is wrapped around too small a radius—too sharp a bend in the cable—it is severely weakened.) Some guy brackets, particularly those in the utility industry, have built-in thimbles.

Cable clamps are awkward to use. The stiff cable resists being passed around the thimble and through the clamps. The U-bolt of the cable clamp must act on the dead-end of the cable necessitating a struggle with the cable to position the clamp properly. When finally in place the clamps can be tightened. This requires tightening two nuts on each clamp.

Strand vises simplify the task. They operate on the wedge principle. The guy cable is passed through the vise and, when tensioned, wedges the cable in the vise grip. The strand vise has a wire bale or loop which passes through the guy bracket. No thimble is needed. Because this bale is usually fixed around the vise, strand vises are suited only for the guy bracket and not for the anchor rod. (The brackets are assembled at the site and the bale of the strand vise inserted into the bracket before it is bolted together. Not so with the anchor rod, where the eye is forged within the rod.)

Less expensive and adaptable to both ends of the guy cable is the preformed cable grip. In use by the utilities since the mid-fifties, they work like a Chinese finger puzzle. After first passing one leg of the helically wound steel strands through the guy bracket or eye of the anchor rod, the grip is wrapped around the guy cable. Tension on the cable pulls the strands tighter together. A fine grit on the inside of the strands insures that the grip has a firm hold on the cable (see fig. 10-6).

Preformed cable grips allow quick adjustment in the guy cable during installation by being easy to remove and to reapply. They are simply unwrapped, the cable pulled taut for example, and then rewrapped. Like cable clamps, these cable grips require the use of thimbles, but their ease of application more than compensates.

Crane Method

The entire tower and wind turbine can be lifted into place with a crane or each can be lifted separately. If you are lifting a complete assembly the crane does not have to be taller than the tower. The combined tower and wind turbine can be lifted at some level below the top. Because the lattice mast is somewhat frail, a nylon sling should be used to spread the lift over the entire cross-section of the tower and not act on one leg alone. Lifting a complete assembly also requires that the wind machine must be

276 • *Wind Energy*

Fig. 10-6. A coffing hoist tensions a guy cable. The free end of the cable is then attached to the turnbuckle with a preformed cable grip by wrapping the helical legs of the grip around the guy cable. The grip works in the same fashion as a Chinese finger puzzle. (Courtesy Preformed Line Products)

fully assembled. This may require lifting the tower off the ground slightly to attach the blades and tail (where used).

If a crane is to be used, bolt all tower sections together making sure that the section with the guy brackets and cable is in the right position. Attach conduit and thread the power cable through it as discussed in the subsequent section on wiring. While the crane holds the tower upright, position it over the base plate. Rest the tower on its pier and connect the guy cables to the anchors and pull them taut. Once this is done the crane can be removed and final adjustments made to the tower.

Gin Pole Method

Bolt two tower sections together and attach temporary guys (3/8-inch polypropylene rope works well). Tip the two sections up onto the pier and tie off the temporary guys to the anchors. They will hold the tower in place until the first guy level is reached. Someone must now climb the tower and bring up the gin pole.

The gin pole is a boom or davit that extends above the tower. It permits tower sections or the wind turbine to be lifted up the tower and set in place without the use of a crane. Gin poles need be nothing more than a long section of pipe strong enough to handle the expected loads and with some means of being attached to the tower. Some gin poles are a little more sophisticated and incorporate a horizontal arm that allows the load to be centered on the tower (see fig. 10-7).

The old Jacobs windchargers were installed with an 11-foot length of 3-inch steel pipe. The modular tower sections of Unarco-Rohn's lattice mast are erected with a 12-foot aluminum pipe. This same gin pole is used by Bergey Windpower Company to lift their 3-meter turbine. But the strength of the gin pole and of its attachment to the tower is not something to be overlooked. If you are planning to use one of your own design, note that well-casing may be strong enough but water pipe may not.

Pulleys are used to direct the hoisting rope over the gin pole to the load. The pulley is either built into the top of the gin pole or attached separately. Never use a gin pole without first routing the hoisting rope through a pulley at the base of the tower. This pulley permits the hoisting crew to stand clear of the tower and be well away from any falling objects. But it also prevents any unnecessary bending of the gin pole. With a base pulley in place, the hoisting tension acts directly on the gin pole from below, not from a distance to the side of the tower. This minimizes bending of the gin pole.

Art and Maxine Cook can testify to the need for a gin pole strong enough for the job at hand. Their mountaintop farm in western Pennsylvania has two wind machines: one, an Aermotor for pumping water, and the other a wind generator for charging batteries.

Art's windcharger developed a bearing noise. He called the manufacturer who sent a technician to replace the heavy, direct-drive generator. Both Art and the technician had previous experience with tower-mounted gin poles (Art has replaced more than one generator since he began using wind power in the mid-seventies). The technician scrounged a gin pole from a local scrap yard. It looked good—right length, right size. The gin pole was bolted to the tower and the noisy generator removed. But raising its replacement proved more difficult. As Art strained on the rope

278 • Wind Energy

Fig. 10-7. Tower installation with gin pole.

at the base of the tower, the machine moved slowly skyward. It was only a few feet off the ground when the gin pole collapsed. The generator fell as Art scurried for cover. Fortunately, no one was hurt and the generator was not damaged.

Similar mishaps have occurred when the attachment of the gin pole to the tower has failed. It is paramount that the gin pole always be firmly attached and that it not move laterally when the load is applied. The manufacturer's recommendations for the materials used in the gin pole and its attachment to the tower should be followed religiously. Otherwise you may learn an expensive lesson in physics.

With the gin pole in place the next tower section is brought up. The third 10-foot section will usually have the lower guy bracket attached. Once the section has been bolted down the guy cables can be strung to the anchors. These cables are tensioned by hand until the three assembled tower sections are perfectly vertical and the tower straight. Vertical alignment can be checked with a level (held vertically) on the lowermost section, with a transit, or with a plumb bob.

The gin pole is then released and moved to the top of the last section. A new section is hoisted up and bolted into place and so on until the tower is completed. After each set of guy cables is attached the tower should be checked for plumb and twist. If the lower sections were aligned properly it is possible to simply sight along the tower to check the alignment of the upper levels. A transit can be used to be sure. The tower must be vertical for proper yawing of the wind turbine.

Normally the guys on small machines can be tensioned by hand. For some wind machines, though, especially those 6-meters and larger, it is necessary to tension the guys to a predetermined level. This imparts to the tower a known stiffness intended to avoid harmonic vibrations. The guys can be tensioned with a coffing hoist (*come-along*) pulling on a cable grip. The tension is measured by a dynamometer in line with the hoist. Follow the manufacturer's directions for attaching the hoist and the amount of tension required.

After the guys are in place, you're ready to raise the wind turbine. Do so only on a calm day. It is also easier to do as much of the assembly on the ground as possible. Use a machine stand near the tower if you have one. This permits you to attach the blades and other components. Put the stand in the bed of a pick-up truck if you need more clearance for the blades.

Attach the hoisting rope and one or two tag lines. The wind turbine may have an eye bolt used for lifting or it may require a special lifting jig. Whatever is used it is important that the hoisting rope lift at the center of gravity (the center of weight). If not, the wind turbine will be a lot more difficult to handle on its way up the tower, and once there you will have a heck of a time mounting it on the tower adapter. The turbine may be easy to move around on the ground but when you are 80 feet in the air there is not much to give you leverage. Everything you do is harder — and more dangerous (see fig. 10-8).

The tag lines (⅜-inch polypropylene rope) keep the machine from banging into the tower and tangling with the guy cables. The tag lines must be longer than the tower is high. For a lift to an 80-foot tower, the tag line will need to be at least 120 feet long or about 1½ times the tower's height. The pull on the tag lines must be moderate (just enough to prevent the machine from hitting the tower), particularly as the turbine nears the top of the tower. It is easy to buckle the gin pole if too much force is used on the tag lines.

When the turbine is in place string the conduit and fish the power cable to the generator. Torque all tower fasteners to the specified value and apply locking nuts where required.

FREE-STANDING TOWERS — ASSEMBLY AND ERECTION

Like the guyed tower, a free-standing truss tower can be erected with a tower-mounted gin pole or with a crane. But only in rare cases such as remote sites in Alaska or in other inaccessible areas are tower-mounted gin poles used. The tower sections are much heavier and more awkward to work with (because sections are 20 feet long as a rule) than those on a guyed tower. Some sections may weigh well over 200 pounds even for a small wind machine. Because of the hazards involved and the sheer toil of assembling a truss tower with a gin pole most installers disregard the cost and call in a crane.

Each individual member on a truss tower is so heavy that by the time the first section is bolted together you are not going to be able to move it anywhere without power equipment. Ideally you would like the crane to simply drive up to the tower, raise it, and in one lift set the tower on its foundation and then leave. You do not want the crane to move sections of the tower around the site because you did not thoroughly plan the assembly. This is particularly important if the wind turbine has been mounted on the tower. The more moving around that is required the greater the likelihood of damaging the turbine.

Crane service is usually billed by the hour. Travel time is included. Some companies have a minimum charge; others don't. Two factors are important in crane rentals: the weight of the lift and the height of the lift. An 80-foot truss tower for a 3-meter turbine will weigh less than 1,500

Fig. 10-8. Wind machine installation with gin pole. Note movable pulley attached to lifting ring on wind machine. The pulley provides a 2:1 mechanical advantage.

pounds; one for a 6-meter turbine will weigh twice that. The height of the lift, though, is not necessarily the height of the tower.

The tower can be raised and then the wind turbine mounted on top in a second lift. For wind systems over 6 meters in diameter this is the common practice. The boom of the crane then must extend several feet above the top of the tower. If you are planning to use a 100-foot tower and erect it in this manner you will need one whale of a crane. But a crane with a shorter boom can do the same job if it raises the wind turbine and tower at one time. This works well with wind machines 3 to 5 meters in diameter. I erected a 60-foot truss tower with a 3-meter turbine this way in less than fifteen minutes (see fig. 10-9).

The lift should be made some distance below the top of the tower yet well above the tower's center of mass. As the tower is raised the weight of the tower keeps the bottom sections on the ground while the upper sections move towards the vertical. If the tower was positioned correctly during assembly the bottom will scoot across the ground to the foundation right where you want it to be. Because the lift is being made below the top of the tower the rotor blades are able to clear the boom as the tower nears the vertical. With this method you can use smaller, less costly cranes such as those used to erect billboards and highway signs rather than heavy cranes designed for industrial work.

With the skillful use of hand trucks, dollies, and come-alongs even the heaviest towers can be fully assembled without power equipment. Once you have decided where the tower should be assembled to make the crane operator's task as simple as possible begin by bolting the lowest — and the heaviest — section together. Truss towers are like giant Erector sets and will be puzzling at first, but after a few of the cross-girts are assembled the rest will become clearer. Tighten the bolts snug but not tight. As you move along you will find that some of the pieces will not fit easily. With the bolts just snug the members can give a little, allowing that bolt to slip through where you did not think it was possible. Erection wrenches (sometimes called spud wrenches) and drift pins are helpful in these situations (see fig. 10-10). An erection wrench has a long tapered shaft that's used by ironworkers (those guys that build skyscrapers) to solve stubborn alignment problems. The shaft is inserted into the holes and with a little muscle it is used to lever the pieces into position. The bolt is popped in and the wrench is then used to tighten the bolt. A bull pin is a similar device. It too has a long tapered shaft, but instead of a wrench on one end it has a striking face. The pin is dropped into the holes needing a little nudge and driven with a hammer until they are aligned. Both of these tools are well suited for aligning holes on the flange plates between each tower section (see fig. 10-11).

Fig. 10-9. Erecting 60-foot truss tower with crane. The completely assembled tower with wind machine attached was hoisted into place in one lift. (Photo by Susan Schillmoeller)

284 • Wind Energy

Fig. 10-10. Assembly tools. *Left,* erection wrench; *right,* drift or bull pin. (Courtesy Klein Tools)

On lightweight towers it is possible to assemble the next 20-foot section near the first and then bolt the two together. This proceeds until the tower is fully assembled. The wind machine is mounted on the tower and the conduit for the power cable installed. You'll find that strapping the conduit to the tower while it is on the ground and fishing through the conductors is much easier than trying to do it after the tower is erected.

Fig. 10-11. Assembling truss tower. Where's this go? Bolting together the cross-girts can be confusing until you get the hang of it. (Photo by Susan Schillmoeller)

Once all the sections have been assembled tighten the bolts to the desired torque.

Because the tower vibrates when the wind turbine is running the nuts on all bolts must be prevented from loosening. The nuts can be of the self-locking type with a nylon insert or a special locking nut can be added after the nut and bolt have been tightened. These are easiest to install while the tower is still on the ground as well. But do not overlook this step. Locking nuts are specified for a reason; towers have failed without them.

The tower is now ready for the lift. Attach the sling so that the stress is distributed onto tower members strong enough to take it. Do not for example wrap the sling around a tower cross-girt. Instead use a tower leg; better yet use two legs. Note that the sling must not slide along the leg as the tower is being lifted.

The crane will slowly set the tower down on the J-bars or flanges in the foundation. Align the holes with the drift pin and judicial use of a crowbar, sliding in the bolts as the holes are aligned. Before removing the sling, level the tower.

There are two ways in which the tower is leveled. On the lighter towers, shims are forced in between the bottom-most flanges. The heavier towers use J-bolts between the foundation and the lower tower section. These J-bolts have adjusting nuts that are used to level the tower. When the tower is level, tighten the mounting bolts and install lock nuts. The sling can now be safely removed and the crane sent on its way.

Truss towers for wind turbines 6 meters and larger are usually erected in 20-foot sections. The first section is placed with the crane and then leveled. The following sections are then added.

Tapered Tube Towers

Free-standing tube towers are usually erected with a crane a section at a time. Most of the towers depend on a slip fit for holding the sections together. The first section is placed on the foundation and leveled with adjusting nuts on the J-bolts. The next section is slipped into the first and so on. Gravity does the rest. The wiring run is then strung inside the tower and the wind machine installed.

Instead of using J-bolts one approach places a short tube section in the foundation. When the first tower section is raised it is slipped into the first that has been embedded in the concrete. This is easier than trying to align several J-bolts but it does demand that the embedded section be perfectly level—there is no adjustment once the concrete has hardened.

Wooden poles can be installed with a crane or with the boom on truck-mounted power augers. The boom was designed for setting utility poles and has a clamp that grasps the pole near the midpoint and tips it upright. The pole is set on a foundation and bolted down or set in an augered hole and concrete added. (Temporary guys are needed to keep the pole vertical until the concrete sets.)

HINGED TOWERS – ASSEMBLY AND ERECTION

All towers, whether guyed or free-standing, can be hinged and tipped upright into the vertical position eliminating the need for a crane. Hinged towers simplify assembly because all tower and wind turbine components can be added while safely on the ground. They simplify service and repair for the same reason. Rather than climbing the tower and manhandling awkward components 80-feet in the air the tower can be lowered and the job done on the ground. By avoiding the need for a crane the cost of installation and service calls is also reduced.

Hinged towers are not without their problems. The hinges add to the cost and complexity of the installation and they introduce a potentially weak link in the tower structure. Some manufacturers refuse to install their wind turbines on hinged towers for this reason. Despite this limitation the advantages of hinged towers dictate that they will be used more frequently.

Hinged towers can be raised by a heavy-duty industrial winch and gin pole or by using a vehicle and gin pole. Bucket trucks or small cranes such as those for servicing highway signs can also be used to pull the tower upright at little cost.

Guyed towers that are hinged use either a three-guy or four-guy layout. There is no lateral restraint on the tower during the lift if only three guys are used. The tower wants to swing from side to side until it is upright and the guys are taut. This lateral motion is prevented by a two-piece gin pole in the shape of an A-frame. The base of the A-frame rests on the concrete anchors used to guy the tower. The apex is connected to the tower by the free guy cable and to the free guy anchor by a block and tackle. The A-frame can be built inexpensively from four sections of lattice mast, but the bases must be hinged to allow the gin pole to move with the tower. The lift is made by drawing the tackle blocks together bringing the A-frame and attached guy cable to the free anchor (see fig. 10-12).

One popular 10-meter machine uses a simpler one-piece gin pole welded to the base of the tower. The tower uses four guys instead of three. Lateral motion is prevented by the two guys at right angles to the lift. As with the A-frame, the free guy cable is routed to the top of the gin pole and the tower raised by drawing together a block and tackle between the gin pole and the free anchor.

Installation and Maintenance • 287

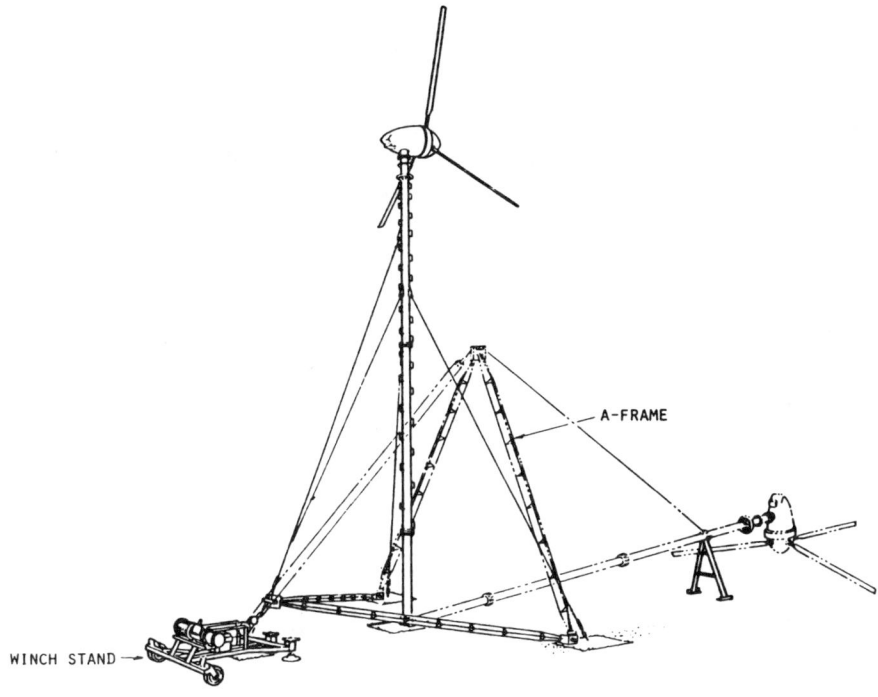

Fig. 10-12. Erecting guyed tower with gin pole and winch. (Courtesy Windworks, Inc.)

Truss towers have been raised in the same way. One turbine manufacturer uses the A-frame gin pole; others use a single pole. The gin pole is either attached to the tower's foundation with a hinge or it stands separately. When standing apart from the tower the gin pole must be guyed to prevent it from moving laterally out from under the load (see fig. 10-13).

Free-standing tube towers have also been erected with a gin pole mounted at the base of the hinged tower. In both the case of the truss tower and the tube tower the hoisting cable is passed over the gin pole to the top of the tower. Once upright the raising cable can be removed or it can be left in place for service calls.

WIRING

The bible on wiring is the National Electrical Code (NEC). Established to prevent electrical fires the NEC is updated and reissued periodically. It is the rule book for what can and can't be done but the final say

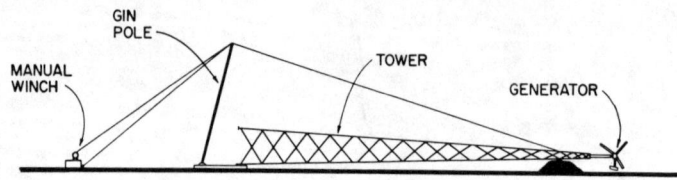

Fig. 10-13. Erecting truss tower with gin pole and winch. (Xavier University, DOE)

in electrical matters is in the hands of your local electrical inspector, fire underwriter, or code enforcement officer.

Licensed electricians in your area will be familiar with the local application of the code. They will not be familiar, however, with wind generators and their specific requirements especially as they relate to interconnection with the utility. This section will be helpful to your electrician and essential if you plan to wire the wind system yourself. In either case, when unsure of your next step check with the electrical inspector *before* you set out to begin work.

Why is the approval of the electrical inspector or fire underwriter so important? Because your fire insurance may be void without it. And in some communities you cannot sell your home without approved wiring.

The electric utility is responsible for all wiring from the nearest transformer to the service drop (where the wires are secured to your house above the kWh meter). Your responsibility begins at this point as does coverage by the NEC.

The wiring from a wind system is much like the power lines from a small generating station. The lines can be strung above ground or buried. Entry of the power lines into the home is accomplished in a manner similar to the utility's service entrance. Let's first look at wiring on the tower and then consider how we are going to deliver the power to the load—your home, barn, or business.

In all cases, the leads from the generator must be connected—*terminated* or *spliced* in the jargon of electricians—to the wires (called *conductors*) running down the tower. These connections must be permanent and weatherproof. How this is accomplished depends on the kind of wind turbine and on the manufacturer.

When the generator is stationary within the tower, as in a vertical axis wind turbine or in some horizontal axis machines with right-angle drives, the generator leads can be directly connected with the conductors on the tower. On more conventional wind turbines, though, there must be a mechanism for transferring power from the moving platform of the wind machine to the stationary tower. Slip rings and brushes usually perform this task.

Some horizontal axis wind machines, however, do not use slip rings. Instead the leads from the generator hang freely through the center of the tower. As the machine yaws, or turns to face the wind, the conductors twist. The cable is permitted to twist several revolutions before it must be unwound. This practice is more common in Europe than in this country.

Good connections between conductors are those that are mechanically tight, electrically insulated, and corrosion resistant. Connections may be made by using a
1. twist-lock plug
2. split-bolt
3. wire nut
4. crimp or compression connector.

Each has its merits.

Weatherproof twist-lock plugs rated for the power output of the generator have been used in the past. They are unsightly and do not offer a reliable connection. They may not be permitted under some building or electrical codes as an impermanent connection.

Split-bolts are easy to use and come in sizes suitable for even the heaviest power cable. The stripped ends of the wires are inserted into the jaws of the bolt and the faces tightened. An insulated boot is needed to cover the connection. Vibrations have been known to loosen split-bolts.

Wire nuts are the most popular connector because of their ready availability and ease of use. Nearly every home handyman is familiar with them. They do have their drawbacks. They too may loosen under vibration and need to be secured with electrical tape. Moreover, they are most practical on solid rather than stranded wire and on the smaller wire sizes. Wire nuts for heavy-gauge wire are hard to find and are more difficult to use.

Crimp or compression connectors are the best all around termination. They are mechanically sound and will not loosen from vibration. Some include an insulated covering but all can be insulated with a boot of shrink tubing slipped over the connector. Their disadvantage is the need for a crimping tool, and in the larger wire sizes this tool is both expensive and awkward to use.

For small-gauge wire, crimp connectors or wire nuts are best. When joining heavy-gauge conductors use either crimp connectors or split-bolts. Cover the connection with shrink tubing or wrap securely with good quality electrical tape.

The connection between the leads from the slip rings and the tower conductors can be made in the open or within a junction box. The conductors are then run down the tower exposed to the elements or protected within conduit ("hard-wired").

Previously, some installers have taken shortcuts and used twist-lock plugs at the top of the tower and then laced the power cable down a tower leg. They would hold the cable in place with electrical tape or with nylon cable-ties. This has proved unsatisfactory. It is unsightly, does not meet building or electrical codes in many communities, and presents a hazard to servicemen, to say nothing about the potential for maintenance problems.

Conductors on the tower should be protected within conduit. Either electrical metallic tubing (EMT) or plastic tubing can be used. I prefer EMT. It offers a better appearance (the galvanizing of the EMT matches that of the tower) and it can be bent with an inexpensive bending tool. Each 10-foot section of EMT is joined by weatherproof compression couplers and the conduit run mounted on the tower with conduit hangers. The hangers should be spaced two per section with one at each end near the coupler. If you are using PVC (polyvinyl chloride) conduit make sure it is rated for electrical use. Because PVC flexes, more conduit hangers must be used. Space them about 3 feet apart (see fig. 10-14).

If the connection between the leads from the slip rings and the tower conductors is made in the open, then a weatherhead will be needed at the top of the conduit string. The weatherhead is the same as those used for the service entrance at your home. The connection is made and slack in the conductors is used to make a drip loop before they are fed into the weatherhead. It is preferred, however, that all connections be made within a weathertight enclosure or junction box. (Most codes insist on it.)

The junction box does not have to be elaborate. An inexpensive bell box is sufficient. The conduit on the tower is coupled directly to the box. Similarly the leads from the slip rings can be fed through a short section of conduit into the box. Manufacturers provide fittings for attaching conduit to the slip ring assembly for this purpose. Once the splices have been made the box can be sealed. The entire conductor run from wind turbine to the base of the tower is then protected. I have found that a junction box designed for explosive environments meets my needs best. It gives plenty of room for the conductors and is sealed with a large threaded cap which is easy to hang on to. The little screws on bell boxes have a way of walking off.

From the junction box to the base of the tower, conduit of ¾-inch diameter or larger is best. You can get by with ½-inch diameter on short runs, but the few pennies saved are not worth the aggravation. Strain reliefs such as a basket grip will be needed to support the conductors within

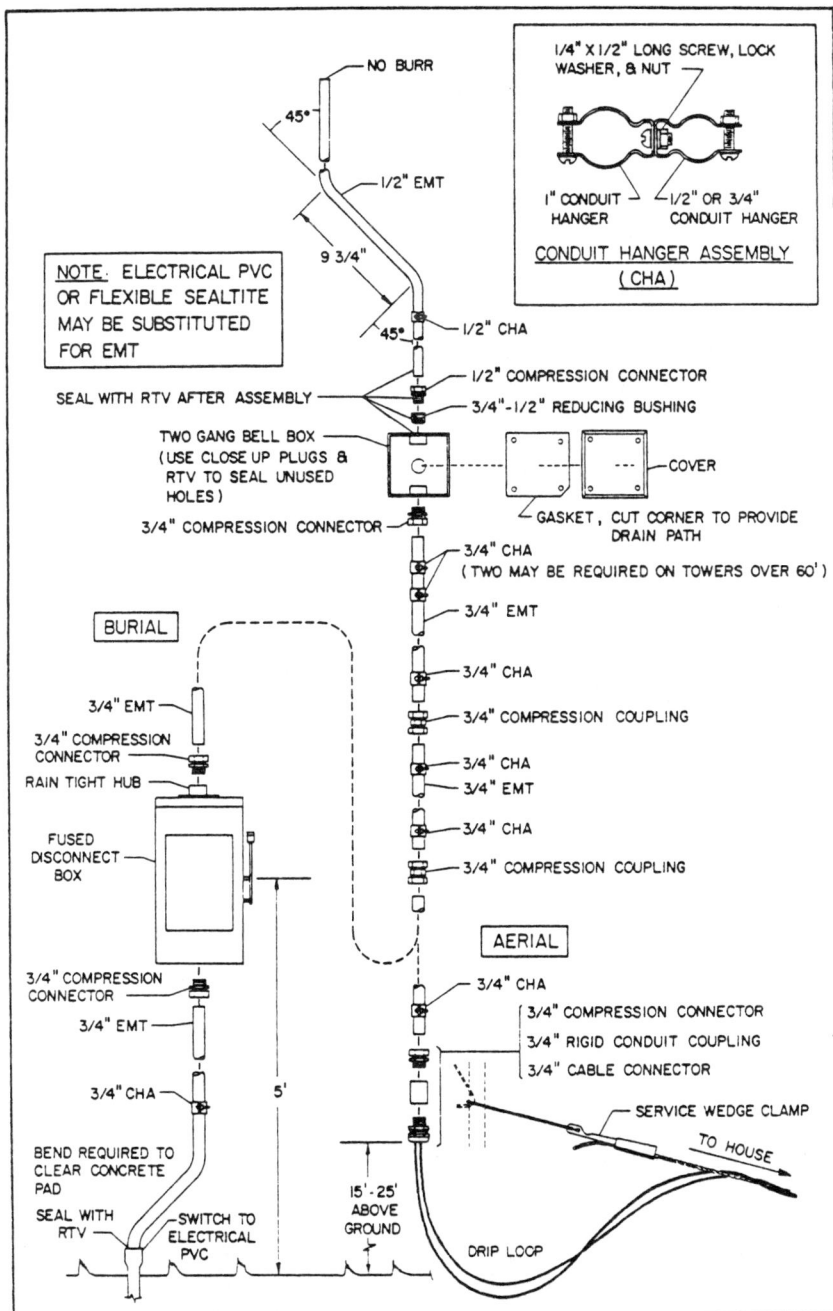

Fig. 10-14. Tower conduit assembly. Power cables from the wind machine should be protected within conduit. (Courtesy Bergey Windpower Co.)

the conduit. Otherwise their weight will pull on the splices and cause them to fail.

Aerial

For aerial runs from the tower to the load, a plastic bushing must be used at the end of the conduit where the wires emerge. This bushing prevents the conduit from nicking the insulation on the conductors.

Aerial runs look easier than they are. If there is any distance at all to cover you will need what is called a messenger cable to support the wire so it will not stretch. This steel or aluminum cable is secured at the tower and at the load with a service wedge clamp, prefabricated cable grip, or wire tie. You will also need conductors rated for exposure to sunlight.

Even with a messenger cable, the length of a single span should not exceed 125 feet to keep the conductors from sagging. The objective in stringing wire is to keep the conductors out of reach. The conductors must be 10 to 12 feet off the ground over residential property, 15 feet over driveways, and 18 feet over roads subject to truck traffic. To achieve these clearances you may need to set poles and attach insulators.

You will need another weatherhead at the load and conduit from the weatherhead to a disconnect switch. The weatherhead should extend above the conductors and be at least 3 feet from windows and doors. It should be mounted so that there is sufficient clearance between the conductors and the roof to meet local code requirements.

Positioning of the weatherhead, or the wind generator's service entrance as it could be called, is determined by where the utility's service enters the house. In many cases you will want to install a disconnect switch in the line between weatherhead and the wind system's control panel (see fig. 10-15). The disconnect switch is included to permit isolation of the wind system from the utility and from the service panel during emergencies. (It also is useful during maintenance of the wind system.)

The disconnect switch is a redundant safety device for interconnected wind systems because they automatically disconnect from the utility line during power outages. The switch gives a utility lineman, though, a positive mechanism for insuring his own safety. By throwing the switch off and inserting a *lock-out* (a metal tag warning that someone is working on the line) through the switch handle, the lineman has done all that he can to protect himself from being electrocuted by the wind generator.

From the utility's viewpoint the lockable disconnect should be located near their kWh meter. This is not a sinister plot by the utilities. Requirements for lockable disconnect switches accessible to repairmen (whether

Installation and Maintenance • *293*

they are utility linemen or private electricians) are a standard safety practice in industry.

Building codes, however, may call for the disconnect switch to be near the service panel and *main disconnect* (the switch between the utility and the service panel) for aiding firefighters. One of the first steps in fighting a fire is to cut all power to the building. Ideally the switches should be

Fig. 10-15. Disconnect switch. Utilities require that a lockable disconnect switch on the service from the wind machine must be easily accessible to their linemen. This switch should be located as near to the utility's service drop and billing meter as practical, recommends AWEA. Note PVC conduit rising out of the ground to meet a junction box which is attached to the disconnect switch. Output from the switch is routed through metal conduit into the basement via the window sill.

located in one place so that firefighters need not fear that an undetected line in the building is live (energized).

The American Wind Energy Association suggests that the wind system's service entrance and the manual disconnect switch be located at or near the utility's service entrance. They also recommend that the conductors from the wind generator must not pass through any buildings before reaching the disconnect switch. In other words, the distance from the wind system's service entrance and the control panel within the building should be minimized. The conductors should run externally to the point on the building nearest the service panel. In most cases this will be where the utility's service also enters the building.

The wind generator's output is then wired to a control panel or synchronous inverter depending on the type of wind system. In no case should DC output ever be connected directly with the service panel. Generators producing DC must incorporate a synchronous inverter before they can be interconnected with utility-supplied AC. Those wind systems producing "dirty" AC, that is, AC of varying voltage and varying frequency, must also use a synchronous inverter.

From the control panel or inverter, conductors are wired to a dedicated circuit in the service panel. Output from the wind system must not be plugged into a wall outlet or wired to a circuit in the building already supplying a load such as a refrigerator or a series of receptacles. Instead, a separate fuse or circuit breaker is "dedicated" to the wind system. In modern buildings or those with upgraded wiring the service panel is larger than needed for existing loads. The panel has "knock-outs" that can be removed for the addition of a new circuit. One or more of these can be used for the wind system circuit. On small wind systems a 30-amp circuit breaker should be more than adequate. For larger systems a 60-amp breaker may be required. For example, a 2.2 kW wind generator producing 110 volts will deliver 20 amps.

Some manufacturers suggest that an outlet just for the wind system should be wired to a dedicated circuit in the service panel. This is the current practice for large appliances such as electric ranges and dryers. Special outlets are mounted on the wall that will accept only a compatible plug. The outlet is then wired to the service panel directly. No other loads are placed on the circuit. The output from the wind generator's control panel or synchronous inverter is wired to a "pig tail" with the plug at the free end. During normal operation the wind system is plugged into the circuit. But during an electrical storm where lightning may be a hazard you pull the plug. Those using such receptacles argue that this gives greater protection to the other circuits in the house than a disconnect switch. Lightning cannot jump the distance between the plug and the receptacle as easily as it can across the terminals of a disconnect switch.

Buried

Overhead power lines increase the likelihood of damage from static electricity and lightning strikes. Falling trees frequently interrupt service on the utility's lines; the same is true of your wind system. These hazards can be avoided by burying the conductors. You can use direct burial cable or you can place the conductors inside PVC conduit.

To bury the run from the tower to the load dig a trench 18-inches deep if you are using PVC conduit, two feet deep if you are using direct burial. Use a backhoe or power trencher. Install a fused disconnect switch at the base of the tower. This minimizes the damage done if someone inadvertently digs up your cables.

Direct burial is easier to install than conduit. The cable is laid directly into the trench. But it is falling out of favor. Conduit allows the conductors to be removed if defective, and there is less likelihood that the conductors will be damaged by sharp rocks as the trench is backfilled.

When using direct burial, thread the conductors through several feet of PVC conduit from the disconnect switch on the tower to the bottom of the trench. This makes a clean installation and protects the cables where they enter the ground. Lay the cable on a bed of gravel or straw at the bottom of the trench. Don't just throw it in and dump the dirt on top. Backfill a little at a time to be sure that no large rocks nick or kink the conductors.

Conduit is now preferred. Ten-foot sections are glued together and laid in the trench. On long runs, pull boxes should be added every 200 feet. They aid fishing the conductors through the conduit and servicing them if problems develop. Special fittings (e.g., flexible conduit) for the rise to the disconnect switch must be used to give the desired alignment.

Conductors

Each manufacturer specifies the type and size of conductors to be used with their wind system for runs of various lengths. But they do not tell you much more. They may not explain, for example, why wire size is important or how to work with wires as thick as your thumb, so we will go over some of that material here.

The conductors used to wire a wind system are insulated. The insulation used is labeled with a multiletter code. Rubber insulation is identified with an R and so on as shown.

R — rubber
RU — latex rubber
T — thermoplastic
H — heat resistant to 167°F
HH — heat resistant to 194°F
W — moisture resistant

Insulation commonly used is a thermoplastic that is heat and moisture resistant (THW). Because certain insulation degrades in sunlight or is susceptible to mildew, conductors intended for aerial runs or direct burial should be rated for such use. Conductors that can be exposed to sunlight such as those with cross-linked polyethylene will be rated for service entrance (SE); those that are fungus and corrosion resistant for direct burial will be labeled UF (underground feeder) or USE (underground service entrance). Conductors not so labeled must be used within conduit.

The quality of the insulation is especially important on wind systems using synchronous inverters. Leakage through the insulation to ground can destroy an inverter in seconds. The insulation must be rated for 600 volts. Though the operating voltage is considerably below this, some wind machines when unloaded can produce up to 600 volts. Weaknesses in the insulation can show. Nicks in the insulation from handling, from sharp conduit edges, or from being stepped on are enough to cause a ground fault and fry an inverter.

Insulation should always be tested before the conductor is put into use. One dealer suggests dipping the whole spool of wire as it comes from the supplier in a tub of water and testing the insulation's resistance with a "megger" or meg-ohm meter. The insulation should have a high resistance. A leaking or faulty insulator has a low resistance in comparison. The insulation should also be checked after the wiring has been completed to detect any faults created during installation.

Conductor Sizing. Wire size is determined by the maximum allowable voltage drop between the wind generator and the load. Voltage drop increases with increasing resistance.

Resistance in a wire is a function of its diameter, length, and whether copper or aluminum is used. Resistance increases with decreasing diameter and with increasing length. Copper offers less resistance than aluminum. Thus the voltage drop across a conductor increases with decreasing wire size and increasing distance between the generator and the load; the voltage drop is greater in aluminum conductors than in copper of the same size.

On interconnected wind systems, voltage drop is critical to the operation of the wind generator or power conditioning equipment. Take the case of a vocational-technical school that installed a 6-meter wind turbine driving an induction generator. The school was responsible for wiring the system into their service panel. Students and their instructor mapped out the conduit run and laid the conductors. When all was finished the dealer flipped the switch and nothing happened. He checked the wind turbine. Everything seemed fine. He checked the students' wiring—okay, too. As they sat and scratched their heads some wise guy suggested they measure the voltage. The problem? Low voltage. The wind turbine's control system sensed a voltage below its disconnect value and would not engage the generator with the line. (The manufacturer incorporated this feature to detect a power outage so the generator would not energize a downed line and kill a lineman.)

The dealer and the school then went back to the books. The wiring was sized according to the specs in the installation manual, or so it seemed. Their mistake was failing to take into account that the panel where the wind system was interconnected with the school's service was several hundred feet from the utility's entrance to the school. The conductors were sized properly for the run from the service panel in the outlying classroom. But the long distance from the utility's service caused an excessive voltage drop by the time the power reached the classroom. They remedied the situation by upgrading the conductors.

The problem with wind systems using synchronous inverters is similar; that is, low voltage disrupts optimum operation of the wind system. Synchronous inverters are used with wind generators whose voltage output varies with wind speed. As wind speed increases and more power can be produced, the inverter increases the load on the generator. The inverter knows when to do this by monitoring the generator's voltage. If the voltage drop from the generator to the inverter is too great the inverter is fooled into thinking that less power is available than may actually be the case. The wind generator is never loaded to its full potential and performance suffers.

Given a wire run of a certain length you have a choice between aluminum or copper conductors in different sizes to limit voltage drop to tolerable levels. Wire size is designated by its American Wire Gauge (AWG) number. The smaller the number the larger the wire. House wiring is usually No. 12 or No. 14 gauge with the heavier gauge wire (No. 12) being specified in some communities. The leads from the generator of a small turbine (3 to 4 meters in diameter) are heavier yet at No. 10 gauge.

Wire size is graded in even numbers down to No. 0 or 1/0, read "one ought." Larger diameters are then designated as No. 00 (2/0) or double ought, No. 000 (3/0) or three ought, to No. 0000 (4/0) or four ought. Larger sizes are available but the designation system changes to the number

298 • Wind Energy

of circular mills. The largest size commonly used for wind systems is 2/0 (double ought) and the smallest size is No. 8 or one size larger than the leads from the generator.

The size used depends on the allowable voltage drop, the distance from the generator to the load, and whether copper or aluminum wire will be used. (See tables 10-6 and 10-7.) For runs over 500 feet aluminum has a definite cost advantage.

From Ohm's law we can derive a simple formula for calculating the maximum wire run permitted by the manufacturer.

$$R = V/I$$

where R is resistance in ohms, V is voltage in volts, and I is current in amps. The total resistance in a wire run is the product of the wire's resistance per foot times the total distance the current must travel—two times the run (L) from the generator to the load.

$$R_T = R \times 2L$$

TABLE 10-6
Distance to Load for 1 Percent Voltage Drop, Copper Wire (ft.)[a]

Approx. Gen. Size (kW)	Maximum Current (amps)	Wire Size (AWG)							
		8	6	4	2	1	0	00	000
At 110 Volts									
1	10	81	130	210	340	430	540	680	860
2	20	40	67	110	170	210	270	340	430
3	30	27	45	71	110	140	180	230	290
4	40	20	34	53	85	110	130	170	210
At 220 Volts									
1	5	320	540	850	1400	1700	2200	2700	3400
2	10	160	270	430	680	850	1100	1400	1700
3	15	110	180	280	450	570	720	900	1100
4	20	81	130	210	340	430	540	680	860
5	25	65	110	170	270	340	430	540	690
6	30	54	90	140	230	280	360	450	570
8	40	41	67	110	170	210	270	340	430
10	50	32	54	85	140	170	220	270	340
15	60	27	45	71	110	140	180	230	290

[a] $L = \dfrac{\text{volt drop}}{2 \ (\text{amps}) \ (\text{ohms/ft.})}$, where L is the one-way distance to the load.

By substituting the equation for total resistance into Ohm's law we can solve for L or the length of the run when V is the permissible voltage drop.

$$L = V/2IR$$

This formula was used to construct tables 6 and 7. In practice, the distance is given and you need to find the right wire size to span it. You can go to the table and find the appropriate wire size if you know the voltage drop permitted or you can calculate the maximum resistance and match this with the resistance of various wire sizes (see table 10-8).

Permitted voltage drops range from one to three percent. Let's work an example of a 200-volt wind generator producing a maximum of 10 amps (2 kW) and an allowable voltage drop of only one percent to the synchronous inverter. Say we found a spool of 2/0 aluminum wire suitable for our use on sale; how far can we get with it? The resistance of 2/0 aluminum conductor is 0.133 ohms per thousand feet.

TABLE 10-7
Distance to Load for 1 Percent Voltage Drop, Aluminum Wire (ft.)[a]

Approx. Gen. Size (kW)	Maximum Current (amps)	Wire Size (AWG)							
		8	6	4	2	1	0	00	000
At 110 Volts									
1	10	51	82	130	200	260	330	410	520
2	20	26	41	65	100	130	160	210	260
3	30	17	27	43	67	87	110	140	170
4	40	13	20	32	52	65	82	100	130
At 220 Volts									
1	5	210	330	520	830	1000	1300	1700	2100
2	10	100	160	260	400	520	650	830	1000
3	15	69	110	170	280	350	440	560	700
4	20	51	82	130	200	260	330	400	520
5	25	41	65	100	170	200	260	330	420
6	30	34	54	86	140	170	220	280	350
8	40	26	41	65	100	130	160	210	260
10	50	21	33	52	83	100	130	170	210
15	60	17	27	43	70	87	110	140	170

[a] $L = \dfrac{\text{volt drop}}{2\,(\text{amps})\,(\text{ohms/ft.})}$, where L is the one-way distance to the load.

TABLE 10-8
DC Resistance Ohm/1000 ft.

Wire Size (AWG)	Copper (Cu)	Aluminum (Al)
10	1.018	1.67
8	0.679	1.07
6	0.410	0.674
4	0.259	0.424
3	0.205	0.336
2	0.162	0.266
1	0.129	0.211
0	0.102	0.168
00	0.0811	0.133
000	0.0642	0.105

$$L = \frac{2 \text{ volt drop}}{2(0.000133 \text{ ohms/ft} \times 10 \text{ amps})}$$

$$= 751 \text{ ft}$$

Note that this distance includes the tower and the run from the wind system's service entrance to the service panel.

Now let's try it the other way. The distance from the generator to the service panel in your home is 500 feet; what size wire is needed?

$$R = \frac{V}{2LI}$$

$$= \frac{2 \text{ volt drop}}{1 \ (500 \text{ ft}) \ (10 \text{ Amps})}$$

$$= \frac{1}{5000}$$

$$= 0.002 \text{ ohms/ft}$$

$$= 0.2 \text{ ohms/1000 ft}$$

From table 10-8 of DC Resistance you can see that No. 2 copper or 1/0 aluminum conductors will be required.

Induction generators are not quite so finicky and the permissible voltage drop is considerably greater (up to 3 percent on some turbines). If you were installing an induction generator you could use wire several sizes smaller than in the preceding example. But always follow the manufacturer's recommendations.

The larger conductors are heavy, stiff, and difficult to work with. The terminal blocks inside inverters, control panels, and disconnect switches were designed to accept much smaller wire than, say, the 2/0 used on a 1,000-foot run. Disconnect switches, for example, are rated by their current-carrying capacity and not by the size of wire they can accept. If you use a switch of the correct current rating it will be too small to make the terminations. On the other hand, if you buy a switch large enough to accept the heavy-gauge conductors used in a long run it is much more costly than what is needed. The solution is to use junction boxes at each end of a long run. The heavy-gauge wire is joined with a wire sized appropriately for the control panel or disconnect switch. It is important to make good splices when doing so to avoid shorts or grounds and to keep the overall resistance of the run to a minimum.

In our example we needed conductors large enough for a 500-foot run. Assume that this includes 100 feet for the tower and from the disconnect switch to the service. It is to our advantage to use smaller wire on the tower and in the junction boxes. By using No. 4 copper on the tower and in the disconnect switch the installation will be much easier. The copper wire can fit within ¾-inch conduit and is less difficult to handle on the tower than the 1/0 aluminum used for the run from the tower to the service entrance. Total resistance and hence voltage drop is kept within the limits.

Small wind machines producing AC are all single-phase. Larger machines, those in the 7-meter range and above, though, are three-phase. Three-phase is more efficient at transmitting energy by reducing the size and number of conductors needed than single-phase. From our earlier discussion remember that voltage drop in a conductor is partly due to the amount of current we are trying to push through it. In a three-phase system, each phase carries only one-third the current of the total generator's output. For example, the current carried by each phase of a three-phase 10 kW generator operating at 220 volts is 15 amps whereas a single-phase generator pushes 45 amps through the conductors to deliver the same power. The wire size required for the three-phase generator can be significantly smaller than that required to carry the 45 amps of the single-phase with the same voltage drop.

Conduit Fill

You can only stuff so much wire into conduit and still meet the requirements of the NEC. The cross-section of a conductor is based not only on the wire size but also the amount of the insulation covering it. For example, ¾-inch conduit will handle three pieces of No. 6 gauge wire with ease but can be used with No. 4 gauge wire only when covered with certain types of insulation. Table 10-9 gives the maximum number of conductors insulated with THHN, THWN, and THW that can be used with several sizes of conduit. Electricians prefer slightly larger conduit than that necessary to meet code requirements to ease the task of pulling the conductors through the conduit. For example, on a 1000-foot wire run we used 2/0 THWN. We pulled it through 1½ inch conduit rather than struggle with the 1¼ inch allowed.

There are several ways to get the wire inside the conduit. For short runs and with a size that is stiff, the wire can be pushed through. On longer runs the wire can be fed through each conduit section before the sections are coupled together. Most often a metal tape, called a fish, is pushed through the conduit. The fish tape has a woven basket or cable grip to which the conductors are attached. The tape is then used to pull the wires through the length of the conduit.

I have found that it is best to fish a line from the bottom up when wiring conduit on the tower. In this way the ground crew must do the pulling from below. All the person on the tower has to do is feed the wire into the conduit. This is considerably easier than pulling the wires up the conduit. You not only have to fight gravity but also the friction in the conduit. When feeding from the top down you can avoid the hassle of first pulling up the conductors by hand. Hang a pulley and hoist line nearby. With the hoist line, the ground crew can pull the conductors up to meet you.

TABLE 10-9
Maximum Number of Wires in Conduit

Wire Size (AWG)	Conduit Size (Inch Diameter)				
	½	¾	1	1¼	1½
	No. Wires Insulated with THHN-THWN, (THW)				
10	6(4)				
8	3	+(3)			
6	1	4(2)			
4	1	2	(3)		
3	1	1	3(2)		
2	1	1	3(2)		
1	—	1	1	3	
0	—	1	1	3(2)	
00	—	1	1	2(1)	3
000	—	1	1	1	3(2)

Surge Protection

Electrical systems are grounded to limit voltage surges due to lightning strikes and to insure the prompt operation of fuses, circuit breakers, and other protective devices. Lightning is only the most obvious of several sources for voltage spikes. Passing clouds can induce voltage spikes as can the rapid opening and closing of switches in the utility's distribution system. In fact voltage spikes from the latter are more frequent than those from lightning.

Lightning is dangerous and unpredictable. It can make a molten mass out of a wind generator in less than a second. We have all seen the damage that lightning can inflict. Though it occurs almost instantaneously—in micro seconds—lightning can be of sufficient voltage to break down insulation and arc over insulators. The effects of lightning are minimized by using a *lightning arrestor*. When a greater than normal voltage exists in a conductor, the lightning arrestor furnishes a path to ground and drains off the excess voltage. After the voltage has returned to normal levels the flow to ground ceases. Lightning arrestors provide limited protection to power lines and other conductors.

Equipment and buildings can also be protected by, in effect, raising the ground level with a static line or lightning rod. The static line on the utility's distribution system and the lightning rods on farm houses are attempts to raise the effective ground line above the structure. Lightning rods, such as those at utility substations, offer a 45-degree cone of protection beneath them. When lightning strikes a lightning rod it passes directly to ground without first passing through the object below. Some wind generators, notably those 10 meters in diameter and larger, sport lightning rods above the nacelle for this reason. Small wind systems rely on thorough grounding of the tower and guy cables to drain off any static charge to minimize the possibility of a direct strike.

Lightning does not always strike the tallest object. There are many documented cases where telecommunications towers have been spared a direct hit while nearby trees have been incinerated. Proper grounding lessens the possibility of a direct strike and minimizes the damage if one does occur.

How often can a strike occur? Some areas are more prone to electrical storms and lightning than others. In areas of the country with thirty-five or more thunderstorms per year there is the chance for a strike or near strike once every two years when the tower is 100 to 150 feet above the surrounding terrain. Much of the country east of the Rockies falls within this zone of incidence. The southern states and areas of the Great Plains have a higher frequency of electrical storms and a greater potential for damage due to lightning.

Do not assume that the tower is electrically grounded simply because it is embedded in the earth. Concrete is an insulator. Towers on concrete piers with concrete anchors are electrically isolated and when hit by lightning will sit and cook like a king-size version of the coil in your toaster.

Ground the tower by driving $5/8$-inch copper-clad (preferred over galvanized steel) ground rods. Eight feet of the ground rod should be in contact with the soil. Drive a ground rod near the mast and near each of the anchors on a guyed tower. Where the rods cannot be driven eight feet into the ground, drive them obliquely (at an angle) or bury them in a trench not less than $2\frac{1}{2}$ feet deep. On free-standing towers, two or more ground rods should be used. Connect the ground rods to the tower or guy cables with No. 4 gauge wire. Do not skimp on the connectors by using ground clamps intended for indoor grounds to water pipes. Use heavy brass or bronze ground clamps (sometimes called acorn clamps) on the ground rods and split-bolts on the guy cables or tower girts.

In areas of high lightning incidence or where the soil is dry and sandy (this soil makes a poor ground) install more ground rods (attach one to each tower leg for example) and use heavier ground wire. Jim Sencenbaugh of Sencenbaugh Wind Electric goes even further. He recommends a ground net for his guyed towers. The tower and each anchor is grounded to its own ground rod. Then all the ground rods are tied together with a buried ground wire. He also electrically bonds each tower section together with a jumper wire to assure a continuous path down the tower to ground. Also eliminate all sharp bends from the ground wire, he suggests, as high voltage prefers to travel in a straight line.

Additional Notes

Electrical codes require that all metal electrical enclosures such as EMT conduit, disconnect switches, control panels, and inverters be grounded so that any fault (short circuit) between a live or hot conductor and the enclosure will be conducted safely to ground. Grounding causes the circuit's protective devices to function—fuses to blow, circuit breakers to trip. This prevents the metal enclosure from becoming energized (hot) and presenting a shock hazard.

Control boxes and switches must be properly secured and all holes or cut-outs not in use must be sealed. There must be a minimum three feet of clearance in front of any panel for servicing. Control panels and inverters with ventilation louvers must be located so as to allow for free air circulation.

All terminals, connectors, and conductors must be compatible. Poor connections and the use of dissimilar materials such as copper and aluminum are a major cause of electrical fires. Aluminum is particularly troublesome. It oxidizes when exposed to the atmosphere and forms a highly resistant crust. This increased resistance causes the connection to heat up under heavy loads and is believed to be responsible for numerous fires. Whenever aluminum is used, terminal blocks, split bolts, and other connectors must be rated for aluminum (Al) or for copper and aluminum (CO/AL, Cu-Al). Aluminum connections should also be coated liberally with an anti-oxidizing compound. One common brand is No-Alox. (Don't overdo it though. These compounds are conductive and could cause a short circuit if you get sloppy.)

OPERATION

You should notify the utility—in writing—several weeks before the installation that you plan to interconnect with their lines. This gives your letter ample time to move through the utility's bureaucracy and get the proper clearances. If the utility requires any special switches or metering, it is better to find out as early as possible to minimize costly modifications once the turbine is installed. Expedite the process by calling first and finding the person responsible for interconnections with small power producers. Address the letter to them.

Include in your notification the following information:
- Brand name and model number of wind generator
- Type of wind system (whether it uses an induction generator or synchronous inverter)
- Maximum power output in kW
- Operating voltage
- Number of phases
- Line drawing of the proposed installation.

The line drawing should show the location of all disconnect switches and protective relays (called contactors) relative to the service panel, the kWh meter, and the utility's service drop. Describe how the wind system functions under both normal and emergency conditions (such as a power outage on the utility's lines). The utility wants to know how the design of the wind system guards against energizing a downed line and endangering their linemen (see fig. 10-16).

Your letter should allay the utility's fears. It should tell them that the wind turbine controls will automatically isolate the wind turbine from the utility's lines whenever:
1. a fault on the wind turbine side of the interconnection is detected,
2. a fault on the utility's line is detected, or
3. abnormal operating voltage or frequency is detected.

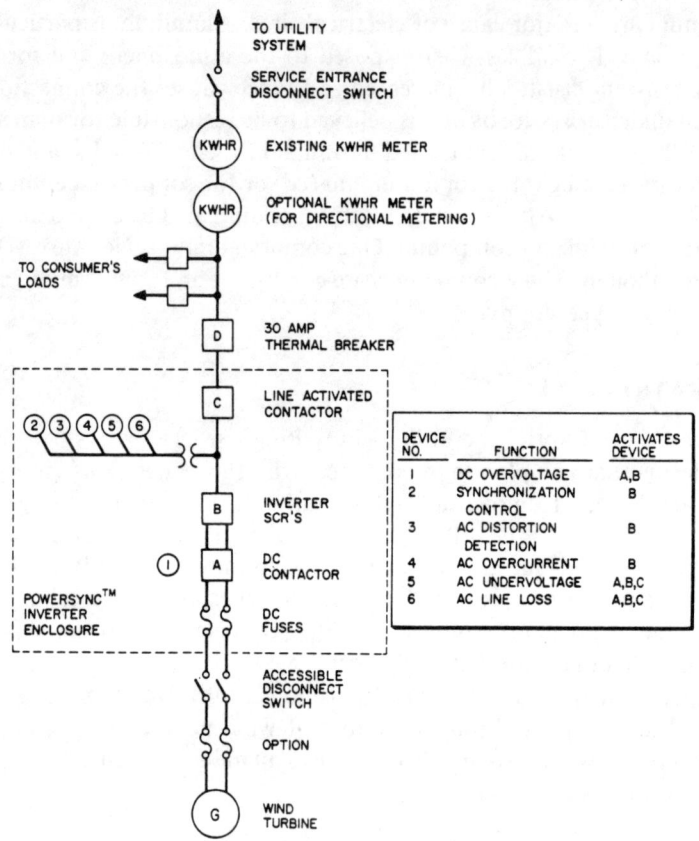

Fig. 10-16. Interconnection schematic. Utilities usually require line drawings such as this before they will permit an interconnection. Such a schematic tells them how the wind system will be interconnected with their lines and what precautions have been taken to protect their linemen. (Courtesy Bergey Windpower Co.)

Additional information on the system's power factor or VAR (volt-ampere-reactance) characteristics, current harmonics, and maximum inrush currents may be helpful to the utility, but is unnecessary to insure a safe interconnection. The turbine's manufacturer will provide this information to you if the utility insists upon it.

The utility is responsible for determining whether the interconnection poses a safety hazard to their linemen and whether it will interfere with their service to other nearby customers. To do so, they will want to inspect the installation before the wind system begins operation. Don't panic; this is a reasonable action. In some cases they will accept the inspection report of the local fire underwriter and give you the go-ahead to begin operation until they can get someone out to your site. But don't tolerate deliberate delays. If they have not responded to your request for an inspection within ten working days, call the state energy office or public utility commission and file a complaint.

Try not to let the thrill of starting your new investment get the better of you. Take your time and look over the entire installation. Go over the manufacturer's check list: all nuts snug, cables secured (where used), wires connected to the proper terminals, and so on. If everything meets your satisfaction, go through the suggested start-up procedure. For example, flip the circuit breaker and disconnect switch "On" and unfurl the tail if you are using an upwind machine; release the brake if it is a downwind turbine. As a rule, you always turn the AC power on first. Then you proceed to turn the wind system on. Avoid starting the machine for the first time in a high wind. If a problem develops, light wind minimizes the potential for damage. The turbine is also easier to bring under control in light winds.

Unlike new cars, there is no need to "break in" the turbine by running it intermittently at different speeds. It is a good idea, though, to be present the first few days and to keep a watchful eye to see that it operates as expected. I've had the misfortune that every time I am ready to start a turbine, there has been no wind. In my case, the small machines I have installed are two or more hours drive from home. I set them up and leave and have never had any problems. On a large machine with a number of complex components, I would prefer to be around for the first few days.

MEASURING PERFORMANCE

If the wind turbine is running properly—no frightening sounds or dangerous vibrations—how can one monitor its performance? Begin by installing an anemometer and checking to see if the turbine starts up and shuts down at the advertised wind speed. You are not going to get the anemometer at exactly the same level as the rotor, so the readings from the anemometer will be somewhat less than the winds actually striking the turbine—but they will be close. This is one simple test.

Another test is to compare the power output from the wind machine at various wind speeds with the manufacturer's performance curve. The control panel on some machines includes a watt meter for this purpose, but on many small machines no meters are provided. You can improvise, though, by using a kWh meter. They are inexpensive and easy to install (see fig. 10-11).

The kWh meter records total energy production in kWh—this is the bottom line in reference to the economic value of a wind system. By measuring the rate at which a kWh is being produced, you can calculate the average power. This is a good way to detect poorly aligned tip brakes or other power-robbing problems before they show up in lower overall energy production.

Within the kWh meter is a metal disc that spins in response to the flow of power. On the meter's face is a number labeled kh. This factor is in watt-hours of energy that pass through the meter per revolution of the disc. Some common values are 1, 1.8, 2, and 3. If kh = 1 the disc must revolve 1,000 times before one kWh registers on the meter (1,000 watt-hours equals 1 kWh).

By measuring the time (T) it takes for the disc to make a revolution, you can calculate the power (P) in watts produced at that instant.

$$P = \text{rev/min} \times kh \text{ (watt-hr/rev)} \times 60 \text{ min/hr}$$

Watch the meter for one minute, say, and count the number of revolutions (there's a black mark on the disc for this purpose) and plug them into the equation. You now have the one-minute average power output. If you don't want to wait around for one minute, time the disc for thirty

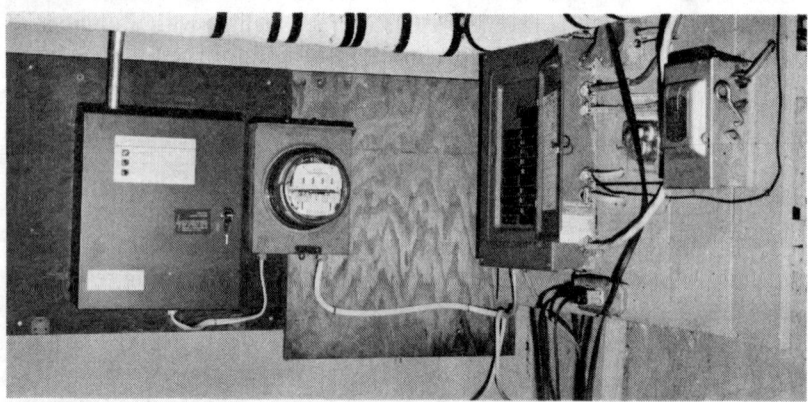

Fig. 10-17. Control box and service panel. The service from the wind machine enters the top of the synchronous inverter on the left. The line-quality output leaves the bottom of the inverter in the plastic insulated cable (Romex) and flows through the kWh meter to a dedicated circuit breaker in the service panel on the right. The arrangement shown is the same on wind systems using induction generators. The control box for the induction wind machine would replace the synchronous inverter in this photo. The kWh meter measures the amount of energy generated by the wind machine.

seconds and double the value. When you want to catch the output during a strong gust, measure the time it takes for one revolution.

$P = \text{rev}/T \text{ in seconds} \times 60 \text{ sec/min} \times \text{kh (watt-hr/rev)} \times 60 \text{ min/hr}$
$= \text{rev}/T \times 3600 \times \text{kh (watt-hr/rev)}$

Example: The timing disc makes 10 revolutions in one minute. kh = 1.8. How much power is being generated?

Solution: 10 rev/min × 1.8 (watt-hrs/rev) × 60 min/hr = 1080 watts = 1.1 kW

Example: Assume you're in a hurry. The disc makes one revolution in 10 seconds. What is the power output?

Solution: $P = 1 \text{ rev}/10 \text{ sec} \times 1.8 \text{ (watt-hrs/rev)} \times 60 \text{ min/hr} \times 60 \text{ sec/min}$
$= 6480 \text{ watts-sec}/10 \text{ sec} = 648 \text{ watts} = 0.65 \text{ kW}$

Now you can compare the average power output with the average wind speed for the same interval. You will quickly find that there is rarely a perfect match between the output curve and the generator's actual performance. Sometimes it will be greater, sometimes less, depending on whether the rotor is coasting after a big gust or is coming up to speed during a gust. You are looking for trends. If power output is consistently below that expected, try to find an explanation.

You can pick up a used meter for $25 (1983) and the socket to go with it for another $25. There are several different designs available. Those with numerical displays are much easier to read than the more common version with the clock faces. They are slightly less accurate but they will do fine. You might find a demand meter helpful. In addition to total energy in kWh, they also register the maximum power in kW that passes through the meter. Demand meters cost two to three times more than regular watt-hour meters and they are only of benefit on the larger installations where power output reaches 10 to 20 kW.

The socket meter bases are preferred. They can be used indoors or out and make a neat installation. To install the meter, mount the socket on the wall, wire it up, and then plug in the meter. That is it. I use a meter on all my installations.

MAINTENANCE

How much maintenance is there? is a question I am often asked. Unfortunately, I don't have a concise answer. It all depends, is the best I can tell them. Some wind machines are marvels of simplicity and they are nearly maintenance free. Others are more complex and the level of regular service required is higher. And there are those that never seem to work right, or for very long, without a major repair. The amount of maintenance required depends on the type of wind machine, its size, and the approach of its designers.

Here are a few guidelines. Rotors with fixed pitch blades require less than those with governors. Machines using direct-drive require less than those using transmissions. Free-wheeling drive trains require less than those where the rotor must be motored up to speed. And those turbines using passive yaw to orient the rotor require less than those using active yaw drives. On the whole, small machines require less maintenance than larger ones.

Most important of all is the designer's attitude. During each step of the design, decisions must be made. If minimizing maintenance is a top priority, it is reflected in the final product. Because I don't like to climb towers, my idea of the correct design philosophy is that espoused by Karl Bergey of Bergey Windpower Company. According to Karl, the maintenance instructions for a wind machine should read: "Once each year, on a windy day, walk out to the base of the tower and look at the mill. If the mill is turning, it's okay." In an age where the vast majority of us are accustomed to automatic chokes and idiot lights on our cars, we should hardly be expected to run up and down the tower in foul weather carrying a grease gun.

Maintenance will not be a concern during the first year or two because the wind system will be under warranty. But, eventually the warranty will expire. What happens then is up to you. You can perform routine maintenance yourself or you can hire the dealer to do it. Most dealers offer maintenance contracts. Price and terms vary but usually for a small fee they will provide emergency service and regular maintenance. If any parts need to be replaced you are billed extra.

Some manufacturers are extending their warranties on major components, but you still have to provide maintenance. Others offer maintenance insurance. You buy a multiyear policy (say two years at a time) from an independent insurer affiliated with the manufacturer. The policy covers all repairs and pays for both parts and labor.

Maintenance is principally a thorough inspection. But it can include tightening bolts, greasing fittings, and changing oil. Occasionally it may also entail cleaning the contacts on the slip rings. If your wind machine has zerk (grease) fittings on any of the bearings, they will have to be greased periodically. The trend, however, is towards sealed bearings and bushings like those on your car that are designed to last the life of the machine without being greased. Oil in transmissions (where used) will have to be changed on a regular schedule. This can be messy. But if it is required, make sure it gets done.

Inspect the turbine at least once each spring (after winter storms) and each fall (in preparation for winter). You can do this in a cursory manner from the ground if you like by using binoculars. Check the rotor for symmetry. See if each blade looks alike. Obviously, if they do not, you have a problem. Watch how the turbine changes direction as the wind shifts. Note if the turbine yaws smoothly or abruptly. Erratic yawing can be due to turbulence. The only treatment is installing a taller tower. It can also be due to the tower being off the vertical. (You can check the plumb of the tower with a level, plumb bob, or transit). While still on terra firma, check that the tower is still properly grounded and that the guy cables (where used) are tensioned correctly. The remainder of the inspection must be made on the tower. Before performing any service, carefully read chapter 11 and take the appropriate precautions. Remember to furl the tail or apply the brake before climbing the tower.

At the top of the tower, check that all bolts on the rotor are snug and that the pitch of all the blades is the same. If any bolts or nuts are missing, replace them immediately. (Usually, if there were any missing, you would have known about it shortly thereafter.) Locknuts are used in the rotor because of the severe vibrations. Locking nuts with a nylon insert can be applied several times before they need to be replaced. (But don't replace them with a regular nut.) While you are at it, check whether the blades track each other — follow the same path. You can do this by measuring the distance from a point on the tower to the tip of each blade. They should all be the same. Check that the tail vane (where used) rests solidly against the nacelle and that it takes the rotor out of the wind when furled. Pull the machine about the top of the tower and check whether it moves freely or binds in one position. If it does, you may need to grease the yaw bearings.

Change the oil, grease the bearings, and adjust the brake as your circumstances require. Check that the bolts securing the tower adapter to the wind turbine and to the top of the tower are snug and that the yaw assembly is level. If not, use the leveling nuts provided on some machines or plumb the tower by adjusting the cable tension. Check that the slip rings are adjusted and their contacts clean. Slip ring brushes can be cleaned with an alcohol based fluid. Buff or lightly sand the contacts as required. Swing the turbine around again and insure that the slip rings make good contact no matter which way the wind blows.

As you are coming down the tower, check that all bolts are snug. Do not overtighten as this will score the galvanized metal coating on the tower causing rapid corrosion. Galvanizing acts somewhat like a lubricant, making it easy to overtorque a bolt.

Check for corrosion and secure connections at all wiring terminations. Use extreme caution anytime you open the control panel, synchronous inverter, or disconnect switch. If the connections look good, leave well enough alone and do not unwrap the insulation. When closing the door on any electrical enclosure, make sure you do not pinch the conductors between the door and the box. Vacuum the control panel or synchronous inverter carefully if you live in a dusty area.

TROUBLESHOOTING

You may be able to make some minor repairs yourself, others will require professional help; only you can be the judge. The following is a guide to give you some idea where to look when trouble develops. When you find a defect in your wind system, first determine if continued operation will cause any harm or create a hazard. If it will, shut the machine down and take corrective action. Make the needed repairs yourself or call in the dealer. If you are unsure, it is best to play it safe. You have a sizable investment to protect. If it is a minor problem, you may have the luxury of letting the machine run until you can get around to fixing it. But do it or hire someone who will.

Problem: Rotor does not turn.
A. Check the anemometer to see that the wind speed is above the start-up speed. If not, wait until the wind speed increases.
B. If the wind speed is above start-up, first check that the tail is fully extended behind the rotor on upwind machines, or that the brake is fully released on downwind machines. Second, if the tail is extended or the brake released, check whether the rotor shaft is frozen. Furl the tail or apply the brake, climb the tower, and spin the rotor by hand.

Problem: Rotor turns slowly in high winds.
A. On rectified alternators, check for shorted varistors or diodes. On induction generators, check for dragging brakes. On rotors with tip brakes, check that the tips are in their proper position and not deployed.
B. Check for correct field excitation. The electronic controls could be loading down the generator prematurely, preventing the rotor from getting up to speed on generators with synchronous inverters.

Problem: Rotor vibrates or is noisy.
A. Furl the tail or apply the brake, then check the rotor for symmetry (that each blade looks alike). Check for missing balance weights, stretched springs or loose bolts. Shake each blade to see that it is securely attached.
B. Check that the pitch (angle of blade to the wind) of each blade is the same.
C. Check whether the blades are chipped or splintered and whether the leading edge covers on the blades are secure.
D. During the winter, check for ice on the blades. Remove any ice (or wait until it melts), but do not beat on the blades or costly damage will result.
E. Check that the bolts mounting the wind turbine are snug. If not, tighten.
F. Check whether the blades track each other.

Problem: Turbine squeaks.
A. Furl the tail or apply the brake, then open the nacelle cover. Check that the bearings on the main shaft are firmly mounted.
B. Lubricate the bearings if needed.
C. Check that the main shaft spins true.

Problem: Wind turbine yaws erratically.
A. Check guy cable tension and plumb of tower.
B. Check that yaw assembly, where used, is level.

Problem: Tower vibrating.
A. Check rotor as described earlier.
B. Check guy cable tension.

Problem: No voltage or power at the control panel.
A. Check that the tail or brake is released and that the wind speed is above 10 mph.
B. If so, check the ampmeter for current. If there is current, the voltmeter or its sensing circuit may be defective. Check the voltmeter by measuring voltage with a multimeter.
C. If there is no current at the ampmeter, check for a fault or open circuit between the generator and the control panel or between the control panel and the service panel. For example, check the slip rings.

Problem: High voltage with no amperage.
A. Check for open circuit.

Problem: Power too low, voltmeter and ampmeter erratic.
A. Furl or brake the rotor and check for loose connectors from the control panel to the wind generator. For example, check the slip rings, tighten loose connections, and clean tarnished terminals as needed.

Problem: Power output low.
A. On rectified alternators check for shorted diodes.
B. Check that the tail is fully unfurled or the brake fully released.
C. Check that tip brakes (where used) are not deployed.
D. Check that the transmission or main shaft is not binding.

This listing is only a sample of troubleshooting hints. The operations and maintenance manual (also called an owner's manual) will have a more thorough and detailed presentation. Always refer to the manual for specific treatment of your machine's problems.

11

Safety

BOULEVARD, CA (UPI) — Terrance Mehrkam, 34, owner of a Hamburg, PA windmill manufacturing company was struck and killed by the blade of one of his own windmills at a "wind farm" in this San Diego County community.

The coroner's office said Mehrkam was struck by one of the blades after falling from a platform . . .

When you are hanging from an 80-foot tower in a strong wind you gain an intimate sense of the danger involved. But it doesn't hit home until someone you know has been killed. I knew Terry. His death shocked me as it did the wind industry. But possibly, his accident and that of Tim McCartney mentioned earlier will serve some purpose by warning the rest of us to be more careful and to take all the precautions necessary so that it doesn't happen again.

Contrary to popular belief, wind systems are not an entirely benign tool of a future solar society. An accident with a wind system can kill or cripple. With a wind system, any number of accidents are possible: falling off a tower, getting caught in the rotor or transmission, being hit by falling debris (worse yet, getting hit by a falling wind turbine), or being electrocuted.

Wind machines are no more dangerous, or pose no more hazards, than many other aspects of modern life. We have all grown to accept the hazards of the electricity and natural gas that flow to our homes. Yet accidents with these common energy sources continue to occur. We continue to drive our cars even when we know that automobile accidents take thousands of lives every year. Common do-it-yourself projects such as painting the eaves or repairing the family car are just as dangerous as working on a wind machine. If I had to rank them, I would say that working on a wind machine is much safer than cutting wood with a chain saw—and more fun. But, treat wind systems with the same respect you would give any machine, and work as though your life's at stake—because it is.

Safety is both a state of mind and a way of life. This is not a new Eastern philosophy. It is a necessity. Foremost in any discussion of safety equipment is that it must be used before it is of any value. A safety belt or hard hat is no good when left on the ground. Likewise, precautions are useless when ignored.

Work on or around the tower poses the greatest hazard because the possibility for accidents and the severity of possible injuries is greatest. This discussion of safety, as a result, will focus first on tower work.

The simplest and most reliable way to avoid accidents is to avoid the hazard. As Ward Slager of Bergey Windpower Company is wont to point out, the objective of wind system designers and manufacturers should be to eliminate working on the tower altogether. Designers, for example, should strive towards complete assembly on the ground and erection of the wind turbine as an entire unit. Furthermore, there should be a minimum of maintenance—preferably none—performed on the turbine once atop the tower. The hinged towers and the new wind generators entering the market are beginning to make this goal a reality. But for most wind systems today, whether during installation or during repairs, work near the top of the tower is still very much a part of the job.

Work Belts

Tower work demands a restraint system that frees the hands for lifting, pulling, and performing other assorted tasks above ground, while keeping you on the tower. To be used consistently and to allow freedom of movement, the restraint's link to the tower—the lanyard—must be easy to release and to reattach. Because standing for long periods on a narrow ladder rung or cross-girt is tiring, it is desirable to use a restraint that also takes up the load on the legs as well as freeing the hands. Protection in case of a fall, though, requires a different kind of restraint: one that safely arrests the fall. Ideally, a single restraint would be used that delivers these three elements. Unfortunately, that is not the case.

In common parlance, the term *safety belt* is a misnomer. The safety belts used on construction sites and throughout industry are more correctly labeled *positioning belts* or, more simply, *work belts*. They are used to free the hands when working in an upright position with the legs carrying all the weight. Work belts are made from a wide strap of leather or nylon webbing buckled around the waist. On each side of the belt are large steel "D" rings. A lanyard links the wearer of the belt to the tower via attachments at the two "D" rings and the tower.

The principle purpose of a safety belt is—as one would expect—to prevent falls. But when a fall does occur, the belt must keep you from hitting the ground. And equally as important, the safety belt itself must not cause injury. Positioning belts are valuable in preventing falls. These belts restrain a person in a hazardous position to reduce the probability of falling. They will also protect against striking the ground (assuming that the lifeline is not too long) in a fall. But they may cause serious injury in the process, because the shock of reaching the end of the line is transmitted to the belt where all the load is placed on the lower back. If the wearer is knocked unconscious, he may even flip upside-down and slip out of the belt completely.

Consider, for example, the popular *lineman's belt* (see fig. 11-1). Originally designed for climbing utility poles, the "lineman's belt is, in fact, a positioning belt," according to William Glynn of SINCO Products, a major manufacturer of work belts. "It's only a work tool." The lineman's belt can give a false sense of security because it was not intended to rescue workers in a fall. (This belt was designed primarily to aid linemen in quickly getting to the work station at the top of the pole.)

SINCO and other belt suppliers (see appendix D) make full body harnesses designed to arrest the most severe free fall without injury. The harnesses include a work belt, leg straps, and a chest harness. The chest harness keeps the wearer upright after a fall and, with the leg straps, distributes the load from the fall more uniformly over the body than a work belt does alone.

Sit Harnesses

When examining fall protection one casts about to find other fields where this problem is a major concern. Mountaineering quickly comes to mind. Sport climbing (wind system installers, in contrast, don't climb

318 • Wind Energy

Fig. 11-1. Lineman's belt. Linemen's belts were designed for climbing utility poles as shown here. (Courtesy Klein Tools)

towers just because they're there) has developed the *sit harness* for long periods of direct aid or technical climbing.

The simplest and least expensive sit harness is the Swami-Belt, a five foot plus sling of nylon webbing wrapped around the buttocks and held in place by a carabiner. A second piece of webbing is often tied around the waist and also attached to the carabiner. In conjunction with a chest harness the Swami-Belt offers the greatest safety at the lowest cost. The Swami-Belt, however, is difficult to use. It is uncomfortable, and tends to slip down the legs when you are moving about.

Sewn sit harnesses are preferable to the Swami-Belt. They are easy to use and, more importantly, comfortable when hanging suspended from the tower for long periods. Sit harnesses are also superior to positioning belts during a fall (the shock is more evenly distributed between the legs, buttocks, and waist). Like positioning belts, sit harnesses allow free use of the hands. Moreover, they can be used to hang from the tower in a near sitting position. This relieves the strain on the legs—there is nothing quite like a cramped calf when you are 80 feet in the air.

OSHA (Occupational Safety and Health Administration) standards on work belts are based principally on strength. Work belts and lifelines must have a minimum breaking strength of 5,400 pounds. Most sit harnesses do not meet this requirement. Depending upon the style and manufacturer, sit harnesses are rated at 3,000 to 4,000 pounds breaking strength. If this is a concern, you have no choice but to stick with traditional work belts and full-body harnesses designed to meet OSHA's regulations.

Fortunately, if neither a full-body harness nor a sit harness is for you, there is a belt that can be used to hang suspended from a positioning line in a half-sitting posture (see fig. 11-2). These Bosun belts (as in Bosun chairs) are work belts that include a strap across the buttocks that can take some of the weight off your legs. They are more comfortable than simple work belts for extended tower work. In either case, work belts and sit harnesses are only part of a fall protection system—without a lifeline they are not a whole lot of help.

Lifelines and Lanyards

Lanyards are short sections of rope with snap hooks attached. They are used as positioning lines or as lifelines. OSHA requires lanyards to "provide a fall of no greater than six feet." Both lanyards and *lifelines* (long lanyards) must have a minimum breaking strength of 5,400 pounds. After catching a fall, OSHA requires that a lifeline or lanyard be discarded.

Lifelines should not only prevent the wearer from hitting the ground

Fig. 11-2. Work belts and lanyard. *Top*, lanyard with snap hooks for use as a safety line or for positioning; *bottom left*, "bosun" work belt, preferable to ordinary work belt because strap can support weight during extended tower work; *bottom right*, work belt with "D" rings for attaching lanyard. Work belts are not "safety" belts though often labeled as such. They are used for positioning. (Courtesy SINCO Products, Inc. and Atlas Safety Equipment Co.)

but they should not cause any injuries as well. The safety line must "give" some in a fall. To minimize the shock to the body when the lifeline or lanyard reaches its limit, nylon rope is used frequently because it can stretch, thus dampening the deceleration. Steel cable, in contrast, will stop a fall cold. Hemp rope is not much better. "Under no circumstances should a steel cable be used [as a lanyard] where the possibility of a free fall exists," warns Atlas Safety Equipment. And avoid hemp whenever possible.

Rope serves several functions during the installation of a wind system. Nylon rope, for example, can be used to lift a load or restrain you in a fall. A safety line should not be used to lift loads. Since use weakens rope, make sure that lifelines or lanyards are at their peak strength. You only get one chance.

Snap Hooks, Carabiners, and Slings

As the name implies, *snap hooks* are shaped like hooks that can be snapped onto a line, tower leg, a cross-girt, or the "D" ring of a work belt. A safety catch (keeper) prevents the hook from releasing accidentally. Snap hooks are easier to attach than *carabiners* (snap links). Riggers and others in the construction industry use snap hooks exclusively. A quick glance always tells you which end is up and how to apply it—not so with carabiners. They also come in several throat sizes—up to 2¼ inch wide for work with reinforcing bars. These *re-bar hooks* are useful for clipping onto tower legs or cross-girts.

Carabiners are used principally in rappelling and direct-aid climbing. Any time a rope must be fastened and unfastened a number of times, a carabiner can make the task easier. Unfortunately, snap links are symmetrical. It is often difficult to tell which end is up. But they are useful for clipping gear onto a work belt or tool belt. Aluminum carabiners have a breaking strength of 4,000 pounds. Steel carabiners are nearly twice as strong. Snap hooks are designed to meet OSHA's breaking strength requirement; aluminum carabiners are not.

Sewn loops of nylon webbing—*slings*—were developed after climbers learned that a great deal of valuable rope was being lost when constructing rope harnesses. Climbers were also seeking a more comfortable harness material. With a carabiner attached, a nylon sling can be used like a lanyard. Or slings can be used to carry equipment up the tower, tie down the rotor, and perform any number of other tasks. Wider and stronger nylon slings are used extensively in rigging. Where an erection jig or fixture is not handy, slings are a good choice for lifting heavy loads such as wind generators or a completely assembled turbine and tower.

Fall Protection Systems

The most dangerous activity when working on a tower is climbing up and descending. It is dangerous because the common practice is simply to scale the tower then secure yourself with a lanyard once you have reached the work station. It is even more dangerous coming down, because you are tired and your timing can be off. This is particularly true in winter when biting winds quickly sap your strength (not to mention the possibility of encountering ice on the tower). Hypothermia (the loss of higher mental functions, and a false sense of confidence) can set in before you know it.

Several manufacturers offer a device designed to mitigate this hazard. They employ a sleeve that slides along a taut cable or rope that runs the length of the tower (see fig. 11-3). You attach your work belt or harness to the sleeve with a snap hook and lanyard. The sleeve rides up the cable as you climb the tower. Should you slip, it locks and arrests the fall. Fall-arresting cables such as these can be found on aircraft beacon lights, lights at sports stadiums, drill rigs, water towers, and on some European wind machines. On the surface this seems to be the ideal fall protection system. But it does have drawbacks. Chief among them is cost: $200 to $300 (1983). Another is that some of the systems do not allow freedom of movement, especially at the top of the tower. (They were designed for climbing a ladder to a protected work platform.)

322 • *Wind Energy*

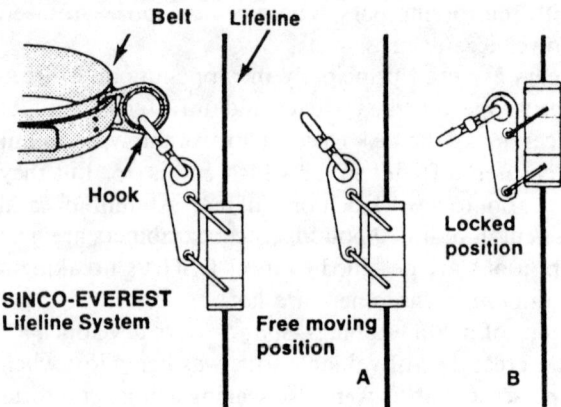

Fig. 11-3. Fall protection system. When the locking device is lower than the work belt as at (*A*), the device slides freely along the lifeline. When the locking device is above the work belt as at (*B*), the device locks onto the lifeline arresting descent. (Courtesy SINCO Products, Inc.)

Wind turbines are naturally the highest point on the tower. To work on them you need to be at their level or above. Cable and sleeve systems cannot extend high enough or they begin to interfere with the wind turbine. This can be alleviated by using a lanyard with the sleeve (thus allowing a degree of movement), or by releasing the lanyard and reattaching it to some point on the wind turbine itself.

In terms of cost and convenience, fall protection systems using rope are superior to those using cable. And those allowing quick release are better than those attached permanently to the lifeline.

Less costly but more time consuming (thus less likely to be used) is a two-lanyard system. When ascending, one lanyard is attached above as far as possible. When it is reached, the second lanyard is attached above and then the first removed and so on. This insures that you are always tied to the tower even when reattaching a lanyard. Though impractical for scaling towers, the two lanyard technique illustrates a good overall safety practice—always keep one lanyard or safety line attached. This is particularly helpful when you need to reposition yourself around the wind turbine. For example, say you need to remove the nacelle cowling. There might be an attachment point on the chassis useful for positioning. But you've got to get there first. A four- to six-foot lanyard clipped at your last position will give you enough slack to reach the turbine while still protecting you in a fall (six feet is far better than sixty).

LeRoy Wagner, an Enertech dealer in western Pennsylvania, has devised a method for belaying a climber when descending after the day's work is done. A lifeline is looped over a tower girt and attached to the work belt or harness. As the climber descends, LeRoy and his installation crew pay-out the line. Once safely on the ground, they then retrieve the line by pulling it back over the girt. It is probably a better idea to run the rope over a carabiner than over the girt (the rough edges of the girt will damage the rope). The carabiner can be left in place for the next maintenance call.

More Tower Tips

Besides work belts and harnesses, there is some other gear that can make tower work both safer and more comfortable. Boots are one, gloves another. Always wear boots with firm, nonslip soles. Your feet tire less and are less likely to slip from a girt or ladder rung than with street shoes. Gloves do more than protect the hands; they help get a better grip — and a good grip is paramount. Leather is best. The galvanizing forms droplets on the steel before it cools. These droplets can be sharp as a knife, cutting through cloth gloves with ease.

Hard hats are essential, though admittedly, they do not qualify as comfortable (particularly in midwinter). They are hard to keep on in a high wind, unless fitted tightly or used with a chin strap. But their value becomes apparent when working around small wind turbines that lack parking brakes, or after a wrench whizzes by your head.

Several small wind machines lack parking brakes because of their added cost and complexity. Even when furled, the rotor may still spin. Those blades may not look like much, but they can sure put a dent in your skull. The blades themselves on small machines are not too heavy, but the generator and transmission contain a lot of inertia when the rotor is turning. This inertia can drive a lightweight blade with damaging force. If the blades strike you, they could send you tumbling down the tower, knock you unconscious, or break an arm. A similar problem is unexpected yawing of the turbine in gusty winds. Just when you think you are clear of all that machinery, the wind changes direction and brings everything swinging your way. It is then that a hard hat, and a safety line for that matter, are truly appreciated. Larger turbines feature a parking brake for the rotor as well as a yaw brake.

This brings us to a cardinal rule: never climb the tower while the turbine is running. Always apply the brake, furl the rotor, or do whatever else is required to shut the unit down first. Sometimes it may be necessary to approach the turbine while it is operating to detect a problem that cannot be found any other way. But never go beyond the area of the rotor disc.

Never work alone, as well. Always have someone act as a spotter — the owner, a neighbor, or a friend — someone who can go for help if you need it. Climb the tower on the windward side where possible. The wind forces you into the tower and not off it. No one should work at the base of the tower while someone is working above. Clear the zone about the base of the tower in case a tool or some lost parts come hurtling to earth.

As for tools, always carry them in a tool belt; don't carry them in your hand. All other items should be hoisted up with a rope once you are safely in place. Take a hand line up with you on your first trip. Then use a nylon or canvas bucket to ferry small parts up and down the tower or to hold parts while you're working.

When working around rotating machinery, whether it be a wind machine or a bicycle, do not wear rings, watches, loose clothing, or long hair (tuck it under your hard hat if need be).

Avoid constructing a tower near utility lines. If any part of the tower, wind turbine, gin pole, or crane comes in contact with them, serious injury or death will result. If you have any doubts about clearance, call the utility company before you start to erect the tower.

Stay clear of the tower if a storm is threatening, especially an electrical storm. A lightning strike anywhere near the tower will energize all metal components. Static buildup before a storm can produce the same effect.

Stay clear of the tower during ice storms or freezing rain. If operating, the rotor will shed ice by throwing it to the ground. Ice projectiles will strike directly below the wind turbine and can be a hazard in the immediate vicinity of the tower. Similarly, it is dangerous to climb the tower when it is covered by ice or in a freezing rain.

Electrical Safety

Wind generators produce high voltages. Use extreme caution anytime you open the control box or the nacelle cowling, or work around the slip rings. Always turn off the power before working around any of the electrical components, and always use insulated tools where possible. Remember, electric shock can kill. But if you are working on the tower, the shock itself may not be the greatest danger. Electric shock can cause you to lose your grip. If the shock doesn't kill you, the fall will!

People sometimes freeze when shocked. They may be unable to let go of the hot wire. Help by pushing them away from the live wire with an insulator such as a broom stick (the insulator is needed to protect you from also being electrocuted). Either push the victim away from the wire or push the live wire away from the victim. If the victim is not breathing when released from the current source, apply mouth to mouth resuscitation.

Avoid an electric shock by checking the circuits with a multimeter before you begin work. You could have thrown the wrong switch by mistake. Take your time and think about what you are doing. Never wear metal jewelry when working around electricity (even with gloves).

Rope and Cable

One of the most essential tools for installing and servicing a wind machine is rope. It can be a woven nylon rope used for a lifeline or a wire rope (cable) for hoisting the turbine. Because of its importance and its many uses, rope has continued to develop over the years until it has reached a state of complexity that is near baffling. To pick the right rope or cable for the job, an understanding of a rope's inner workings is helpful.

Rope is made from fibers. These can be natural or synthetic. Hemp, manila, cotton, and sisal are examples of natural rope. (Manila is the most common these days.) Synthetic rope is made of nylon, polyethylene, polypropylene, or polyester. The fibers are twisted to form yarns and strands wound from the yarn. Several strands are then combined to form rope.

As opposed to rope, the strands of steel cable are wound from wires instead of yarn. The wire is made from various grades of steel depending upon their use. For example, stainless steel may be used when corrosion protection is desired.

Europeans have developed a jacketed rope that has become a climber's standard accessory. In the *kernmantle construction* a core (kern) of braided or twisted strands is covered by a protective, braided sheath (mantle). No strands are exposed at the surface. Consequently, there is less opportunity for debris to work between the strands and damage the rope. The sheath can take a lot of abuse while still preserving the integrity of the core.

Manufacturers indicate the breaking strength of rope or cable on the spool. You should look for this when you purchase rope and cable because it will partially determine the rope you choose for a particular job. But breaking strength is not what you might think it is. The *breaking strength* (or *tensile strength*) of rope is the linear tension at which a rope breaks. So far so good. A rope is considered to break, however, when one segment of the rope breaks—not the entire rope. And breaking strength applies only to new ropes; it decreases with age and use. Breaking strengths, furthermore, may vary between manufacturers because of the various testing methods employed, so always check the figures closely.

Breaking strengths also vary between American and European manufacturers. American manufacturers list the average breaking strength; European manufacturers list the minimum breaking strength which can be 10 percent to 15 percent less than the average breaking strength. The accompanying table is a general guide to the breaking strength of new rope (see table 11-1).

TABLE 11-1
Breaking Strengths of Various Types of Rope (lbs.)

Rope Diameter (in.)	Manila	Nylon	Polypropylene	Polyester
3/8	1350	3725	2350	2700
1/2	2650	6080	3990	6080
9/16	3450	7600	4845	7600
5/8	4400	9800	5890	9500
3/4	5400	13490	8075	11875
13/16	6500	16150	9405	14725
7/8	7700	19000	10925	17100
1	9000	23750	13300	20900

When choosing the type and size of rope for a specific job, buy it according to its working strength not its breaking strength. *Working strength* allows a reserve—a margin of safety—for the effects from falling loads and knots. The force on the rope can greatly exceed the weight of the load when the lift is uneven or the load drops suddenly. Knots, on the other hand, significantly reduce breaking strength. Depending on the knot they can lower breaking strength from 45 percent to 75 percent. It's imperative to consider the weakening effects of knots when you use rope. As a result of these effects, working strength is one-fifth of the breaking strength. For example, if you are planning to lift a 500-pound wind machine, a rope with a breaking strength of 2,500 pounds should be used. Though a 1/2-inch manila rope would be adequate for the task, a 5/8-inch manila rope would be a better choice. It not only provides a wide margin of safety, but it's also thick enough that you can get a good grip on it. If you are working with used and partially worn rope, figure working strength as one-tenth of its breaking strength to be safe.

In general, nylon has twice the strength of manila rope. Nylon stretches; manila does not. Polyethylene stretches similar to nylon but is made from a coarser fiber and is more resistant to abrasion. Polyester is more resistant to sunlight than the other synthetics, but like manila, it does not stretch. These features determine which rope is better suited for the job. Nylon is preferred, for example, over manila for lifeline because it stretches.

Like rope, not all steel cable is created equal. For example, guy strand is stiff and heavy, it is hard to work with, and comes in several different grades, from common or utility grade with the lowest breaking strength, to extra high strength at the other extreme (see table 11-2). Guy strand is also galvanized. Other cable may be of stainless steel as in aircraft cable, or simply untreated steel. Winch cable, in contrast, is flexible so it can be wound around the spool of a winch or over a pulley.

The construction of steel cable is given by the number of strands and the wires per strand. Almost all cable has at least six to eight strands. For example, a six-by-seven cable has six strands and seven wires per strand. What you use depends on what you want to do with the cable. There is no sense in using the more expensive winch cable for guys; guy strand works better at lower cost. For a hoisting line, winch cable is best.

TABLE 11-2
Breaking Strength of Steel Cable

Size	Winch Cable	Guy Strand	
		Common	EHS
$1/8$	1760		
$7/32$	5600		
$1/4$	7000	1900	6650
$5/16$	9800	3200	11200
$3/8$	14800	4250	15400
$7/16$		5700	20800

Inspect rope regularly for signs of fatigue and abrasion. Color changes from deterioration and chemicals may not be readily apparent on the surface; examine the inner fibers for color change. Natural ropes usually show clear signs of fatigue and wear. They become limp, surface fibers become soft and fuzzy, and the inner fibers rot. Synthetic ropes will become limp and soft with frayed surface fibers. Rope should not be used for more than five years, even when properly cared for.

When using steel cable for the hoisting line, check for nicks or kinks that may weaken the cable. You can keep the cable from fraying by wrapping tape around the ends. When cutting the cable, keep a grip on both ends so it does not recoil and whip around. The cable should not be pulled over a sharp radius or tied in a knot like a rope. Cable should be looped over a thimble and secured with Crosby clamps wherever it is attached. Remember that the U-bolts of the Crosby clamps bear on the dead-end (the free end) of the cable.

Pulleys

Pulleys serve two purposes: changing the direction of the force applied to a hoisting rope, and gaining mechanical advantage. If a pulley is fixed, say on a gin pole at the top of a tower, it allows you to lift a load from the security of ground level rather than finding yourself in the absurd and dangerous position of carrying up tower and wind turbine components. (Though carrying a heavy generator up a tower may seem ludicrous, I've seen engineers do it; but they never did it again.) You can carry this a step further and attach a *snatch block* (pulley with a hook) at the base of the tower. The lift can now be made at a safe distance from the tower by pulling horizontally on the hoisting rope. Friction in the pulleys slightly increases the force required, but it is well worth it. Pulleys make the job easier and safer.

Fixed pulleys alone provide no mechanical advantage. To do so there must also be a pulley that moves with the load. The number of lines suspending the movable pulley or block determines the advantage gained. If the movable pulley hangs from two lines, the force needed to lift the load is reduced by half. To gain this advantage, though, twice as much rope must be used. Work equals force times distance. To perform the same amount of work as before when we reduce the force by half, we must double the distance through which it acts.

In practice, hoisting a small wind machine by hand requires at least one pulley in the movable block. This block is attached to the lifting ring on the wind turbine. The hoisting rope is routed through a pulley at the base of the tower, over a pulley on the gin pole, and through the movable block. The fixed end is attached near the top of the gin pole. With this arrangement, a 75-pound force will be needed to lift a 150-pound wind machine, and a minimum of 300 feet of rope will be required on a 100-foot tower. (For a 2:1 advantage, we have three lines draped down the tower when the load is on the ground.) This is a manageable lift for two people. A heavier load either requires more help or another pulley in the fixed block on the gin pole. When the fixed end of the hoisting line is routed through the second pulley in the fixed block, and then attached to the movable block, a 3:1 advantage is gained, but another 100 feet of hoisting rope are needed.

Winches and Hoists

Heavy loads, such as complete tower assemblies and wind machines weighing several hundred pounds, can be lifted with vehicles or winches. After using a truck to lower many pre-REA windchargers, I can sum up this approach—it's tricky. Whenever a vehicle is used, whether a truck, tractor, or an auto, you lose a degree of control over the lift no matter how well the driver feathers the clutch. I could afford the risk of damaging the wind turbine when I was salvaging abandoned equipment—and we sometimes did. Not so when installing a new and costly wind turbine I want to last for twenty years.

Winches are preferred for raising wind turbines and heavy tower components, but most winches have limited spool capacity. They can hold only 50 to 100 feet of cable (truck-mounted winches usually have less than 25 feet of cable). For 100-foot towers, the winch must have at least a 100-foot spool capacity, more if a block and tackle are used. These winches are found in industrial supply catalogs along with the handy capstan winch.

The capstan winch drives a spool or capstan. The hoisting rope is looped over the spool a few turns, but instead of winding the rope onto the spool directly, the spool merely aids in pulling the rope along. You have to take up the slack. Battery powered versions with load capacities up to 2,000 pounds can be bought for less than $300 (1983).

Winches should have a remote control switch so you can stand clear of the winch in use. They should also have a brake that locks the spool in either direction, and it should engage if power is lost. You can mount the winch on a special stand for heavy lifts, or on the bed of a pickup truck for light lifts.

Loss Prevention

As soon as possible after the tower has been erected, prominently post warning signs at eye level on the tower. These signs should exclaim DANGER: HIGH VOLTAGE or DANGER: AUTHORIZED PERSONNEL ONLY.

If there is the possibility that the tower will become an "attractive nuisance" and be scaled by thrill seekers, children, or vandals, install anti-climb guards. These can be purchased from the tower supplier, or can be made from 8- to 12-foot-long wood planks, wide enough to cover each face of the tower. Protect manual controls at the base of the tower by removing winch handles or chaining them down. You can also prevent unauthorized access to the tower and its manual controls by surrounding it with a chain-link fence.

Because the attachments of the guy cables on guyed towers are so tempting to vandals, mushroom the threads of bolts to prevent the nuts from being removed. Install a safety cable through the turnbuckles (see fig. 11-4). This prevents both vandals and normal vibrations from loosening the turnbuckles and releasing the guy cable.

Fig. 11-4. Safety cable. Looping a scrap piece of guy cable in a figure eight through the equalizer plate, the turnbuckle, and the guy cable as shown, prevents the turnbuckle from loosening.

Vandals are not the only concern. Errant vehicles, cows, and horses can all damage the tower. Avoid collisions with the tower by installing metal or concrete posts at each tower leg or guy anchor where this danger exists. The posts can be obscured by shrubs if that's your desire, or they can be called out from their surroundings with florescent paint.

Avoid placing guy anchors in traveled ways, but where you must, consider planting low shrubs or bushes around the anchors. People tend to detour around hedgerows rather than go charging through them. Shrubs can also soften the line between the tower and the anchor, or the appearance of a fence. Slip florescent guy guards over the guy cables to make them more visible.

Fence out horses and cattle. They love to scratch their backs on the edges of the tower. Don't give them the chance.

Appendix A

Air Density Corrections for Temperature

**Air Density Corrections
(after Park and Schwind)**

Temperature	0	20	40	60	80	100	120
Adjustment	1.13	1.08	1.04	1.0	0.96	0.93	0.90

Appendix B

Capital Recovery and Compound Amount Factors

TABLE B-1
Capital Recovery Factor

Interest Rate
(Percent)

Year	4	5	6	7	8	9	10	11	12	13	14	15	16
1	1.04	1.05	1.06	1.07	1.08	1.09	1.10	1.11	1.12	1.13	1.14	1.15	1.16
2	0.53	0.54	0.55	0.55	0.56	0.57	0.58	0.58	0.59	0.60	0.61	0.62	0.62
3	0.36	0.37	0.37	0.38	0.39	0.40	0.40	0.41	0.42	0.42	0.43	0.44	0.45
4	0.28	0.28	0.29	0.30	0.30	0.31	0.32	0.32	0.33	0.34	0.34	0.35	0.36
5	0.22	0.23	0.24	0.24	0.25	0.26	0.26	0.27	0.28	0.28	0.29	0.30	0.31
6	0.19	0.20	0.20	0.21	0.22	0.22	0.23	0.24	0.24	0.25	0.26	0.26	0.27
7	0.17	0.17	0.18	0.19	0.19	0.20	0.21	0.21	0.22	0.23	0.23	0.24	0.25
8	0.15	0.15	0.16	0.17	0.17	0.18	0.19	0.19	0.20	0.21	0.22	0.22	0.23
9	0.13	0.14	0.15	0.15	0.16	0.17	0.17	0.18	0.19	0.19	0.20	0.21	0.22
10	0.12	0.13	0.14	0.14	0.15	0.16	0.16	0.17	0.18	0.18	0.19	0.20	0.21
12	0.11	0.11	0.12	0.13	0.13	0.14	0.15	0.15	0.16	0.17	0.18	0.18	0.19
15	0.09	0.10	0.10	0.11	0.12	0.12	0.13	0.14	0.15	0.15	0.16	0.17	0.18
20	0.07	0.08	0.09	0.09	0.10	0.11	0.12	0.13	0.13	0.14	0.15	0.16	0.17
25	0.06	0.07	0.08	0.09	0.09	0.10	0.11	0.12	0.13	0.14	0.15	0.15	0.16
30	0.06	0.07	0.07	0.08	0.09	0.10	0.11	0.12	0.12	0.13	0.14	0.15	0.16

TABLE B-2
Compound Amount Factor

Escalation Rate
(Percent)

Year	4	5	6	7	8	9	10	11	12	13	14	15	16
1	1.04	1.05	1.06	1.07	1.08	1.09	1.10	1.11	1.12	1.13	1.14	1.15	1.16
2	1.08	1.10	1.12	1.14	1.17	1.19	1.21	1.23	1.25	1.28	1.30	1.32	1.35
3	1.12	1.16	1.19	1.23	1.26	1.30	1.33	1.37	1.40	1.44	1.48	1.52	1.56
4	1.17	1.22	1.26	1.31	1.36	1.41	1.46	1.52	1.57	1.63	1.69	1.75	1.81
5	1.22	1.28	1.34	1.40	1.47	1.54	1.61	1.69	1.76	1.84	1.93	2.01	2.10
6	1.27	1.34	1.42	1.50	1.59	1.68	1.77	1.87	1.97	2.08	2.19	2.31	2.44
7	1.32	1.41	1.50	1.61	1.71	1.83	1.95	2.08	2.21	2.35	2.50	2.66	2.83
8	1.37	1.48	1.59	1.72	1.85	1.99	2.14	2.30	2.48	2.66	2.85	3.06	3.28
9	1.42	1.55	1.69	1.84	2.00	2.17	2.36	2.56	2.77	3.00	3.25	3.52	3.80
10	1.48	1.63	1.79	1.97	2.16	2.37	2.59	2.84	3.11	3.39	3.71	4.05	4.41
12	1.60	1.80	2.01	2.25	2.52	2.81	3.14	3.50	3.90	4.33	4.82	5.35	5.94
15	1.80	2.08	2.40	2.76	3.17	3.64	4.18	4.78	5.47	6.25	7.14	8.14	9.27
20	2.19	2.65	3.21	3.87	4.66	5.60	6.73	8.06	9.65	11.52	13.74	16.37	19.46
25	2.67	3.39	4.29	5.43	6.85	8.62	10.83	13.59	17.00	21.23	26.46	32.92	40.87
30	3.24	4.32	5.74	7.61	10.06	13.27	17.45	22.89	29.96	39.12	50.95	66.21	85.85

Appendix C

Wind Data

REGIONAL MAPS OF AVERAGE ANNUAL WIND POWER

Battelle's Pacific Northwest Laboratory conducted a national assessment of wind resources for the Department of Energy. The results of this study are presented in twelve regional atlases. Several states are grouped into each region (see fig. C-1).

The Wind Energy Resource Atlas for your region is the first place to go if you want information on wind speed and power at your fingertips. The atlases are available from the National Technical Information Service, 5285 Port Royal Road, Springfield, VA 22151. The price varies with each atlas, but the usual cost is less than ten dollars.

The atlases contain information on each state, as well as a compilation for the entire region. The data for each state is presented in a consistent format including a description of the geography, topography, and the climate of the state. Following the text is a series of maps, tables, and graphs, including not only information on the average annual wind power, but also data on the area distribution of each power class and on the average seasonal wind power. The graphs for selected stations may be

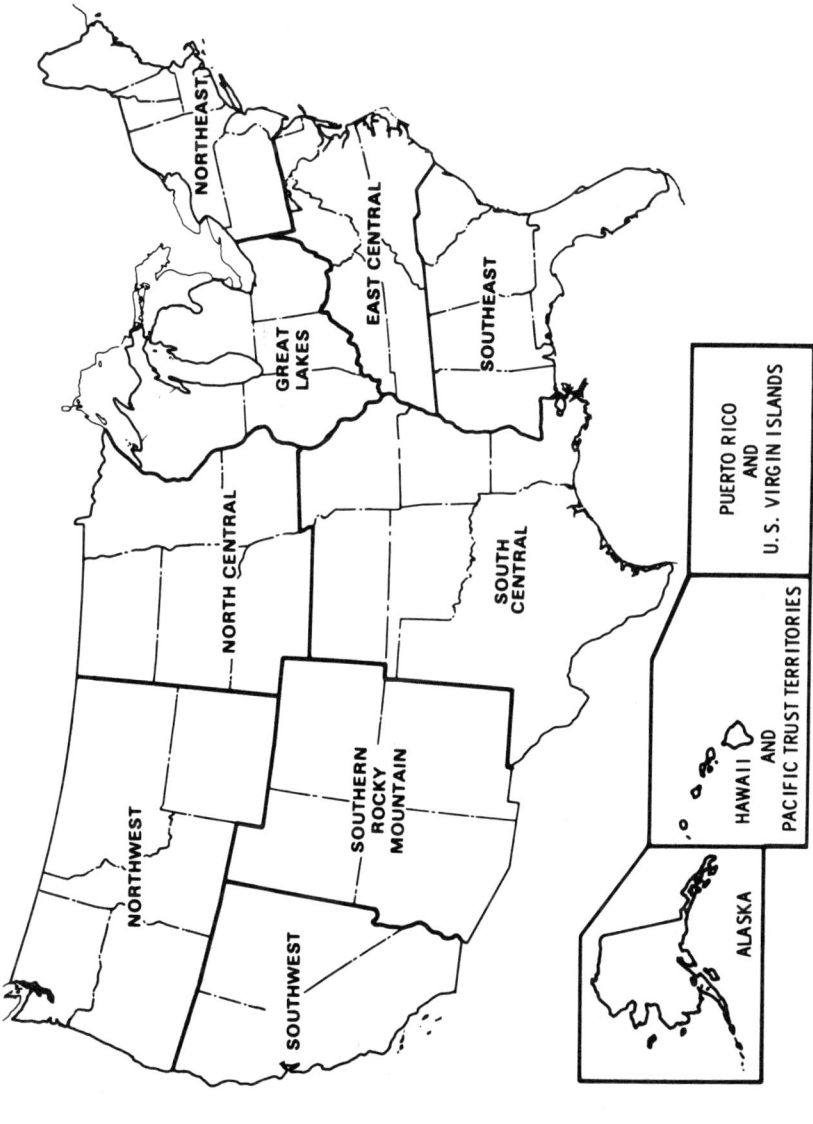

Fig. C-1. Geographic divisions for regional wind energy atlases. Locate your region and find your approximate position on the wind power maps for that region. Check the tables accompanying the wind power maps for the long-term recording station nearest you and compare the average annual wind power density with that indicated on the wind power maps. (Battelle PNL, DOE)

particularly useful. The stations were chosen because they had several years of detailed records. The text describes each station, its location and surrounding terrain, and how the summarized data in the graphs may represent local or regional wind patterns. Following the regional maps of wind power are tables of the average speed and power for selected stations.

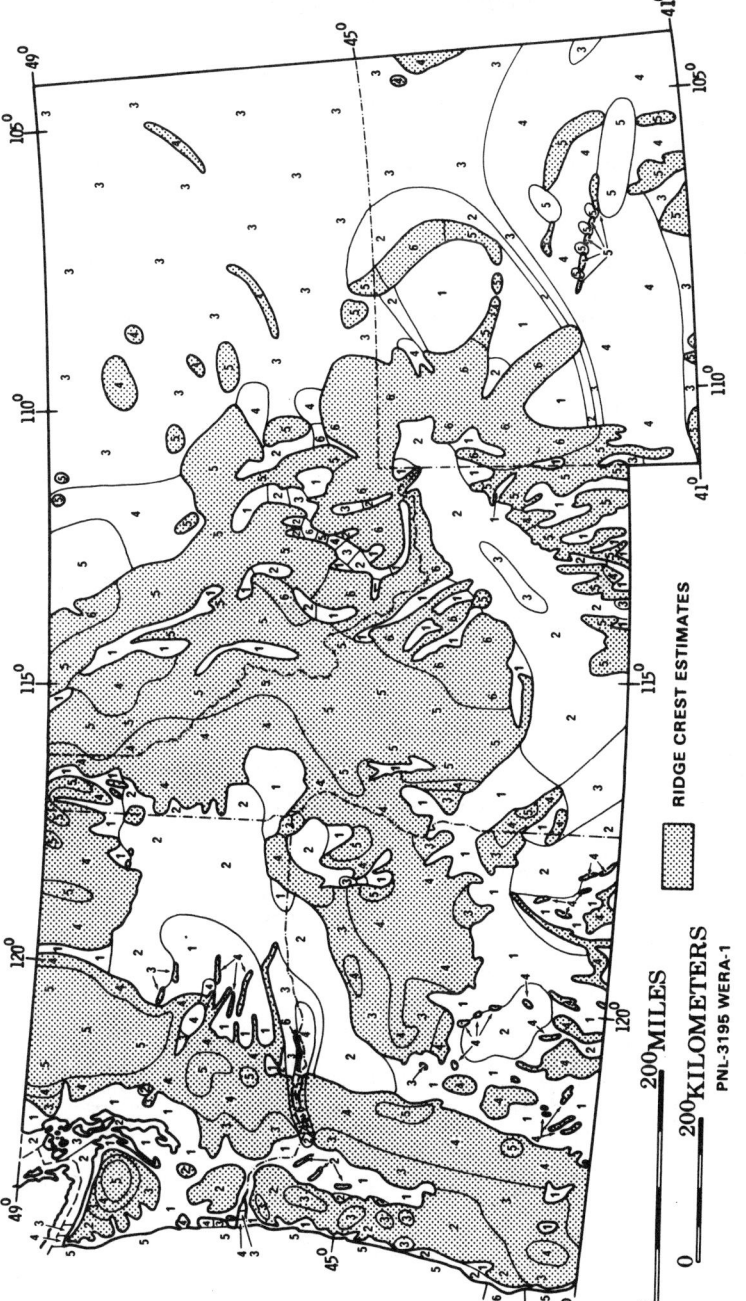

Fig. C-2-1. Average annual wind power in the Northwest.

THE NORTHWEST REGION

Washington

Station	Station Name[a]	Average Annual Speed (m/s)	(mph)	Average Annual Power (W/m²)
Dallesport	The Dalles (Oregon) Airport	4.0	9.0	126
Ellensburg	Ellensburg Bowers Airport	4.5	10.1	215
Hoquiam	Hoquiam Bowerman Airport	5.5	12.3	201
North Head	North Head WBO	6.2	14.0	349
Pasco	Pasco NAS	3.3	7.4	83
Quillayute	Quillayute Airport	3.3	7.4	49
Seattle-Tacoma	Sea-Tac International Airport	4.2	9.4	83
Spokane	Fairchild AFB	3.3	7.4	60
Spokane	Spokane International Airport	4.5	10.1	103
Stampede Pass	Stampede Pass WBO	5.3	11.9	148
Stevenson	Stevenson WBO	7.1	16.1	525
Tatoosh Island	Tatoosh Island WBO	6.1	13.6	335
Whidbey Island	Whidbey Island NAS	4.2	9.4	132

Oregon

Station	Station Name[a]	Average Annual Speed (m/s)	(mph)	Average Annual Power (W/m²)
Eugene	Mahlon Sweet Field	3.8	8.5	66
La Grande	La Grande Airport	4.5	10.0	151
Meacham	Meacham Airport	2.9	6.5	30
Newport	Newport CAA	4.5	10.0	141
North Bend	North Bend Airport	4.8	10.8	136
Ontario	Ontario Airport	3.2	7.2	63
Pendleton	Pendleton Airport	4.6	10.3	126
Portland	Portland International Airport	3.9	8.7	89
Rome	Rome State Airport	3.8	8.5	82
Sexton SMT.	Sexton Summit WBO	4.8	10.8	129
Troutdale	Troutdale Airport	3.9	8.7	116

Idaho

Station	Station Name[a]	Average Annual Speed (m/s)	(mph)	Average Annual Power (W/m²)
Boise	Boise Airport	4.2	9.4	83
Burley	Burley Airport	4.1	9.2	104
Dubois	Dubois Airport	3.7	8.3	80
Gooding	Gooding Airport	4.6	10.3	111
Idaho Falls	Idaho Falls Fanning Field	4.4	9.9	114
Idaho Falls	SPL, 67 km (42 mi) NW of Idaho Falls	3.2	7.2	77
Idaho Falls	SPL, 73 km (46 mi) W of Idaho Falls	3.5	7.8	111
Malad City	Malad City Airport	2.8	6.3	33
Mullan Pass	Mullan Pass CAA	4.4	9.9	95
Pocatello	Pocatello Airport	4.8	10.8	138
Strevell	Strevell CAA	5.1	11.4	150

Montana

Station	Station Name[a]	Average Annual Speed (m/s)	(mph)	Average Annual Power (W/m²)
Billings	Billings Logan International Airport	5.5	12.3	156
Bozeman	Bozeman Airport	3.5	7.8	63
Butte	Silver Bow County Airport	3.4	7.6	70
Cut Bank	Cut Bank Airport	6.1	13.7	284
Dillon	Beaverhead County Airport	4.5	10.1	103
Glasgow	Glasgow AFB	5.0	11.2	165
Glasgow	Glasgow International Airport	5.4	12.1	174
Glasgow	Glasgow WBO (City)	3.8	8.5	76
Great Falls	Malmstrom AFB	4.4	9.9	136
Great Falls	Great Falls International Airport	5.7	12.8	217
Havre	Havre City-County Airport	5.1	11.4	161
Havre	Havre WBO (City)	3.6	8.1	57
Lewiston	Lewiston Airport	4.9	11.0	137
Livingston	Livingston Airport	6.5	14.6	403
Miles City	Miles City Airport	4.5	10.1	108
Superior	Superior Airport	2.2	4.9	13
Whitehall	Whitehall CAA	6.0	13.4	340

Wyoming

Station	Station Name[a]	Average Annual Speed (m/s)	(mph)	Average Annual Power (W/m²)
Casper	Natrona County International Airport	6.2	13.9	245
Cheyenne	Cheyenne Airport	6.0	13.4	211
Douglas	Converse County Airport	5.3	11.9	223
Ft. Bridger	Fort Bridger CAA	6.1	13.7	242
Laramie	Laramie CAA	5.7	12.8	192
Moorcroft	Moorcroft CAA	4.3	9.6	129
Rawlins	Rawlins Airport	5.3	11.9	201
Rock Springs	Rock Springs Airport	5.3	11.9	167
Sinclair	Sinclair CAA	6.7	15.0	424

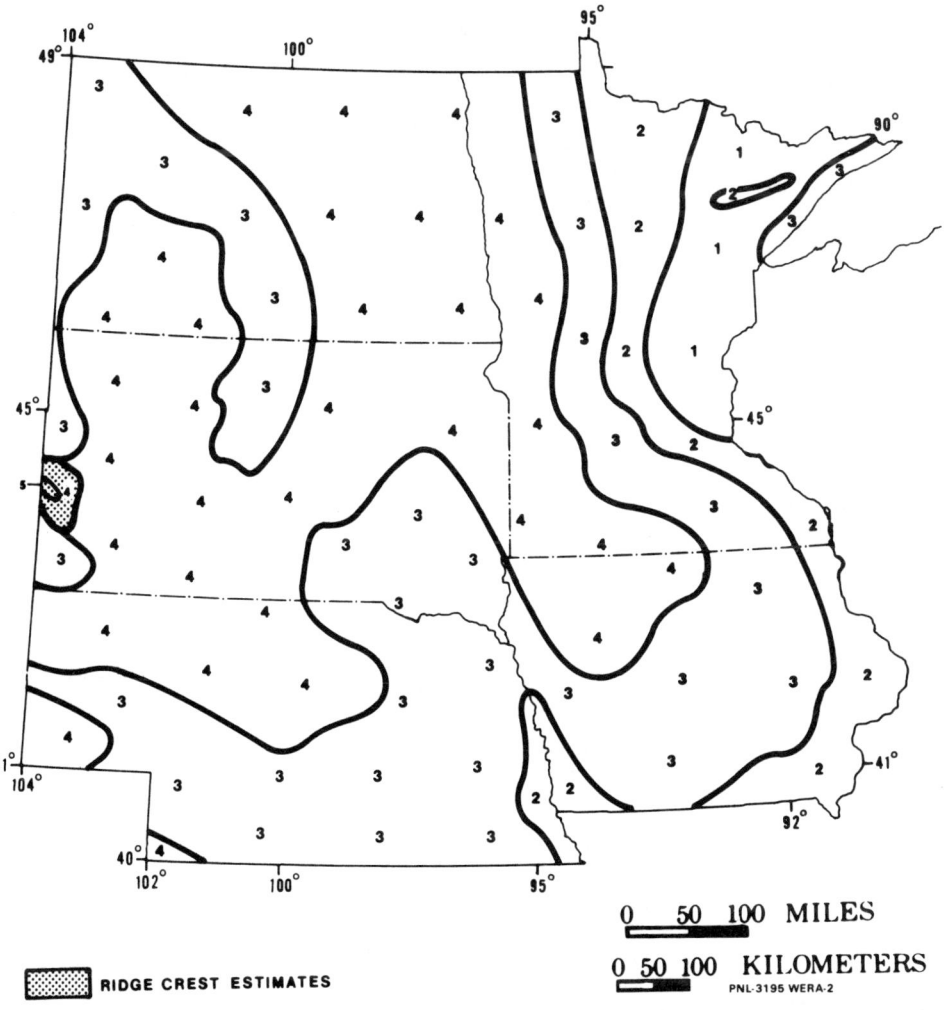

Fig. C-2-2. Average annual wind power in the North Central Region.

THE NORTH CENTRAL REGION

North Dakota

Station	Station Name[a]	Average Annual Speed (m/s)	(mph)	Average Annual Power (W/m²)
Bismarck	Bismarck Airport	4.8	10.8	154
Bismarck	Bismarck Airport	4.7	10.5	150
Dickinson	Dickinson FAA	6.2	13.9	259
Fargo	Fargo, Hector Airport	5.9	13.2	224
Fargo	Fargo, Hector Airport	5.6	12.5	203
Grand Forks	Grand Forks FAA	5.7	12.8	256
Grand Forks	Grand Forks AFB	5.1	11.4	194
Jamestown	Jamestown FAA	5.5	12.3	238
Minot AFB	Minot AFB	5.0	11.2	178
Minot AFB	Minot AFB	5.1	11.4	192
Minot	Minot International Airport	5.8	13.0	234
Minot	Minot International Airport	6.0	13.4	216
Pembina	Pembina CAA	5.8	13.0	243
Williston	Williston WBO[c]	3.6	8.1	56
Williston	Williston, Sloulin International Airport	4.8	10.8	144

South Dakota

Station	Station Name[a]	Average Annual Speed (m/s)	(mph)	Average Annual Power (W/m²)
Aberdeen	Aberdeen Airport	5.4	12.1	215
Aberdeen	Aberdeen Airport	5.4	12.1	184
Huron	Huron Airport	5.3	11.9	182
Huron	Huron Airport	5.5	12.3	192
Philip	Philip Airport	5.2	11.6	202
Pierre	Pierre Airport	4.7	10.5	162
Pierre	Pierre Airport	5.6	12.5	200
Rapid City	Old Rapid City Municipal Airport	5.9	13.2	297
Rapid City	Rapid City Regional Airport	5.0	11.2	185
Rapid City	Rapid City Regional Airport	5.4	12.1	237
Rapid City	Rapid City/Ellsworth AFB	4.8	10.8	221
Sioux Falls[b]	Sioux Falls/Foss Field	4.7	10.5	138
Sioux Falls	Sioux Falls/Foss Field	5.5	12.3	185
Watertown	Watertown Airport	5.3	11.9	213

Kansas

Station	Station Name[a]	Average Annual Speed (m/s)	(mph)	Average Annual Power (W/m^2)
Anthony	Anthony CAA	5.2	11.6	180
Chanute	Martin Johnson Airport	5.5	12.3	164
Concordia	Riosser Airport	5.8	13.0	193
Dodge City	Municipal Airport	6.2	13.9	226
Emporia	Emporia	5.5	12.3	193
Ft. Leavenworth	Sherman Field AAF	3.1	6.9	62
Ft. Riley	Marshall AAF	3.9	8.7	122
Garden City	Garden City CAA	5.3	11.9	224
Goodland	Renner Field Airport	6.2	13.9	218
Hill City	Municipal Airport	5.0	11.2	176
Hutchinson	Hutchinson NAS	5.7	12.8	210
Lebo	Lebo CAA	5.1	11.4	170
Olathe	Olathe NAS	4.4	9.9	117
Russell	Municipal Airport	6.1	13.7	210
Salina	Salina Airport	5.6	12.5	219
Salina	Salina AFB	4.8	10.8	166
Topeka	Billard Airport	4.5	10.1	117
Wichita	Municipal Airport	5.6	12.5	190

Minnesota

Station	Station Name[a]	Average Annual Speed (m/s)	(mph)	Average Annual Power (W/m^2)
Alexandria	Alexandria FAA	5.4	12.1	191
Duluth	Duluth Airport	5.2	11.6	162
Duluth	Duluth Airport	5.0	11.2	134
International Falls	International Falls Airport	4.2	9.4	105
International Falls	International Falls Airport	4.3	9.6	96
Minneapolis	Minneapolis-St. Paul International Airport	4.9	11.0	144
Minneapolis	Minneapolis-St. Paul International Airport	4.8	10.8	122
Redwood Falls	Redwood Falls FAA	5.0	11.2	174
Rochester	Rochester WBAS	4.1	9.2	94
Rochester[b]	Rochester Airport	6.1	13.7	231
St. Cloud[c]	St. Cloud Airport	3.6	8.1	74
St. Cloud[d]	St. Cloud Airport	3.9	8.7	74
St. Paul	St. Paul Downtown Airport	4.6	10.3	117
Willmar	Willmar CAA	5.6	12.5	275

Iowa

Station	Station Name[a]	Average Annual Speed (m/s)	(mph)	Average Annual Power (W/m²)
Burlington	Burlington Airport	4.7	10.5	118
Burlington	Burlington Airport	4.8	10.8	122
Des Moines	Des Moines Airport	4.8	10.8	117
Des Moines	Des Moines Airport	5.0	11.2	142
Dubuque[b]	Dubuque Airport	5.3	11.9	175
Dubuque[c]	Dubuque Airport	5.5	12.3	170
Iowa City	Iowa City SAWR	4.1	9.2	117
Lamoni	Lamoni CAA	5.2	11.6	176
Mason City	Mason City Airport	5.9	13.2	245
Mason City	Mason City Airport	5.3	11.9	171
Ottumwa	Ottumwa Airport	5.2	11.6	166
Sioux City	Sioux City Airport	5.1	11.4	161
Sioux City	Sioux City Airport	5.1	11.4	160
Waterloo	Waterloo Airport	5.1	11.4	163

Wind Data • 347

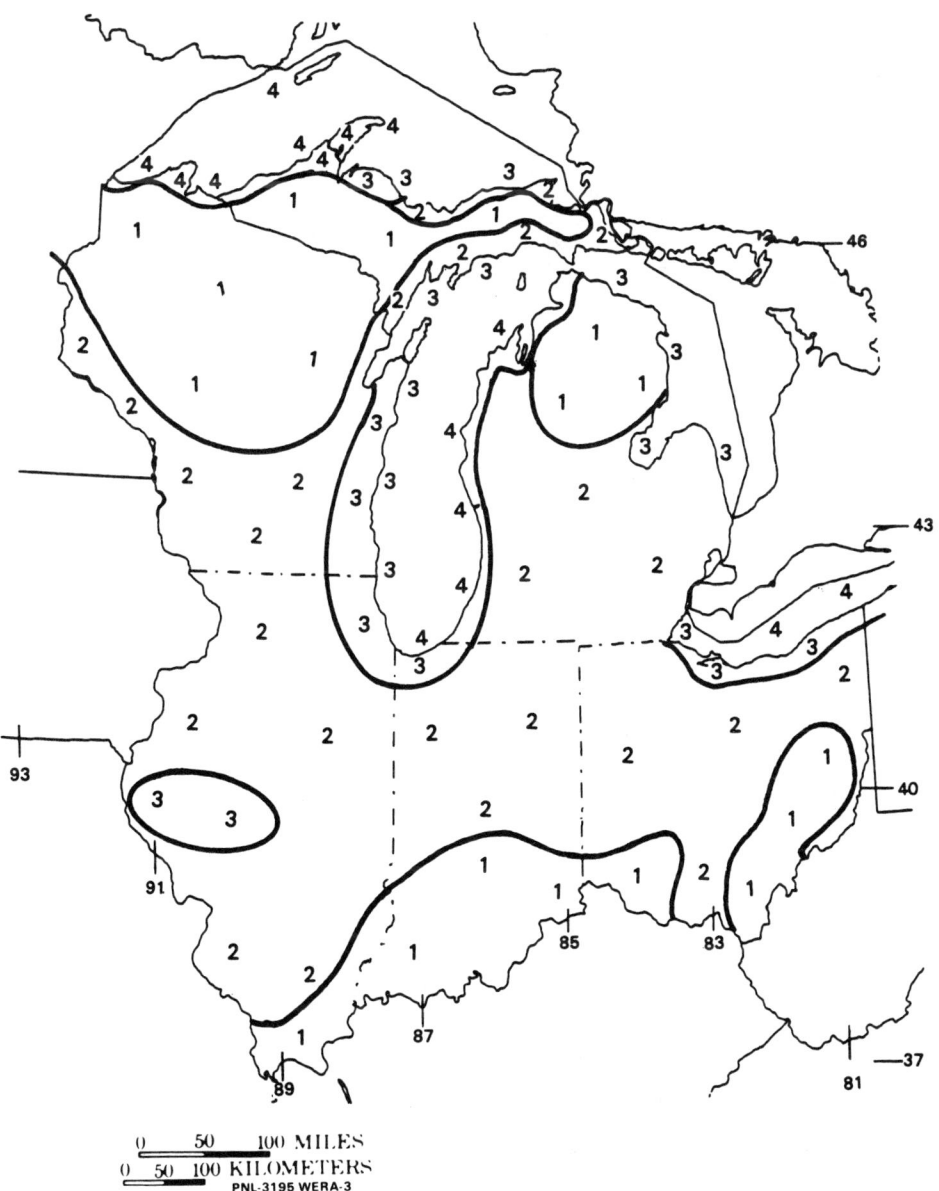

Fig. C-2-3. Average annual wind power in the Great Lakes Region.

THE GREAT LAKES REGION

Wisconsin

Station	Station Name[a]	Average Annual Speed (m/s)	(mph)	Average Annual Power (W/m²)
Camp Douglas	Volk Field ANG	3.5	7.8	74
Eau Claire	Municipal Airport	4.0	9.0	96
Eau Claire	Municipal Airport	4.4	9.9	108
Grantsburg,		3.8	8.5	73
Green Bay	Austin Straubel Airport	4.5	10.1	122
Green Bay	Austin Straubel Airport	4.7	10.5	126
La Crosse	Municipal Airport WBAS	4.3	9.6	107
La Crosse	Municipal Airport WBAS	3.7	8.3	76
Lone Rock	Tri-County Airport WBAS	3.7	8.3	86
Madison	Truax Field WBAS	5.0	8.3	182
Madison	Truax Field WBAS	4.5	11.2	111
Milwaukee	Mitchell Field WBAS	4.9	10.1	138
			11.0	
Milwaukee	Mitchell Field WBAS	5.4	12.1	168
Wausau	Municipal Airport	3.9	8.7	80

Illinois

Station	Station Name[a]	Average Annual Speed (m/s)	(mph)	Average Annual Power (W/m²)
Belleville	Scott AFB	3.2	7.2	72
Bradford	Rinkenberger	4.7	10.5	122
Chicago	Midway Airport WBAS	4.3	9.6	95
Chicago	Midway Airport WBAS	5.1	11.4	146
Chicago	O'Hare International Airport WBAS	4.8	10.8	127
Effingham	CAA	4.9	11.0	118
Glenview	NAS	4.0	9.0	102
Joliet	Municipal Airport WBAS	4.6	10.3	142
Moline	Quad-City Airport WBAS	4.3	9.6	108
Moline	Quad-City Airport WBAS	4.7	10.5	131
Peoria	Greater Peoria Airport	4.8	10.8	124
Quincy	Baldwin Field CAA	4.9	11.0	151
Rantoul	Chanute AFB	4.3	9.6	113
Rockford	Greater Rockford Airport	4.9	11.0	134
Springfield	Capital Airport WBAS	5.1	11.4	166
Springfield	Capital Airport WBAS	5.3	11.9	162
Vandalia		4.6	10.3	126

Indiana

Station	Station Name[a]	Average Annual Speed (m/s)	(mph)	Average Annual Power (W/m²)
Bunker Hill	NAS	4.1	9.2	107
Columbus	Bakalar AFB	3.3	7.4	54
Evansville	Dress Memorial Airport	3.7	8.3	71
Evansville	Dress Memorial Airport	3.7	8.3	72
Ft. Wayne	Baer Field	4.7	10.5	118
Ft. Wayne	Baer Field	4.9	11.0	145
Goshen	CAA	4.2	9.4	99
Indianapolis	Weir Cook WBAS	4.5	10.1	110
Indianapolis	Weir Cook WBAS	4.3	9.6	94
South Bend	St. Joseph County Airport WBAS	4.5	10.1	103
South Bend	St. Joseph County Airport WBAS	4.9	11.0	133
Terre Haute	Hulman Field CAA	4.1	9.2	106
West Lafayette	CAA	4.8	10.8	135

Michigan

Station	Station Name[a]	Average Annual Speed (m/s)	(mph)	Average Annual Power (W/m²)
Alpena	Phelps Collins Field WBAS	4.1	9.2	84
Cadillac	Municipal Airport CAA	4.7	10.5	165
Detroit City	Detroit City Airport WBAS	4.7	10.5	103
Detroit/Wayne	Detroit Metro-Wayne	4.9	11.0	137
Flint	Bishop Airport WBAS	4.8	10.8	126
Gladwin	Municipal Airport	3.2	7.2	59
Grand Marais	Coast Guard	4.4	10.0	103
Grand Rapids	Kent County Airport	4.8	10.8	129
Gwinn	KI/Sawyer AFB	4.3	9.6	123
Houghton	Houghton Airport CAA	4.8	10.8	139
Jackson	Reynolds Airport	4.4	9.9	122
Lansing	Capital City Airport	4.6	10.3	126
Mt. Clemens	Selfridge AFB	4.3	9.6	114
Muskegon	Muskegon County Airport	5.2	11.6	166
Oscoda	Wurtsmith AFB	4.0	9.0	95
Saginaw	Tri-City Airport CAA	4.8	10.8	148
Sault St. Marie	Municipal Airport WBAS	4.2	9.4	94
Traverse City	Cherry Cap CAA	4.2	9.4	91

Ohio

Station	Station Name[a]	Average Annual Speed (m/s)	(mph)	Average Annual Power (W/m²)
Akron/Canton	Akron/Canton Airport	4.6	10.3	103
Cleveland	Hopkins International Airport	4.5	10.1	110
Cleveland	Hopkins International Airport	4.9	11.0	132
Columbus/Lockbourne	Lockbourne AFB	3.5	7.8	75
Columbus	Pt. Columbus Airport	4.3	9.6	99
Dayton/Cox	JM Cox/Dayton	4.7	10.5	119
Dayton/Cox	JM Cox/Dayton	4.6	10.3	107
Dayton/Wright	Wright-Patterson AFB	3.4	7.6	77
Findlay	CAA	4.5	10.1	106
Mansfield	LAHM Municipal Airport	5.1	11.4	155
Mansfield	LAHM Municipal Airport	5.3	11.9	151
Perry	CAA	5.3	11.9	188
Toledo	Express Airport WBAS	4.3	9.6	87
Toledo	WBAS	4.6	10.3	132
Wilmington	Clinton County AFB	3.6	8.1	74
Youngstown	Municipal Airport WBAS	4.8	10.8	116
Zanesville	Municipal Airport CAA	4.1	9.2	115

Fig. C-2-4. Average annual wind power in the Northeast.

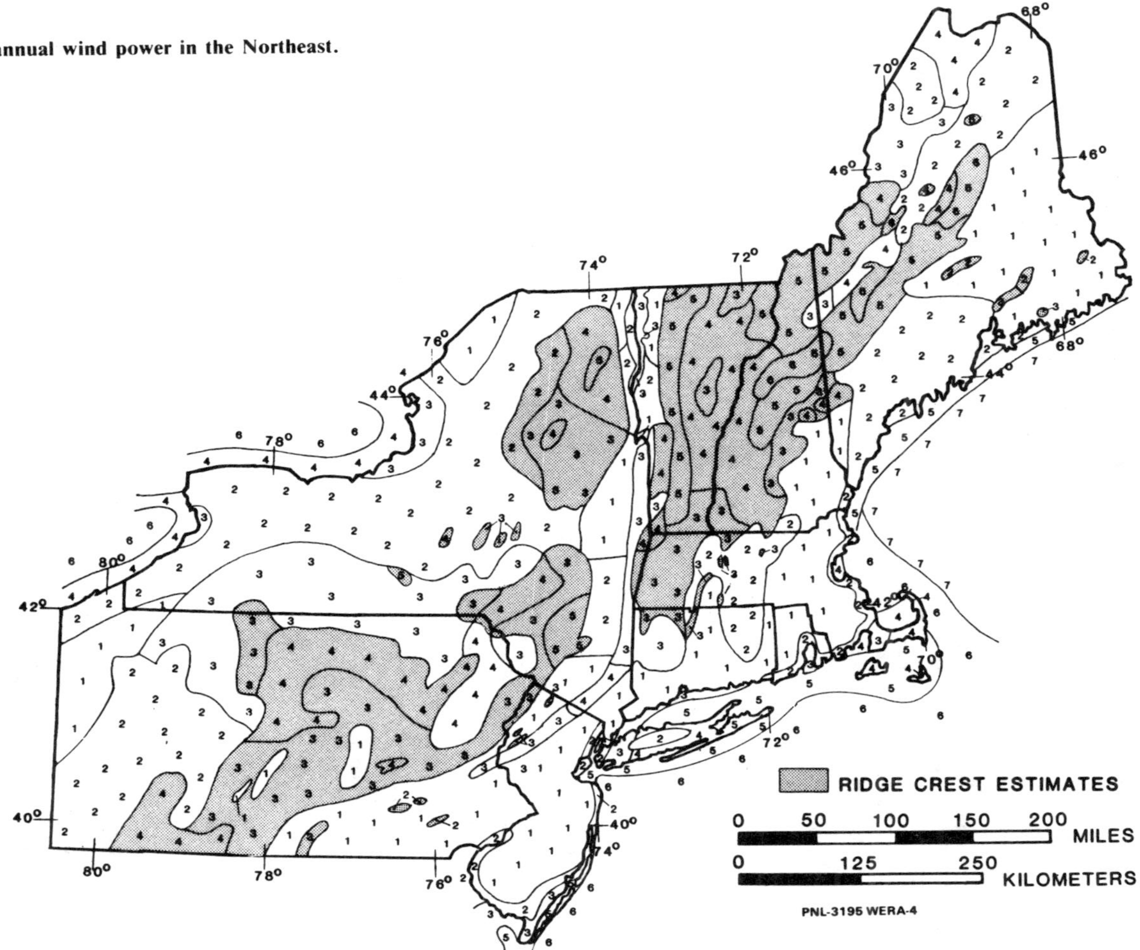

THE NORTHEAST REGION

New York

Station	Station Name[a]	Average Annual Speed (m/s)	(mph)	Average Annual Power (W/m²)
Albany	Albany County Airport	3.8	8.5	79
Albany	Albany County Airport	4.4	9.9	117
Bear Mountain	WBO	5.3	11.9	253
Binghamton	Broome County Airport	4.9	11.0	111
Buffalo	Greater Buffalo International Airport	5.2	11.6	161
Buffalo	Greater Buffalo International Airport	5.6	12.5	196
Dansville	Dansville Municipal Airport	3.7	8.3	59
Dunkirk	FAA	5.3	11.9	209
Elmira	Chemung County Airport	2.6	5.8	38
Geneva	Sampson AFB	5.0	11.2	199
Glens Falls	Warren County Airport	3.5	7.8	71
Hempstead	Mitchell AFB	4.2	9.4	117
JFK	JFK Airport	5.1	11.4	144
JFK	JFK Airport	5.8	13.0	206
LaGuardia	LaGuardia Airport	5.7	12.8	170
Massena	Richards Airport	4.3	9.6	112
Massena	Richards Airport	3.8	8.5	79
Newburgh	Stewart AFB	3.3	7.4	74
NY Shoals	TT4	7.0	15.7	478
Plattsburgh	Plattsburgh AFB	3.4	7.6	77
Rochester	Rochester-Monroe County Airport	4.8	10.8	129
Rochester	Rochester-Monroe County Airport	4.9	11.0	138
Rome	Rome/Griffis AFB	3.2	7.2	72
Syracuse	Syracuse/Hancock Airport	4.0	9.0	79
Syracuse	Syracuse/Hancock Airport	4.7	10.5	130
Utica	Utica/Oneida County Airport	5.0	11.2	166
Watertown	Watertown International Airport	5.0	11.2	223
Watertown	Watertown International Airport	4.7	10.5	133
Westhampton Beach	Suffolk County AFB	4.6	10.3	139
White Plains	Westchester County Airport	3.7	8.3	91

Pennsylvania

Station	Station Name[a]	Average Annual Speed (m/s)	(mph)	Average Annual Power (W/m²)
Allentown	Bethlehem Eastern Airport	4.0	9.0	110
Altoona	Blair County Airport	3.8	8.5	94
Bradford	Bradford Regional Airport	3.1	6.9	41
Bradford	Bradford Regional Airport	3.9	8.7	67
Brookville	Brookville Airport	3.6	8.1	65
Curwensville	Curwensville WBO	4.3	9.6	77
Erie	Erie Airport	4.6	10.3	111
Erie	Erie Airport	5.6	12.5	180
Johnstown	Cambia County Airport	4.8	10.8	122
Philadelphia	Philadelphia International Airport	4.8	10.8	119
Philadelphia	Philadelphia NE Airport	3.6	8.1	68
Philipsburg	Philipsburg Mid State Airport	3.5	7.8	68
Pittsburgh	Allegheny County Airport	4.3	9.6	86
Pittsburgh	Greater Pittsburgh Airport	4.6	10.3	107
Selinsgrove	Selinsgrove Airport	3.0	6.7	53
Wilkes Barre	Wilkes Barre/Scranton Airport	4.1	9.2	71
Williamsport	Lycoming County Airport	3.8	8.5	82

Maine

Station	Station Name[a]	Average Annual Speed (m/s)	(mph)	Average Annual Power (W/m²)
Augusta	Augusta Airport	4.4	9.9	117
Bangor	Dow AFB	3.7	8.3	96
Brunswick	Brunswick Naval Station	3.5	7.8	76
Caribou	Caribou Municipal Airport	5.0	11.2	183
Limestone	Loring AFB	3.8	8.5	89
Millinocket	Millinocket Municipal Airport	3.1	6.9	54
Portland	Portland International Jet Port	3.9	8.7	76
Portland	Portland International Jet Port	4.2	9.4	93

New Hampshire and Vermont

Station	Station Name[a]	Average Annual Speed (m/s)	(mph)	Average Annual Power (W/m²)
Concord, NH	Concord Municipal Airport	2.8	6.3	42
Concord, NH	Concord Municipal Airport	3.4	7.6	72
Lebanon, NH	Lebanon Regional Airport	2.1	4.7	32
Manchester, NH	Grenier Field	2.6	5.8	48
Portsmouth, NH	Pease AFB	3.9	8.7	98
Burlington, VT	Burlington International Airport	4.0	9.0	87
Burlington, VT	Burlington International Airport	4.2	9.4	87
Montpelier, VT	Montpelier Ed. F. Knapp Airport	3.5	7.8	85

Connecticut and Rhode Island

Station	Station Name[a]	Average Annual Speed (m/s)	(mph)	Average Annual Power (W/m²)
Bridgeport, CT	Bridgeport Airport	4.2	9.4	116
Bridgeport, CT	Bridgeport Airport	5.1	11.4	145
Hartford, CT	Brainard Field	3.6	8.1	73
New Haven, CT	Tweed Airport	3.4	7.6	48
New London, CT	New London/ Groton Airport	4.0	9.0	96
Windsor Locks, CT	Bradley International	3.9	8.7	77
Block Island, RI	WBO	7.6	17.0	537
Providence, RI	T.F. Green State Airport	5.4	12.1	178
Providence, RI	T.F. Green State	5.1	11.4	143
Quonset Point RI	Quonset Point Naval Station	4.5	10.1	124

Massachusetts

Station	Station Name[a]	Average Annual Speed (m/s)	(mph)	Average Annual Power (W/m²)
Bedford	L.G. Hanscom Field	3.6	8.1	84
Boston	Logan Airport	5.9	13.2	216
Boston	Logan Airport	5.9	13.2	207
Chicopee Falls	Westover AFB	3.1	6.9	68
Falmouth	Otis AFB	5.0	11.2	198
George's Shoals	Texas Tower 2	7.2	16.1	441
Nantucket	Nantucket Municipal Airport	6.5	14.6	297
Nantucket Shoals	Texas Tower 3	6.8	15.2	398
S. Weymouth	S. Weymouth Navy Station	3.7	8.3	81
Westfield	Barnes Airport	3.4	7.6	57
Worcester	Worcester Airport	6.1	13.7	278
Worcester	Worcester Airport	4.8	10.8	128

New Jersey

Station	Station Name[a]	Average Annual Speed (m/s)	(mph)	Average Annual Power (W/m²)
Atlantic City	Atlantic City Airport	5.0	11.2	152
Fort Dix	McGuire AFB	3.4	7.6	75
Lakehurst	Lakehurst Naval	3.9	8.7	79
Millville	Millville Municipal Airport	3.7	8.3	89
Newark	Newark International Airport	4.3	9.6	94
Newark	Newark International Airport	5.0	11.2	128
Teterboro	Teterboro Airport	3.9	8.7	100

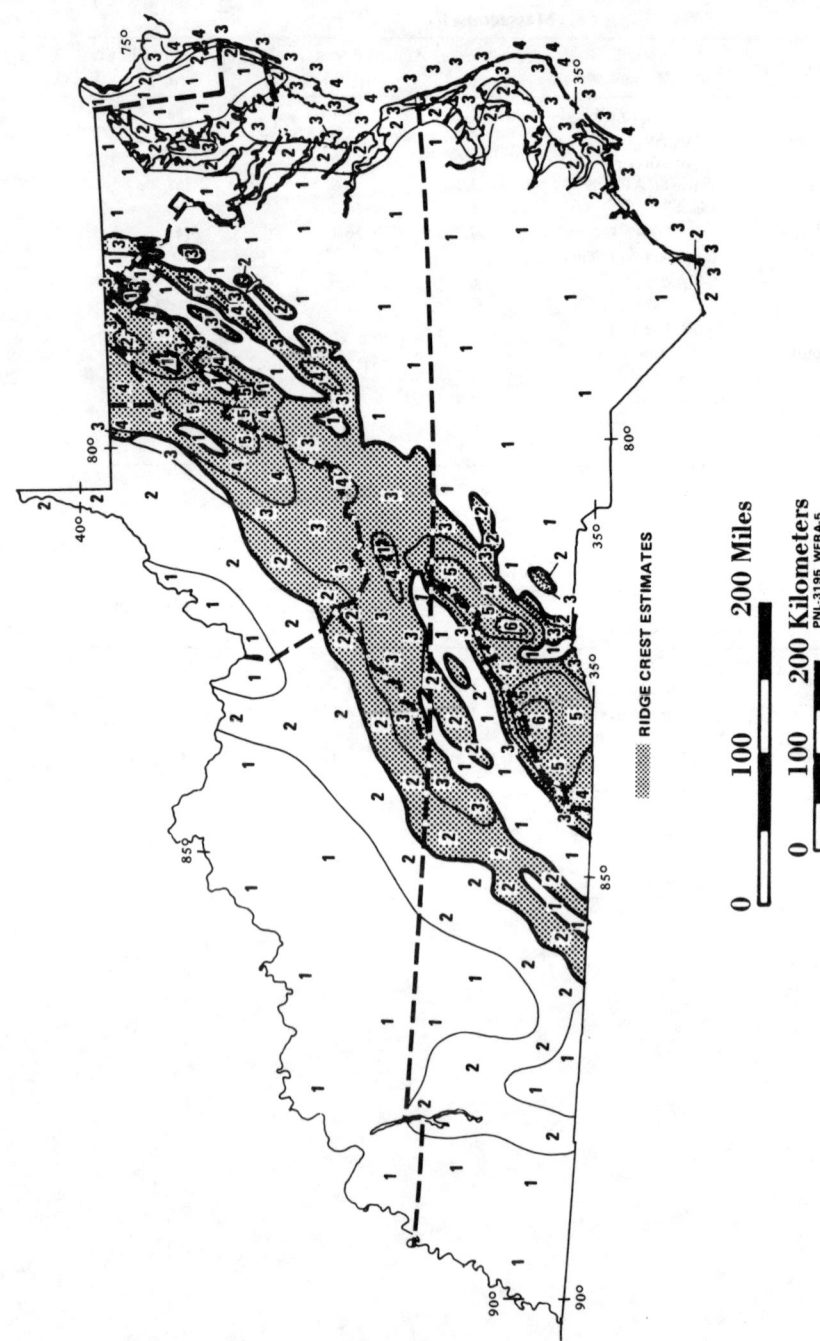

Fig. C-2-5. Average annual wind power in the East Central Region.

THE EAST CENTRAL REGION

Tennessee

Station	Station Name[a]	Average Annual Speed (m/s)	(mph)	Average Annual Power (W/m^2)
Bristol	Bristol Tri-City Airport	2.6	5.8	39
Chattanooga	Chattanooga Lovell Field	2.9	6.5	46
Dyersburg	Dyersburg AAF	3.6	8.1	68
Jackson	Jackson FSS	3.6	8.1	60
Knoxville	Knoxville McGhee-Tyson Airport	3.2	7.2	50
Memphis	Memphis Naval Air Station	3.6	8.1	87
Memphis	Memphis International Airport	4.2	9.4	90
Nashville	Nashville Berry Field	4.0	9.0	76
Smithville	Smithville CAA	3.3	7.4	52
Smyrna	Smyrna Sewart AFB	2.3	5.2	39
Tullahoma	Tullahoma Northern Field	2.1	4.7	57

Kentucky

Station	Station Name[a]	Average Annual Speed (m/s)	(mph)	Average Annual Power (W/m^2)
Bowling Green	Bowling Green City-County Airport	3.1	6.9	60
Corbin	Corbin CAA	2.3	5.2	29
Covington/Cin.	Greater Cincinnati Airport	4.2	9.4	88
Fort Campbell	Fort Campbell AFB	2.7	6.0	46
Fort Knox	Fort Knox Godman AAF	3.1	6.9	61
Lexington	Lexington Blue Grass Field	4.1	9.2	76
London	London Municipal Airport	2.7	6.0	35
Louisville	Louisville Standiford Field	4.0	9.0	75
Paducah	Paducah Barkley Airport	3.6	8.1	75

West Virginia

Station	Station Name[a]	Average Annual Speed (m/s)	(mph)	Average Annual Power (W/m²)
Beckley	Beckley-Raleigh Memorial County Airport	4.0	9.0	75
Charleston	Charleston Kanawha Airport	2.8	6.3	38
Elkins	Elkins-Randolph County Airport	3.1	6.9	54
Huntington	Huntington Chesapeake Airport	1.4	3.1	17
Huntington/TS	Huntington Tri-State Airport	3.1	6.9	42
Martinsburg	Martinsburg Municipal Airport	2.9	6.5	70
Morgantown	Morgantown Municipal Airport	3.2	7.2	48
Parkersburg	Parkersburg WBO	2.9	6.5	45
Petersburg	Petersburg WBO	2.6	5.8	31
Wheeling	Wheeling Ohio County Airport	3.8	8.5	80

Delaware and Maryland

Station	Station Name[a]	Average Annual Speed (m/s)	(mph)	Average Annual Power (W/m²)
Dover, DE	Dover AFB	3.9	8.7	102
Wilmington, DE	Greater Wilmington/ New Castle Airport	4.5	10.1	112
Aberdeen, MD	Aberdeen Phillips Field	3.6	8.1	86
Annapolis, MD	Annapolis Naval Air Facility	3.5	7.8	62
Baltimore, MD	Baltimore Dunkirk Airport	4.8	10.8	150
Baltimore, MD	Baltimore Friendship International Airport	4.2	9.4	94
Frederick, MD	Camp Detrick AF	2.7	6.0	70
Fort Meade, MD	Fort Meade Tipton AAF	2.4	5.4	46
Patuxent River, MD	Patuxent River Naval Air Station	3.8	8.5	86
Salisbury, MD	Salisbury Wicomico County Airport	3.8	8.5	84
Wash/Andrews, MD	Washington, D.C. Andrews AFB	4.0	9.0	118
Wash/Bolling, MD	Washington, D.C. Bolling Field	2.9	6.5	56

Virginia

Station	Station Name[a]	Average Annual Speed (m/s)	(mph)	Average Annual Power (W/m²)
Blackstone	Blackstone CAA	3.5	7.8	55
Charlottesville	Charlottesville Albemarle Airport	2.9	6.5	35
Chincoteague	Chincoteague Naval Air Station	4.8	10.8	126
Danville	Danville Municipal Airport	3.2	7.2	47
Davison	Davison AAF	2.1	4.7	42
Fort Eustis	Fort Eustis Felker AAF	3.2	7.2	50
Front Royal	Front Royal CAA	3.6	8.1	89
Gordonsville	Gordonsville CAA	2.9	6.5	49
Hampton	Hampton Langley AFB	4.2	9.4	123
Lynchburg	Lynchburg Municipal Airport	3.4	7.6	51
Norfolk	Norfolk Naval Air Station	4.2	9.4	110
Norfolk	Norfolk Regional Airport	4.9	11.0	138
Oceana	Oceana Naval Air Station	3.5	7.8	77
Pulaski	Pulaski New River Airport	4.0	9.0	87
Quantico	Quantico Marine Corps Air Station	2.8	6.3	43
Richmond	Richmond Byrd Field	3.5	7.8	53
Roanoke	Roanoke Woodrum Field	3.9	8.7	78
Urbanna	Urbanna Aviation Reports	3.5	7.8	49
Wallops Island	Wallops Island WBO	5.0	11.2	155
Wash/Dulles	Washington, D.C. Dulles Int. Airport	3.6	8.1	79
Wash/National	Washington, D.C. National Airport	4.3	9.6	95

North Carolina

Station	Station Name[a]	Average Annual Speed (m/s)	(mph)	Average Annual Power (W/m²)
Asheville	Asheville Municipal Airport	3.8	8.5	86
Cape Hatteras	Cape Hatteras WBO	5.1	11.4	137
Charlotte	Charlotte Douglas Municipal Airport	3.5	7.8	54
Cherry Point	Cherry Point Marine Corps Air Station	3.3	7.4	63
Elizabeth City	Elizabeth City FSS	4.2	9.4	81
Fayetteville	Fayetteville Pope AFB	2.4	5.4	40
Fort Bragg	Fort Bragg Simmons Field	2.2	4.9	28
Goldsboro/S/J	Seymour-Johnson AFB	2.8	6.3	49
Greensboro	Greensboro-High Point Airport	3.5	7.8	55
Hatteras	Hatteras WBO	5.3	11.9	165
Hickory	Hickory Municipal Airport	2.6	5.8	35
Jacksonville	New River Marine Corps Air Field	2.8	6.3	46
Lumberton	Lumberton CAA	2.7	6.1	34
New Bern	New Bern FSS	3.7	8.3	56
Raleigh/R-D	Raleigh-Durham Airport	3.9	8.7	68
Rocky Mount	Rocky Mount Municipal Airport	3.1	6.9	48
Wilmington	Wilmington-New Hanover Airport	3.9	8.7	75
Winston-Salem	Smith Reynolds Airport	3.6	8.1	78

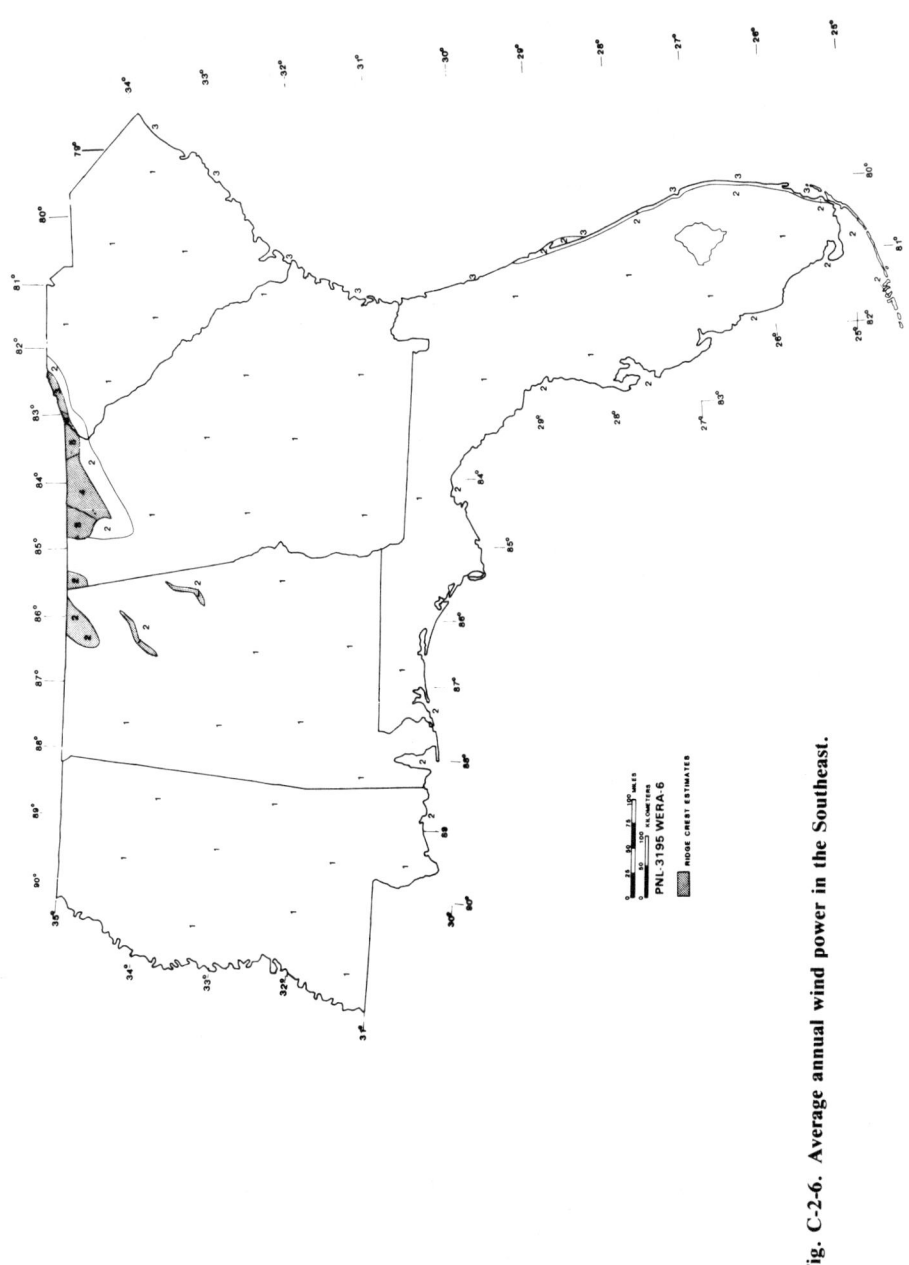

Fig. C-2-6. Average annual wind power in the Southeast.

THE SOUTHEAST REGION

Mississippi

Station	Station Name[a]	Average Annual Speed (m/s)	(mph)	Average Annual Power (W/m²)
Biloxi	Biloxi Airport[b]	3.5	7.8	70
Columbus	Columbus AFB	2.5	5.6	43
Greenville	Greenville AFB	3.0	6.7	51
Greenwood	Greenwood Airport	4.3	9.6	110
Jackson	A.C. Thompson Field	3.6	8.1	67
McComb	McComb Airport	3.1	6.9	44
Meridian	Key Field	2.8	6.3	42

Alabama

Station	Station Name[a]	Average Annual Speed (m/s)	(mph)	Average Annual Power (W/m²)
Anniston	Anniston/Calhoun County Airport	2.8	6.3	45
Birmingham	Birmingham Municipal Airport	3.1	6.9	50
Dothan	Dothan Airport	3.4	7.6	57
Evergreen	Evergreen Airport	2.5	5.6	33
Cairns	Ft. Rucker Field	2.6	5.8	37
Huntville	Huntville/Madison County Airport	3.7	8.3	74
Mobile	Bates Field	4.0	8.9	78
Mobile	Brookley AFB	3.3	7.4	58
Montgomery	Dannelly Field	3.2	7.2	50
Muscle Shoals	Muscle Shoals Airport	3.3	7.4	57
Selma	Craig AFB	2.6	5.8	34
Tuscaloosa	Tuscaloosa Municipal Airport	2.5	5.6	33

Georgia

Station	Station Name[a]	Average Annual Speed (m/s)	(mph)	Average Annual Power (W/m²)
Albany	Albany NAS	2.3	5.2	30
Alma	Bacon County Airport	2.9	6.5	45
Athens	Athens Municipal Airport	3.4	7.6	51
Atlanta	Hartsfield Airport	4.0	9.0	69
Augusta	Bush Field	3.2	7.2	51
Brunswick	Glynco NAS	2.8	6.3	41
Columbus	Columbus Metropolitan Airport	3.2	7.2	46
Macon	Lewis B. Wilson Airport	3.4	7.6	56
Marietta	Dobbins AFB	2.5	5.6	40
Rome[b]	R.B. Russel Field	2.8	6.3	40
Savannah	Savannah Municipal Airport	3.6	8.1	61
Valdosta	Valdosta Municipal Airport	3.2	7.2	46
Warner Robins	Robins AFB	2.5	5.6	42

South Carolina

Station	Station Name[a]	Average Annual Speed (m/s)	(mph)	Average Annual Power (W/m²)
Anderson	Anderson County Airport	3.4	7.6	68
Beaufort	Marine Corps Air Station	2.8	6.3	43
Charleston	Charleston International Airport	4.1	9.2	83
Columbia	Columbia Airport	3.2	7.2	48
Eastover	McEntire Air National Guard Base	2.7	6.0	42
Florence	City-County Airport	3.7	8.3	57
Greenville	Greenville-Spartanburg Jet Age Airport	3.2	7.2	45
Myrtle Beach	Myrtle Beach AFB	3.3	7.4	62
Spartanburg	Downtown Memorial Airport	3.4	7.6	60
Sumter	Shaw AFB	2.8	6.3	45

Florida

Station	Station Name[a]	Average Annual Speed (m/s)	(mph)	Average Annual Power (W/m^2)
Avon Park[b]	Avon Park AFB	2.4	5.4	31
Cape Kennedy	Cape Kennedy AFB	3.4	7.6	52
Cocoa Beach	Patrick AFB	4.5	10.0	113
Daytona Beach	Daytona Regional Airport	4.0	8.9	83
Ft. Myers	Page Field	3.9	8.7	74
Homestead	Homestead AFB	3.5	7.8	73
Jacksonville	Jacksonville International Airport	3.4	7.6	64
Key West	Key West International Airport	5.2	11.6	132
Miami	Miami International	4.4	9.9	95
Milton	Whiting NAS	2.7	6.0	40
Orlando	Herndon Field	4.2	9.4	84
Panama City	Tyndall AFB	3.7	8.3	77
Pensacola[b]	Saufley Field	3.6	8.1	72
Tallahassee	Tallahassee Municipal Airport	3.0	6.7	43
Tampa	Tampa International Airport	3.9	8.7	64
Valparaiso	Eglin Field	3.3	7.4	54
West Palm Beach	Palm Beach International Airport	4.5	10.1	112

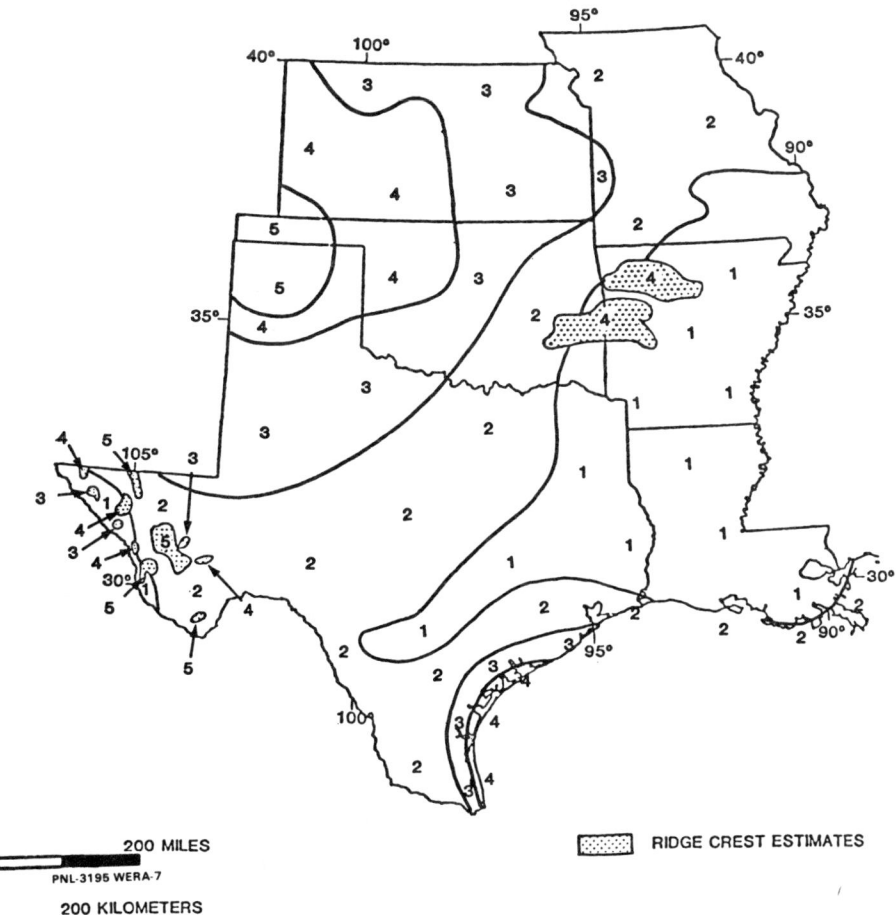

Fig. C-2-7. Average annual wind power in the South Central Region.

THE SOUTH CENTRAL REGION

Nebraska

Station	Station Name[a]	Average Annual Speed (m/s)	(mph)	Average Annual Power (W/m²)
Chadron	Chadron Airport	5.1	11.4	204
Grand Island	Grand Island Airport	5.3	11.9	156
Grand Island	Grand Island Airport	5.6	12.5	189
Hayes Center	Hayes Center	5.2	11.6	187
Imperial	Imperial Airport	4.7	10.5	146
Lexington	Lexington Airport	5.4	12.1	182
Lincoln	Lincoln AFB	4.1	9.2	130
Lincoln	Lincoln Airport	5.0	11.2	164
Norfolk[b]	Stefan Memorial Airport	5.2	11.6	174
North Platte	Lee Bird Airport	4.9	11.0	157
North Platte	Lee Bird Airport	4.6	10.3	131
Omaha	Eppley Field	4.6	10.3	121
Omaha	Eppley Field	4.7	10.5	137
Omaha/Offutt	Offutt AFB	4.0	9.0	115
Scottsbluff	Scottsbluff Airport	4.8	10.8	136
Scottsbluff	Scottsbluff Airport	5.0	11.2	152
Sidney	Sidney FSS	5.5	12.3	184
Valentine[b]	Miller Field	4.0	9.0	104

Oklahoma

Station	Station Name[a]	Average Annual Speed (m/s)	(mph)	Average Annual Power (W/m²)
Altus	Altus AFB	4.7	10.5	159
Ardmore	Ardmore CAA	4.4	9.9	118
Clinton	Clinton-Sherman AFB	5.8	13.0	235
Enid	Vance AFB	5.6	12.5	226
Enid	Enid Airport	4.5	10.0	118
Ft. Sill	Ft. Sill ASC	4.3	9.6	134
Gage	Gage Airport	5.7	12.8	216
Gene Autrey	Ardmore AFB	3.3	7.4	73
Hobart	Hobart Airport	5.8	13.0	215
Oklahoma City	Oklahoma City Airport	6.0	13.4	244
Oklahoma City	Will Rogers Airport	5.5	12.3	179
Ponca City	Ponca City Airport	5.7	12.8	190
Tulsa	Tulsa Airport	5.0	11.2	141

Texas

Station	Station Name[a]	Average Annual Speed (m/s)	Average Annual Speed (mph)	Average Annual Power (W/m²)
Abilene	Municipal Airport	5.5	12.3	166
Abilene/Dyess	Dyess AFB	4.4	9.9	119
Alice	Alice Airport	5.0	11.2	159
Amarillo	English Field Airport	6.6	14.8	252
Amarillo	English Field Airport	5.5	12.3	156
Austin	Mueller Airport	4.4	9.9	99
Beeville	Chase Field NAAS	3.9	8.7	94
Big Spring	Big Spring SAWR	5.2	11.6	162
Brownsville	Rio Grande International Airport	5.4	12.1	178
Bryan	Bryan AFB	3.4	7.6	64
Childress	Childress Airport	5.2	11.6	159
Clarendon	Clarendon Airport	5.6	12.5	186
Corpus Christi	Corpus Christi NAS	5.7	12.3	200
Corpus Christi	Corpus Christi Airport	5.7	12.8	194
Cotulla	Municipal Airport	4.3	9.6	102
Dalhart	Dalhart Airport	6.1	13.7	251
Dallas	Dallas NAS	4.0	9.0	90
Dallas/Love	Love Field WBAS	5.4	12.1	152
Del Rio	International Airport	4.7	10.5	99
El Paso/Biggs	Biggs AFB	3.4	7.6	86
El Paso	International Airport	3.6	8.1	63
Ft. Hood	Gray AFB	5.0	11.2	146
Ft. Worth	Carswell AFB	4.3	9.6	125
Ft. Worth	Greater SW International Airport	4.7	10.5	121
Galveston	Scholes Field Airport	5.4	12.1	165
Guadalupe Pass	Guadalupe Pass CAA	8.2	18.4	695
Harlingen	Harlingen AFB	4.3	9.6	118
Houston	Intercontinental Airport	3.6	8.1	62
Houston/Hobby	Hobby Airport	4.7	10.5	126
Junction	Junction Airport	3.2	7.2	51
Kingsville	Kingsville NAAS	4.1	9.2	104
Laredo	Municipal Airport	5.2	11.6	141
Longview	Longview Airport	4.3	9.6	85
Lubbock	West Texas AT	5.4	12.1	162
Lufkin	Angelina County Airport	3.5	7.8	56
Marfa	Marfa Airport	4.6	10.3	132
Matagorda Is.	Matagorda Is.[b] Air Field	5.5	12.3	178
Midland	Mid-Odessa RAT	5.5	12.3	158
Mineral Wells	Municipal Airport	4.8	10.8	127
Palacios	Palacios Airport	4.9	11.0	123
Port Arthur	Jefferson County Airport	4.6	10.3	111
Port Isabel	Port Isabel[b] NAAS	6.2	13.9	254
Salt Flat	Salt Flat CAA	3.6	8.1	98
San Angelo	Mathis Field Airport	4.8	10.8	119
San Antonio	International Airport	4.5	10.1	92
Sherman	Perrin AFB	4.7	10.5	137
Tyler	Tyler Airport	4.3	9.6	103
Victoria	Victoria Airport	4.8	10.8	126
Waco	Municipal Airport	5.0	11.2	139
Waco/Connally	James Connally AFB	4.0	9.0	100
Wichita Falls	Wichita Falls Airport	5.6	12.5	191
Wink	Winkler County Airport		9.6	111

Missouri

Station	Station Name[a]	Average Annual Speed (m/s)	(mph)	Average Annual Power (W/m²)
Advance	Advance CAA	3.7	8.3	74
Butler	Butler CAA	4.9	11.0	159
Chillicothe	Chillicothe CAA	4.8	10.8	127
Columbia	Municipal Airport	4.4	9.9	93
Farmington	Farmington CAA	3.7	8.3	69
Ft. Leonardwood	Ft. Leonardwood AAF	3.7	8.3	86
Grandview	Richards-Gebaur AFB	4.1	9.2	92
Joplin	Joplin Airport	5.2	11.6	159
Kansas City	Municipal Airport	4.6	10.3	99
Kirksville	Cannon Memorial Airport	4.6	10.3	104
Malden	Malden CAA	4.1	9.2	95
Marshall	Marshall CAA	4.1	9.2	88
New Florence	New Florence CAA	5.2	11.6	129
St. Joseph	St. Joseph Airport	4.4	9.9	120
St. Louis	Lambert Field Airport	4.6	10.3	114
Springfield	Municipal Airport	4.7	10.5	109
Tarkio	Tarkio CAA	3.5	7.8	99
Vichy	Vichy Airport	4.2	9.4	77

Arkansas

Station	Station Name[a]	Average Annual Speed (m/s)	(mph)	Average Annual Power (W/m²)
Blytheville	Blytheville AFB	3.4	7.6	78
El Dorado	Goodwin Airport	3.1	6.9	48
Fayetteville	Fayetteville Airport	3.4	7.6	78
Flippin	Flippin CAA	2.7	6.0	42
Fort Smith	Municipal Airport	3.5	7.8	54
Harrison	Harrison Airport	4.2	9.4	82
Jacksonville	Little Rock AFB	2.8	6.3	46
Little Rock	Adams Field Airport	3.8	8.5	66
Pine Bluff	Grider Field Airport	3.3	7.4	59
Texarkana	Webb Field Airport	3.5	7.8	58
Texarkana	Webb Field Airport	4.0	9.0	88
Walnut Ridge	Municipal Airport	3.3	7.4	64

Louisiana

Station	Station Name[a]	Average Annual Speed (m/s)	(mph)	Average Annual Power (W/m^2)
Alexandria	Esler Field Airport	3.0	6.7	52
Baton Rouge	Ryan Field Airport	3.7	8.3	67
Boothville	Boothville WBO	4.2	9.4	96
Burrwood	Burrwood SPL	4.8	10.8	139
Ft. Polk[b]	Ft. Polk AAF	3.0	6.7	45
Lafayette	Municipal Airport	3.4	7.6	59
Lake Charles	Lake Charles Airport	4.1	9.2	87
Lake Charles	Lake Charles Airport	3.8	8.5	88
Monroe	Monroe Airport	3.4	7.6	65
New Iberia	New Iberia Airport	1.9	4.3	17
New Orleans	Moisant Int'l. Airport	3.8	8.5	72
New Orleans	New Orleans NAS	2.7	6.0	45
Shreveport	Municipal Airport	3.7	8.3	62
Shreveport	Barksdale AFB	2.9	6.5	52

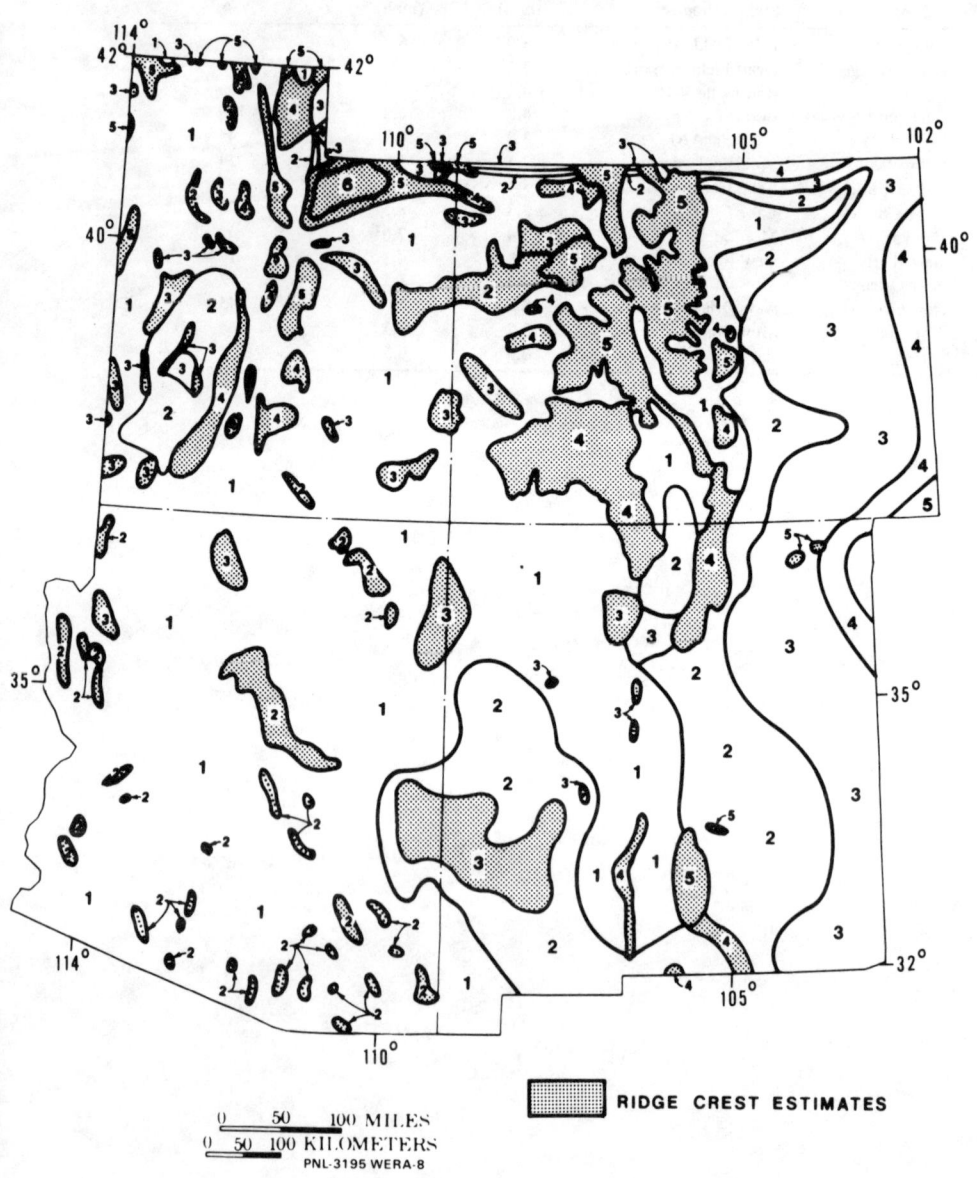

Fig. C-2-8. Average annual wind power in the Southern Rocky Mountain Region.

THE SOUTHERN ROCKY MOUNTAIN REGION

Utah

Station	Station Name[a]	Average Annual Speed (m/s)	(mph)	Average Annual Power (W/m²)
Bryce Canyon	Bryce Canyon Airport	3.3	7.4	53
Cedar City	Cedar City Airport	4.0	9.0	100
Delta	Delta Airport	3.8	8.5	99
Dugway Proving Ground	Michaels AFB	2.6	5.8	56
Fairfield	Fairfield CAA	3.4	7.6	81
Hanksville	Hanksville Airport	2.7	6.0	56
Lucin	Lucin CAA	4.1	9.2	90
Milford[b]	Milford Airport	4.0	9.0	111
		5.6[b]	12.5	197[b]
Ogden	Ogden CAA	4.0	9.0	79
Ogden	Hill AFB	3.9	8.7	89
Salt Lake City	Salt Lake City International Airport	3.8	8.5	64
		4.2	9.4	90
St. George	St. George CAA	3.1	6.9	56
Wendover	Wendover Airport	3.3	7.8	72

Arizona

Station	Station Name[a]	Average Annual Speed (m/s)	(mph)	Average Annual Power (W/m²)
Chandler	Williams AFB	2.4	5.4	27
Flagstaff	Flagstaff Airport	3.5	7.8	52
Fort Huachuca	Libby AAB	3.0	6.7	45
Gila Bend	Gila Bend Airport	3.7	8.3	57
Payson	Payson Airport	2.5	5.6	25
Phoenix	Skyharbor International Airport	2.3	5.2	26
Phoenix[b]	Litchfield NAF	2.3	5.2	24
Phoenix	Luke AFB	2.2	4.9	27
Prescott	Prescott Airport	4.0	9.0	69
Tucson	Tucson International Airport	3.1	6.9	48
		4.3	9.6	87
Tucson	Davis-Monthan AFB	2.9	6.5	54
Winslow	Winslow Airport	3.6	8.1	80
		4.2	9.4	111
Yuma	Yuma International Airport	3.7	8.3	70
		3.7	8.3	68

Colorado

Station	Station Name[a]	Average Annual Speed (m/s)	(mph)	Average Annual Power (W/m²)
Akron	Akron CAA	6.0	13.4	234
Alamosa[b]	Alamosa Airport	4.4	9.9	131
Aurora	Buckley Field	3.7	8.3	65
Colorado Springs	Peterson Field	4.8	10.8	112
Denver[b,c]	Aurora NAS	3.7	8.3	58
Denver	Stapleton International Airport	4.5	10.1	93
Eagle	Eagle County Airport	2.4	5.4	40
Fort Carson[b]	Butts AFB	4.6	10.3	151
Grand Junction	Grand Junction Airport	3.7	8.3	62
La Junta	La Junta Airport	3.9	8.7	97
Pueblo	Pueblo Memorial Airport	3.6	8.1	90
		4.5	10.1	129
Trinidad	Los Animas County Airport	4.5	10.1	122
U.S. Air Force Academy	USAF Academy	4.4	9.9	101

New Mexico

Station	Station Name[a]	Average Annual Speed (m/s)	(mph)	Average Annual Power (W/m²)
Acomita	Acomita Airport	4.6	10.3	97
Alamogordo	Holloman AFB	3.2	7.2	57
Albuquerque	Albuquerque International Airport	3.8	8.5	76
		4.1	9.2	85
Carlsbad	Carlsbad Airport	5.0	11.2	193
Clayton	Clayton Airport	6.6	14.8	280
Clovis	Cannon AFB	5.0	11.2	130
Clovis[b]	Clovis Airport	4.9	11.0	158
Columbus	Hacienda Sur Luna Airport	4.0	9.0	101
El Morro[c]	El Morro CAA	2.9[c]	6.5	50[c]
Farmington	Farmington Airport	3.2	7.2	54
		3.9[d]	8.7	89[d]
Gallup	Gallup Airport	3.3	7.4	82
Hobbs	Hobbs Airport	5.3	11.9	163
Las Cruces	White Sands AF	3.1	6.9	92
Las Vegas	Las Vegas Airport	5.1	11.4	168
Melrose[e]	Melrose Air Force Range	5.4	12.1	165
Otto	Otto CAA	3.4	7.6	82
Raton	Crews Airport	4.3	9.6	151
Roswell	Roswell WBAS (city)	4.7	10.5	148
Roswell	Walker AFB	4.0	9.0	106
Roswell	Roswell FAA	4.2	9.4	91
Santa Fe	Santa Fe Airport	5.3	11.9	158
Silver City[b]	Grant County Airport	5.4	12.1	150
Truth or Consequences	Truth or Consequences Airport	4.6	10.3	129
Tucumcari	Tucumcari Airport	5.3	11.9	168
Zuni	Black Rock Airport	4.3	9.6	98

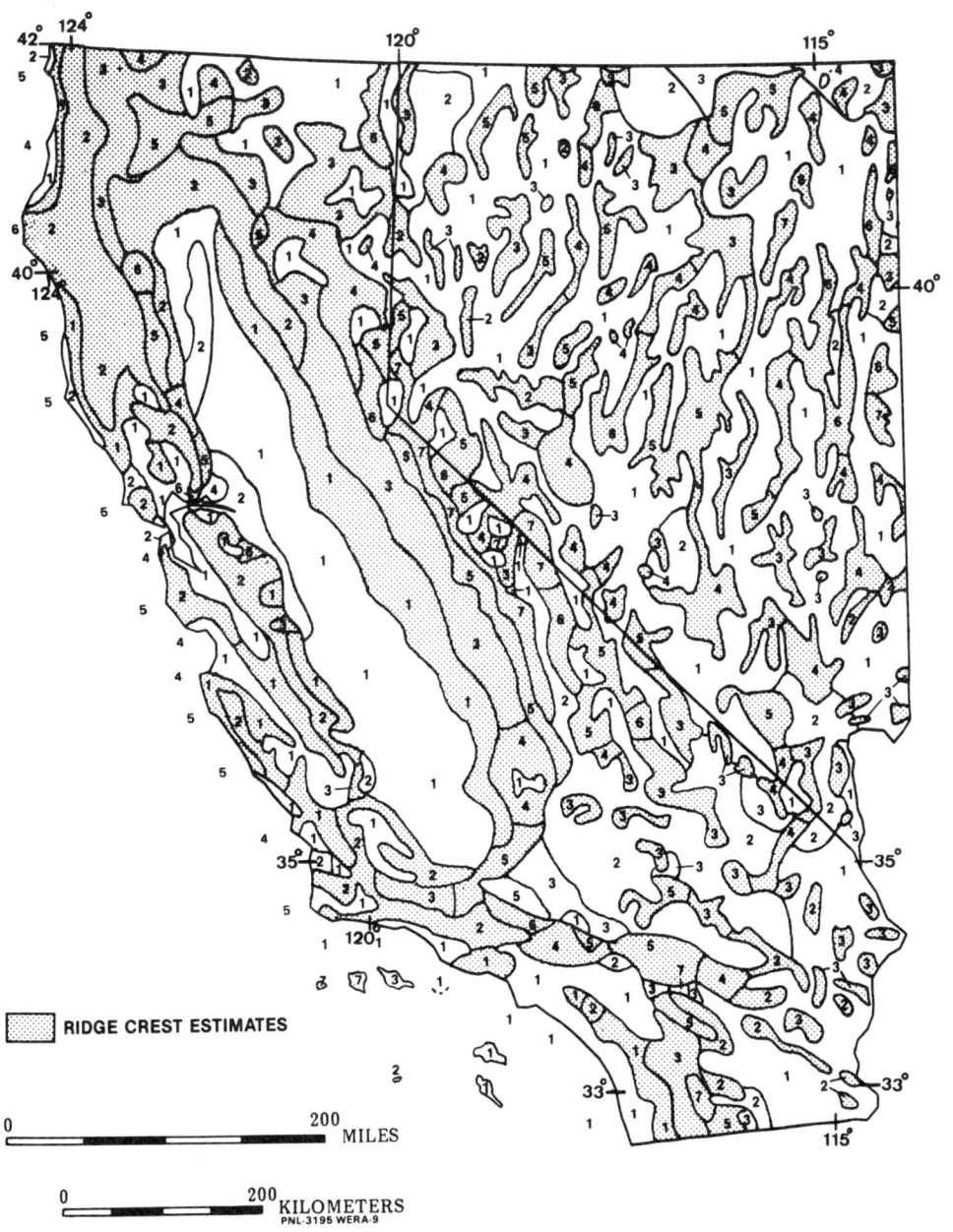

Fig. C-2-9. Average annual wind power in the Southwest.

THE SOUTHWEST REGION

Northern California

Station	Station Name[a]	Average Annual Speed (m/s)	(mph)	Average Annual Power (W/m^2)
Arcata	Arcata Airport	3.0	6.7	59
Marysville	Beale AFB	2.9	6.5	68
Bishop	Bishop Airport	4.4	9.9	101
Blue Canyon	Blue Canyon Airport	3.1	6.9	43
Crescent City	Crescent City Airport	4.4	9.9	178
Donner Summit	Donner Summit CAA	6.4	14.3	401
Fairfield	Travis AFB	5.2	11.6	209
Merced	Castle AFB	3.1	6.9	50
Montague	Siskiyou County Airport	3.5	7.8	88
Mt. Shasta City	Mt. Shasta WBO	2.9	6.5	30
Oakland	Oakland Airport	4.2	9.4	89
Red Bluff	Red Bluff Airport	4.1	9.2	103
Sacramento	Sacramento Airport	3.7	8.3	72
San Francisco	San Francisco International Airport	5.2	11.7	177
San Rafael	Hamilton AFB	2.7	6.1	52
Stockton	Stockton Airport	4.2	9.4	84
Sunnyvale	Moffet NAS	2.6	5.8	45
Ukiah	Ukiah Airport	1.8	4.0	28

Southern California

Station	Station Name[a]	Average Annual Speed (m/s)	(mph)	Average Annual Power (W/m^2)
Bakersfield	Bakersfield Airport	3.4	7.6	45
China Lake	China Lake NAF	3.4	7.6	107
Daggett	Daggett Airport	5.4	12.1	197
El Centro	El Centro NAAS	3.8	8.5	92
Fresno	Fresno Air Terminal	3.2	7.2	37
In-Ko-Pah Gorge	Desert View Tower	7.9	17.7	547
Los Angeles	Los Angeles International Airport	3.9	8.7	64
Mt. Laguna	Mt. Laguna CAA	9.8	22.0	1579
Needles	Needles Airport	3.8	8.5	98
Palmdale	Palmdale Airport	5.1	11.4	188
Paso Robles	Paso Robles Airport	3.0	6.7	64
Riverside	March AFB	2.2	4.9	40
Salinas	Salinas Airport	4.0	9.0	89
Sandberg	Sandberg WBO	5.9	13.2	209
San Diego	San Diego International Airport	3.5	7.8	41
San Nicholas Is.	San Nicholas NF	5.2	11.7	199
Thermal	Thermal Airport	3.9	8.7	74
Vandenberg	Vandenberg AFB	3.4	7.6	81

Nevada

Station	Station Name[a]	Average Annual Speed (m/s)	(mph)	Average Annual Power (W/m^2)
Battle Mtn.	Battle Mtn. Airport	4.1	9.2	94
Elko	Elko Airport	2.9	6.5	45
Ely	Ely Airport	4.7	10.5	106
Fallon	Fallon NAAS	2.4	5.4	36
Las Vegas	McCarran International Airport	4.6	10.3	124
Lovelock	Lovelock Airport	3.0	6.7	55
Pequop Smt.	SPL BPA Anemometer	6.6	15.0	326
Reno	Reno Int'l. Airport	3.1	6.9	66
Tonapah	Tonapah Airport	4.5	10.1	99
Winnemucca	Winnemucca Airport	3.9	8.7	63
Yucca Flat	Yucca Flat WBO	3.3	7.4	71

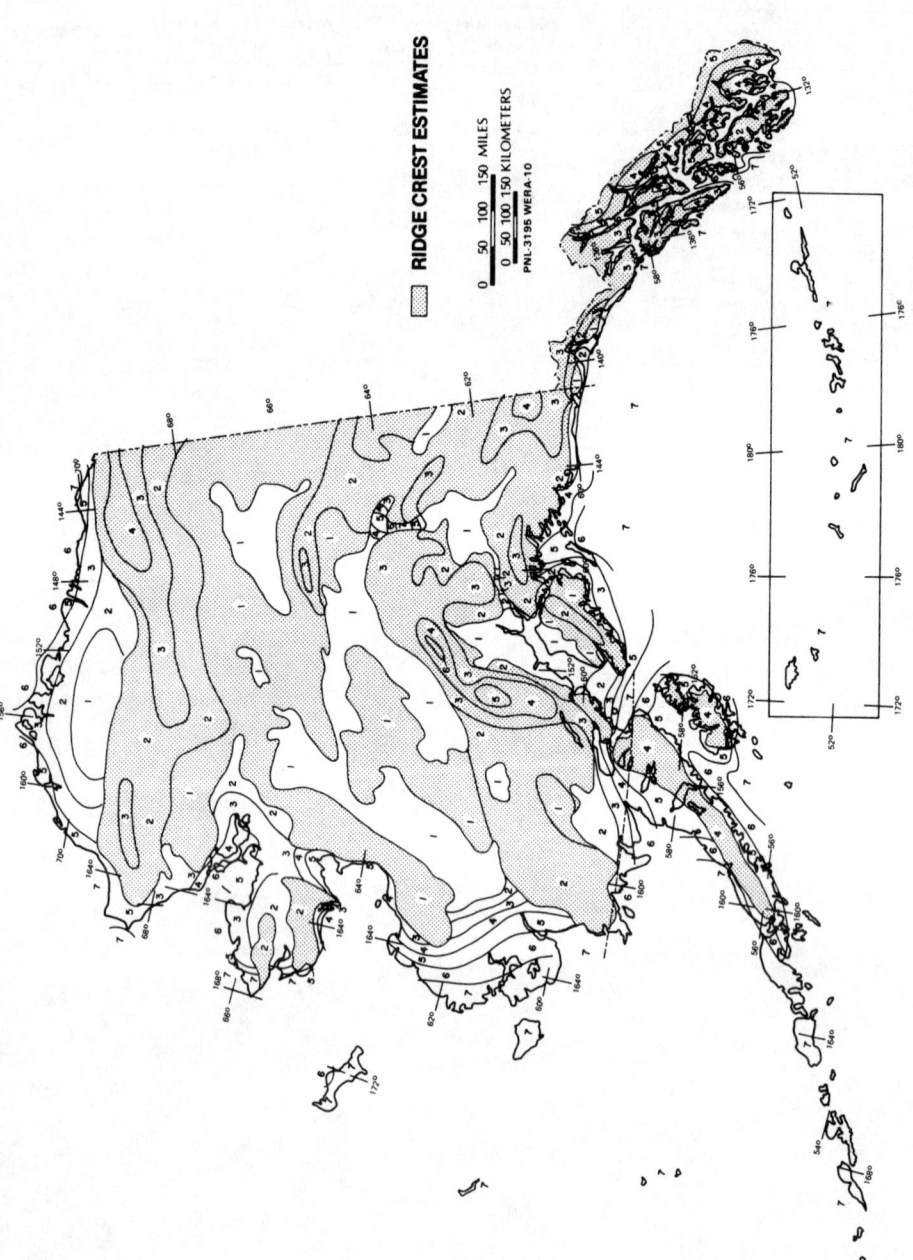

Fig. C-2-10. Average annual wind power in Alaska.

ALASKA
Northern Alaska

Station	Station Name[a]	Average Annual Speed (m/s)	(mph)	Average Annual Power (W/m^2)
Barrow	Wiley Post-Will Rogers Airport	5.4	12.1	198
Barter Island	Barter Island Airport	5.8	13.0	416
Bettles	Bettles CAA	3.0	6.7	40
Cape Lisburne	Cape Lisburne AFS	6.0	13.4	404
Fairbanks EIE	Eielson AFB	1.6	3.6	19
Fairbanks INT	Fairbanks International Airport	2.5	5.6	26
Flayman Island	Flayman Island Air Field	5.7	12.8	334
Fort Yukon	Fort Yukon CAA	3.8	8.5	78
Galena	Galena AFS	2.4	5.4	35
Indian Mountain	Indian Mountain AFS	2.9	6.5	83
Kotzebue	Ralph Wein Memorial Airport	5.8	13.0	317
Lonely Point	Lonely Point Air Field	5.0	11.2	246
Moses Point	Moses Point AAF	5.6	12.5	262
Nenana	Nenana AAF	2.8	6.3	41
Nome	Nome Airport	4.9	11.0	179
Oliktak	Oliktok Air Field	5.9	13.2	366
Point Lay	Point Lay Air Field	6.1	14.0	405
Tanana	Tanana FSS	2.5	5.6	42
Tin City	Tin City AFS	8.4	18.8	772
Umiat	Umiat Airport	3.3	7.4	89
Wainwright	Wainwright Air Field	5.1	11.4	277

Southcentral Alaska

Station	Station Name[a]	Average Annual Speed (m/s)	(mph)	Average Annual Power (W/m²)
Anchorage/ELM	Elmendorf AFB	2.1	4.7	27
Anchorage/INT	Anchorage International Airport	3.4	7.6	62
Aniak	Aniak CAA	3.1	6.9	51
Bethel	Bethel WBAS	6.1	13.7	258
Big Delta	Big Delta CAA	4.3	9.6	227
Cape Romanzof	Cape Romanzof AFS	6.6	15.0	514
Cordova	Cordova WBAS	2.3	5.2	37
Farewell	Farewell CAA	3.8	8.5	129
Gulkana	Gulkana CAA	3.1	6.9	79
Homer	Homer CAA	3.0	6.7	54
Iliamna	Iliamna CAA	4.6	10.3	153
Kenai	Kenai CAA	3.4	7.6	71
L. Minchumina	Minchumina CAA	2.3	5.2	41
McGrath	McGrath WBAS	2.3	5.2	29
Middleton Island	Middleton Island CAA	6.3	14.1	404
Northeast CP	Northeast Cape AFS	6.4	14.3	439
Northway	Northway FAA	1.8	4.0	25
Skwentna	Skwentna CAA	1.4	3.1	14
Sparrevohn	Sparrevohn AFS	2.7	6.1	79
Summit	Summit CAA	4.2	9.4	113
Talkeetna	Talkeetna CAA	2.2	4.9	29
Unalakleet	Unalakleet CAA	5.5	12.3	270
Yakataga	Yakataga CAA	3.7	8.3	111

Southeastern Alaska

Station	Station Name[a]	Average Annual Speed (m/s)	(mph)	Average Annual Power (W/m²)
Annette Island	Annette Airport	4.8	10.8	158
Gustavus	Gustavus FAA	3.2	7.2	69
Haines	Haines CAA	4.1	9.2	121
Juneau	Juneau Airport	4.0	9.0	115
Petersburg	Petersburg Airport	2.6	5.8	36
Sitka	Sitka FAA	3.0	6.7	81
Yakutat	Yakutat WBAS State Airport	3.4	7.6	81

Southwestern Alaska

Station	Station Name[a]	Average Annual Speed (m/s)	(mph)	Average Annual Power (W/m^2)
Adak	Adak Naval Station	5.7	12.8	287
Amchitka	Amchitka Island AFB	9.6	21.5	1186
ASI Tanaga	ASI Tanaga Naval Station	8.5	19.0	1025
Attu	Attu Naval Station	5.3	11.9	325
Cape Newenham	Cape Newenham AFS	5.8	13.0	360
Cold Bay	Cold Bay WBAS	8.0	18.0	626
Driftwood Bay	Driftwood Bay AFS	4.9	11.0	228
King Salmon	King Salmon WBAS	5.2	11.6	202
Kodiak	Kodiak Naval Air Facility	4.8	10.7	233
Nikolski	Nikolski AFS	7.3	16.4	516
Port Moller	Port Moller AFS	4.4	10.0	166
St. Paul Island	St. Paul Island WBAS	7.2	16.1	410
Shemya	Shemya WBAS	9.4	21.1	1001

380 • Appendix C

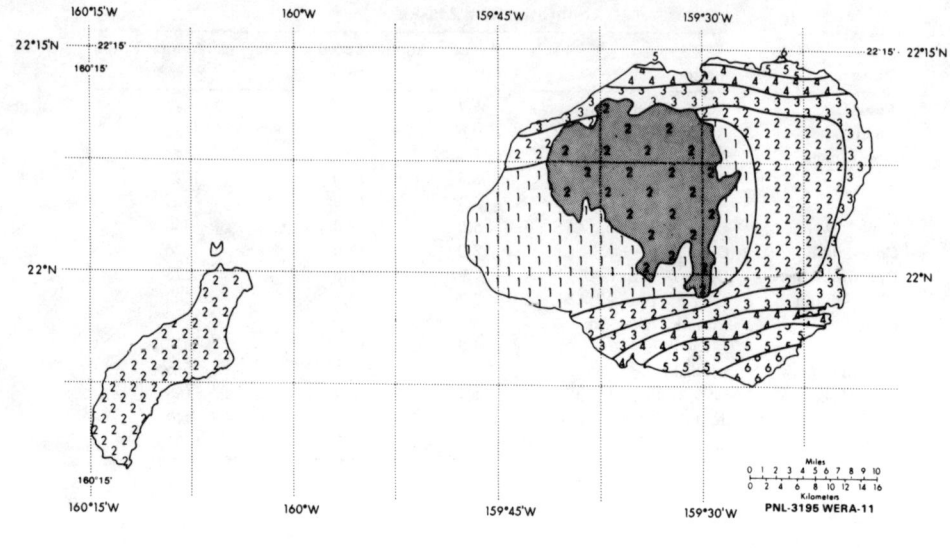

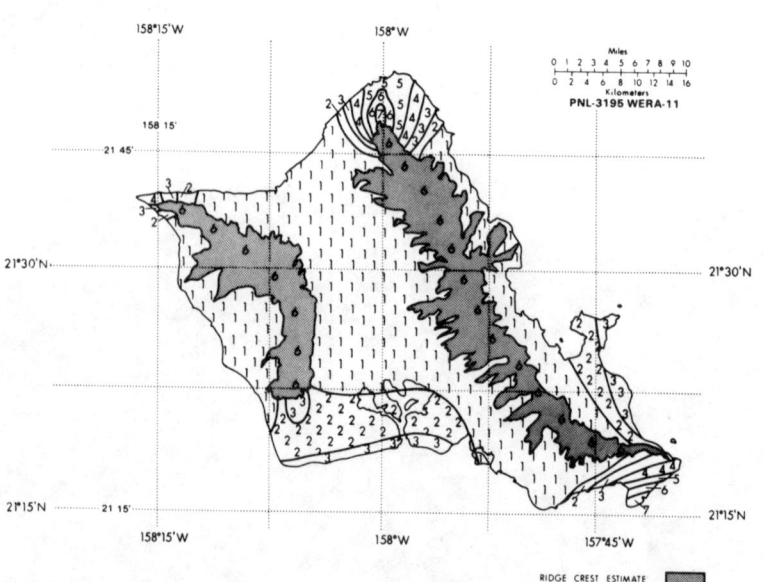

Fig. C-2-11. Kauai County (*top*) and Honolulu County (*bottom*) average annual wind power.

Wind Data • 381

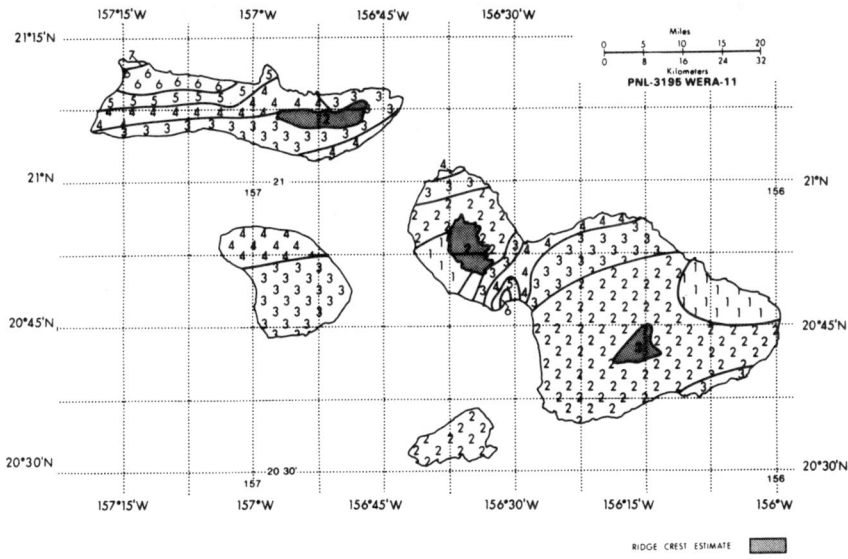

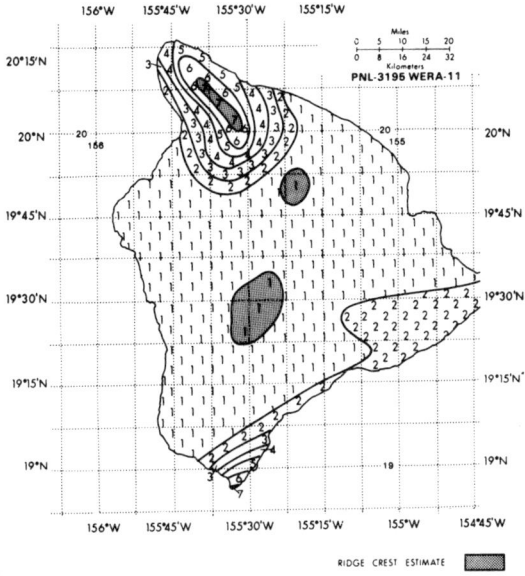

Fig. C-2-12. Maui County (*top*) and Hawaii County (*bottom*) average annual wind power.

HAWAII AND PACIFIC NORTHWEST TERRITORIES
Hawaii

County	Station Name[a]	Average Annual Speed (m/s)	(mph)	Average Annual Power (W/m²)
Honolulu	Barbers Point NAS	4.0	9.0	77
Kauai	Barking Sands Naval Facility	2.3	5.2	36
NW Hawaii (Honolulu)[c]	French Frigate Shoals, NAS	6.7	15.0	289
Hawaii	Hilo Airport	3.1	6.9	33
Honolulu	Honolulu Airport	5.6	12.5	167
Honolulu	Kaena Point Tower, UH Site	5.9	13.2	233
Hawaii	Kahua Ranch UH Site	11.4	25.5	1569
Honolulu	Kahuku Hill LLL Site	8.6	19.2	563
Maui	Kahului Airport	5.8	13.0	211
Honolulu	Kaneohe Bay MCAS	5.1	11.4	138
Honolulu	Koko Head UH Site	7.9	17.7	406
Kauai	Lihue Airport	5.6	12.5	148
Maui	Molokai Airport UH Site	6.0	13.4	210
Kauai	Port Allen CAA	6.3	14.1	224
Maui	Puunene, CAA	4.5	10.1	121
Honolulu	Tantalus Tower UH Site	7.4	15.6	472
Hawaii	Waimea Airport UH Site	7.9	18.0	515
Honolulu	Wahiawa Wheeler AFB	3.1	6.9	58

Pacific Islands

Island Group	Station Name[a]	Average Annual Speed (m/s)	(mph)	Average Annual Power (W/m²)
Marshall	Enewetak AAF	7.9	18.0	405
Guam	Agana NAS	4.0	9.0	76
Guam	Andersen AFB	4.9	11.0	121
Guam	Orote AAF	3.4	7.6	55
Guam	Taguac WSMO	4.9	11.0	112
Detached	Johnston Atoll Airfield	7.2	16.1	281
Caroline	Koror Airport	3.1	7.0	37
Marshall	Kwajalein Airfield	6.3	14.1	226
Marshall	Majuro Airport	4.9	11.0	116
Detached	Midway Island NAS	4.8	10.8	126
Samoa	Pago Pago Airport	4.9	11.0	132
Caroline	Ponape Airport	3.3	7.4	39
Mariana	Saipan NF	3.2	7.2	48
Caroline	Truk Airport	4.6	10.3	99
Detached	Wake Island Airfield	6.7	15.0	271
Caroline	Yap Airport	3.9	8.7	62

[a] AAB—Army Air Base; AAF—Army Air Field; AF—Air Field; AFB—Air Force Base; CAA—Civil Aeronautics Administration Facility; FLD—Field; NAS—Naval Air Station; WBAS—Weather Bureau Airport Station; WBO—Weather Bureau Office.
[b] Station had less than eighteen observations per day.
[c] Station had less than one year of digitized records.
[d] Station data for this period not included in graphs in atlases.
[e] Station had less than twelve observations per day.

Appendix D

Further Sources of Information

INSTITUTIONS

Alternative Energy Institute
West Texas State University
Box 248
Canyon, TX
(806) 656-3904

American Wind Energy Association
1050 17th Street, NW
Washington, DC 20036
(202) 775-8910

Atlantic Wind Test Site
Institute of Man and Resources
P.O. Box 189
Tignish, Prince Edward Island
Canada C0B 2B0
(902) 882-2746

Rocky Flats Small Wind Systems
  Test Center
P.O. Box 464
Golden, CO 80401
(303) 441-1351

Solar Energy Research Institute
1617 Cole Boulevard
Golden, CO 80401
(303) 231-1158

Test Plant for Small Windmills
Riso National Laboratory
DK-4000 Roskilde
Denmark

WIND MACHINE MANUFACTURERS

For a list of American manufacturers write the American Wind Energy Association or the Rocky Flats Small Wind Systems Test Center. For a comprehensive directory of over 300 manufacturers worldwide write:

Wind Books, P.O. Box 14, Rockville Centre, NY 11571

Because the industry is always changing, published directories are out of date by the time they are printed. Moreover, only a few of the manufacturers listed in such directories are producing what I call commercial products; that is, they have wind machines operating in the field that work reliably. Many products in such lists are just off the drawing board; others are too large or too small to be of value for homeowners and small businesses, and a few are—there's no other word for it—just junk.

Rather than reprint such a directory here I've chosen to list the addresses of only a few manufacturers. These are firms that are mentioned in the text. They all have been in business several years, and they all have a reputation for honoring their warranty commitments. The wind machines they produce are suitable for remote applications or for supplementing power from the utility (see table E-1). Their wind machines are available for purchase by homeowners and small businesses. In most cases the models listed have been in use for several years. Their performance has been tested and the results are part of the public record. This listing is not an endorsement. (I do have my preferences among them, though.) But they are reputable manufacturers who stand behind their products.

There are also several companies producing reliable wind machines for wind farm applications. They emphasize sales of more than one unit per customer so I have not included them in this list. If you want only one wind machine, you may still get help from them if your application is unique. For information contact AWEA. The companies in the following list cater to the residential and small commercial market and are accustomed to clients seeking a single wind machine.

Aerolite, Inc.
P.O. Box 576
South Dartmouth, MA 02748
(617) 993-9999

Bergey Windpower Co.
2001 Priestley Ave.
Norman, OK 73069
(405) 364-4214

Jay Carter Enterprises
P.O. Box 684
Burkburnett, TX 76354
(817) 569-2238

Enertech Corp.
P.O. Box 420
Norwich, VT 05055
(802) 649-1145

Northwind Power Co.
Box 556
Moretown, VT 05660
(802) 496-2955

Sencenbaugh Wind Electric Co.
P.O. Box 11174
Palo Alto, CA 94306
(415) 964-1593

TABLE E-1
Wind Machine Characteristics

Model	Rotor Size[a] and Type	Generator Capacity and Type	Application[b]	Rotor Control System
Aerolite, Inc.				
7 meter	7-meter, downwind	10 kW induction generator	SP	Brake, tip brakes
Bergey Windpower Co.				
1000	3-meter, upwind	1.2 kW permanent-magnet alternator	BC, WH, SP	Self-furling
Excel	7-meter, upwind	10 kW permanent-magnet alternator	BC, WH, SP	Self-furling
Jay Carter Enterprises				
25	10-meter, downwind	25 kW induction generator	SP	Pitch to stall, brake
Enertech, Corp.				
1800	4-meter, downwind	2.2 kW induction generator	SP	Brake, tip brakes
21/5	6-meter, downwind	5 kW induction generator	SP	Brake, tip brakes
E-44	13-meter, downwind	20 to 40 kW induction generator	SP	Brake, tip brakes
Northwind Power Co.				
HR 2	5-meter, upwind	3 kW Lundel alternator	BC	Variable axis (tip back)
Sencenbaugh Wind Electric Co.				
500	2.5-meter, upwind	0.6 kW alternator	BC	Self-furling
500 HDS	2.5-meter, upwind	0.6 kW alternator	BC	Variable axis (tip back)
1000	4-meter, upwind	1.2 kW alternator	BC	Self-furling
Windworks, Inc.				
Windworker 10	10-meter, downwind	9 kW permanent-magnet alternator	SP	Pitch to feather

[a] Approximate rotor diameter
[b] BC = Battery charging
WH = Water heating
SP = Supplemental power interconnected with the utility

Windworks, Inc.
Rt. 3, Box 44A
Mukwonago, WI 53149
(414) 363-4088

TOWERS

As a rule, tower manufacturers avoid sales to individual homeowners. They prefer to deal either with the local wind system dealer or with the manufacturer of the wind machine directly. By handling sales in this way, they can assure themselves that the tower is matched properly to the wind machine and installed properly. The following manufacturers are listed here for your reference.

American Pole Structures
Combustion Engineering
P.O. Box 40020
Houston, TX 77040
   Tapered tube towers

ITT-Meyer Industries
Highway 55
Red Wing, MN 55066
(612) 388-4755
   Tapered tube towers

Koppers Co.
Forest Products Div.
10 Rooney Circle
W. Orange, NJ 07052
(201) 736-9150
   Wood poles

Pi-Rod
P.O. Box 128
Plymouth, IN 46563
(219) 936-4221
   Guyed lattice and truss towers

Solargy Corp.
17914 E. Warren Avenue
Detroit, MI 48224
(313) 881-5510
   Telescoping tube towers

Unarco-Rohn
6718 West Plank Road
Peoria, IL 61656
(309) 697-4400
   Guyed lattice and truss towers

Valmont Industries, Inc.
Valley, NE 68064
(402) 359-2201
   Tapered tube towers

ANEMOMETERS

Aeolian Kinetics
P.O. Box 100
Providence, RI 02901
(401) 421-5033

Approach Fish, Inc.
Rt. 1, Box 620 B
Ringgold, VA 24586
(604) 793-2828

Dwyer Instruments
P.O. Box 373
Michigan City, IN 46360
(219) 872-9141

Helion
Box 445
Brownsville, CA 95919
(916) 675-2478

Maximum
42 South Avenue
Natick, MA 01760
(617) 785-0113

Natural Power
Francestown Turnpike
New Boston, NH 03070
(603) 487-5512

NRG Systems
Box 714
Bristol, VT 05443

Omnidata International
P.O. Box 3489
Logan, UT 84321

Parkway Energy Products
22 Parkway Road, Suite 2
Brookline, MA 02146

Parkway Energy Products
22 Parkway Road, Suite 2
Brookline, MA 02146

Second Wind
2000 Massachusetts Avenue
Cambridge, MA 02140
(617) 497-1580

Sign X Laboratories
Stetson Road
Brooklyn, CT 06234
(203) 774-5233

Simerl Instruments
238 West Street
Annapolis, MD 21401

Solar Plexus
3230 St. Matthews Drive
Sacramento, CA 95821

M. C. Stewart
Ashburnham, MA 01430

Wind Engineering Consultants
3421 Adams Avenue
San Diego, CA 92116

ANCHORS

A.B. Chance Co.
210 N. Allen Street
Centralia, MO 65240

Everstick Anchor
Iowa Malleable Iron Co.
9th and Kirkwood Avenue
Fairfield, IA 52556

Joslyn Manufacturing and
  Supply Co.
2 N. Riverside Plaza
Chicago, IL 60606

GUYING HARDWARE

Joslyn Manufacturing and
  Supply Co.
2 N. Riverside Plaza
Chicago, IL 60606

Preformed Line Products Co.
P.O. Box 91129
Cleveland, OH 44101

TOOLS AND WORKBELTS

Atlas Safety Equipment Co.
1 Johnson Avenue
Matawan, NJ 07747
(201) 583-6200
   Work belts and lanyards

Klein Tools, Inc.
7200 McCormick Road
Chicago, IL 60645
   Work belts, lanyards,
     erection tools

SINCO Products, Inc.
P.O. Box 361
East Hampton, CT 06245
   Work belts and lanyards

PLANS

Low Energy Systems
3 Larkfield Gardens
Dublin 6
Ireland
   Plans for sail wing VAWT

National Centre for
   Alternative Technology
Machynlleth
Powys, Wales
England
   Plans for Cretan-sail
     wind machine

REBUILT WINDCHARGERS

See the classified advertisements in:

Alternative Sources of Energy
107 S. Central Avenue
Milaca, MN 56353

Mother Earth News
105 Stoney Mountain Road
Hendersonville, NC 28791

SPECIALTY BLADES

Aeroglass Energy Systems
1078 Hillcrest Drive
Xenia, OH 45385
(513) 376-2150

Santa Rosa Machined Props
P.O. Box 1541
Santa Rosa, CA 95402
(707) 545-8634

Sebastopol Air Power
610 Jonson Street
Sebastopol, CA 95472
(707) 823-8357

Windstead Workshop
RFD 1, Box 180
Battle Lake, MN 56515

Windwright
Rt. 4, Box 345
Neosho, MO 64850
(417) 451-5495

SYNCHRONOUS INVERTERS

Acheval Wind Electronics
361 Aiken Street
Lowell, MA 01854
(617) 453-0874

Gemini Synchronous Inverters
Windworks
Box 44A, Rt. 3
Mukwonago, WI 53149
(414) 363-4088

SURPLUS AIRCRAFT GENERATORS

Airborne Sales Co.
P.O. Box 2727
Culver City, CA 90230

Surplus Center 1000
P.O. Box 82209
Lincoln, NE 68501

WINCHES

Surplus Center
P.O. Box 82209
Lincoln NE 68501
   Capstan and furling winches, motors, etc.

KWH METERS

Hialeah Meter Co.
441 W. 28th Street
Hialeah, FL 33010

Notes

CHAPTER 4. MEASURING THE WIND

1. The correct term is V for velocity. Speed and velocity have separate and distinct meanings. However, they will be used synonymously here.
2. To convert W/m² to W/ft² use the following formula.

$$\frac{W/m^2}{10.76} = 0.093 \times W/m^2 = W/ft^2$$

3. One knot = 1.15 mph.
4. One meter per second (mps) = 2.24 mph.
5. If accuracy is one of your strong suits then you will prefer a more sophisticated method for finding the midpoint of the speed class.

$$\text{Midpoint of Speed Class} = \left(\frac{(S_1)^3 + (S_2)^3}{2}\right)^{1/3}$$

where S_1 and S_2 are the upper and lower end of the speed range and 1/3 signifies the cube root. This is not something you are able to do in your head. You will need a calculator that can perform cube roots. For the 8 mph to 12 mph interval this formula for the geometric midpoint of the speed class will give

$$\text{Midpoint} = \frac{(8)^3 + (12)^3}{2} = 10.4 \text{ mph}$$

The result is 4 percent greater than that obtained from the simple arithmetic average. This difference translates into a 12 percent difference in power because speed is cubed in the power equation; the average of the cubes is greater than the cube of the average. Using the simplified approach given above will result in power estimates about 10 percent lower than those which actually could be expected.

6. Sierra Club Books, 530 Bush Street, San Francisco, CA 94108. (An excellent source book on alternative energy.)

7. Available by subscription from NCC, Federal Building, Asheville, NC 28801.

8. The value is higher than that shown in figure 4-7 and in most other published sources. The average speed at the height of the anemometer, in this example 20 feet, is 9.4 mph. At 10 meters (33 feet) the average speed is 10.3 mph.

CHAPTER 5. ESTIMATING OUTPUT AND PAYBACK

1. Commercial users of electricity do pay a "demand" charge for the instantaneous amount of power they demand—use—at any one time.

2. The instantaneous efficiencies in table 5-1 were compiled from manufacturer's power curves. Overall conversion efficiencies were calculated from their estimates of annual energy output (AEO) using a Rayleigh speed distribution. Actual performance of these wind machines may be better or worse depending on the site.

3. Purchases of a traditional American farm windmill (e.g., for stock watering) are excluded from the commercial energy credits but qualify for the residential energy credits.

4. It can also be supplied with a 40-kW generator.

Bibliography

GENERAL INTEREST

Leckie, Jim et al. 1975. *More Other Homes and Garbage.* Sierra Club Books, 550 Bush St., San Francisco, CA 94108. One of the best overall books on alternative energy available. Runs the gamut from wind to biconversion.

Marier, Donald. 1981. *Wind Power for the Homeowner.* Rodale Press, Emmaus, PA 18049. A readable introduction to wind energy; intended for the homeowner.

Park, Jack, and Schwind, Dick. 1977. *Wind Power for Farms, Homes, and Small Industry.* Wind Books, P.O. Box 14, Rockville Center, NY 11571. Packed with useful information. Available from National Technical Information Service.

WIND RESOURCES AND SITING

Battelle PNL. 1980a. *A Siting Handbook for Small Wind Energy Conversion Systems.* Wind Books, P.O. Box 14, Rockville Center, NY 11571. Useful when siting your own wind turbine. Available from National Technical Information Service.

394 • Bibliography

———. 1980b. *Wind Energy Resource Atlas*. Available from NTIS. PNL 3195 WERA—
1. The Northwest Region
2. The North Central Region
3. The Great Lakes Region
4. The Northeast Region
5. The East Central Region
6. The Southeast Region
7. The South Central Region
8. The Southern Rocky Mountain Region
9. The Southwest Region
10. The Alaska Region
11. Hawaii and Pacific Northwest Territories
12. Puerto Rico and U.S. Virgin Islands

DESIGNING OR BUILDING YOUR OWN

Hackleman, Michael. 1974. *Wind and Windspinners*. Earthmind, 5246 Boyer Rd., Mariposa, CA 95338. Describes how to build your own Savonius rotor; in down-to-earth language.

———. 1975. *The Home-Built, Wind-Generated Electricity Handbook*. Earthmind, 5246 Boyer Rd., Mariposa, CA 95338. Explains how to find, lower, and rebuild pre-Rea wind generators, plus much more.

Park, Jack. *Simplified Wind Power Systems for Experimenters*. Helion, Inc., Box 445, Brownsville, CA 95919. Contains essential information to build your own wind generator.

———. 1981. *The Wind Power Book*. Cheshire Books, 514 Bryant St., Palo Alto, CA 94301. The best overall book on the technology and design of wind machines. An excellent reference for experimenters and professionals alike.

INSTALLATION

Bergey Windpower Co. *BWC 1000 Windpower Generator Installation Manual*. 2001 Priestley Ave., Norman OK 73069. No doubt the best installation manual in the business today. Many details on installing Rohn guyed towers as well as the use of screw, expanding, and rock anchors.

Energy Task Force. September 1977. *Windmill Power for City People* (CSA Pamphlet 6143-8). Superintendent of Documents, U.S. Government Printing Office, Washington, D.C. 20402. Useful notes on anchor loads and the problems associated with installing a wind machine on a roof top.

Enertech Corp. *Enertech Installation and Operation* (for Series 1800 or Series 4000). P.O. Box 420, Norwich, VT 05055.

REMOTE WIND SYSTEMS

Paul, Terrence D. *How to Design an Independent Power System.* Best Energy Systems for Tomorrow, Inc., P.O. Box 280, Necedah, WI 54646. *The* book for planning an independent power system whether it be with a diesel generator or a wind machine. Straightforward and to the point. It's worth twice the price.

PERIODICALS

Alternative Sources of Energy, 107 S. Central Ave., Milaca, MN 56353. Bi-monthly. Occasional features on wind energy written for a general audience.

Renewable Energy News, Box 4869, STNE, Ottawa, Ontario, Canada K1S 5J1. Regular features on wind energy development in United States and Canada. Occasional features on developments overseas.

Solar Age, Church Hill, Harrisville, NH 03450. Monthly. Occasional features on wind; geared to a professional audience.

Wind Energy Report, P.O. Box 14, Rockville Center, NY 11571. Monthly. Comprehensive and in-depth coverage of wind energy development worldwide. WER keeps tabs on the pulse of the industry.

Wind Industry News Digest by Alternative Sources of Energy, Inc., 107 S. Central Ave., Milaca, MN 56353. Monthly. News items and tidbits on developments in the American wind industry.

Index

Accumulators, 101-2
Adjusted annual value, 138-46
Adjusted total cost, 138-46
Air density, 76-77, 332
Airfoils, 163-64
Alternators, 188-92
American Wind Energy Association
   guidelines, 60, 134, 205, 208, 221, 249,
   250-51, *293*, 294
Anchor rods, 270, 275
Anchors, 261-63, 265-70
   concrete, 265-66
   expanding, 266-68
   holding capacity, 261-62, 267, 270
   rock, 270
   screw, 268-70
Anemometer towers, 107-11
Anemometers, 100-107, 121
   height of, 89, 93, 97-98
   for performance monitoring, 307
Angle of attack, *164*, 165
Annual energy output (AEO), 125-28,
   130-32, 134, 251
   comparison table, 135
   for conventional wind machines, 127-28
   from corrected speed distribution, 131
   formula for, 126
   manufacturer's estimates of, 134
   from site survey, 94
Appliances, cost per kWh, 35
Augering, 265-69, 272

Battery charging wind systems, 24-27
Betz limit, 121, 167-68
Blades, 158-60, 173-79
   articulating, 170, 172, *173*
   aspect ratio, 170
   cascade effect, 162
   cost of, 169
   design, 19, 166, 169
   downwind-sweeping, 153, *156*
   failure of, 47-48
   fixed-pitch, 165
   hinged, *181*, 198
   lift/drag ratio of, 165-66
   safety hazards of, 323
   swept area of, 77-78
   tapered-twisted, 166, 168, *202, 217*
   width of, 171, 172
Blade area, 166
Blade attachment, 179-81
Blade pitch, 165-66
Blade pitch, variable, *159, 200*, 201, *202*
Blade speed, 165, 166, 169
Blade tips, 166, *178, 196*, 204-7
Brakes, 171, 203-7, *212*, 252
Building permit, 45-46, 52-53
Buying (a wind system), 117, 248-57
   cost-effectiveness, 13-14, 32-33,
     148-49
   cost vs. energy needs, 57, 96
   cost vs. size, 38, 57, 227, 248
   terms of sale, 254-55
   vs. cost of site wind monitoring, 96-97

Cable, 215, 218, 325-27
Cable, buy. *See* Guy cable
Cable, hoisting, 287
Cable, safety, 329-30
Cable grip, 275, *276*
Calculations, rounding off, 19-20
Capacitors, 60, 61, 63
Capital recovery factor, 142, 334
Carabiners, 320
Chance system (of soil grading), 262
Cogging, 188-89
Compound amount factor, 143, 335
Concrete work, 263-65
Conductor run, 43-44, 288-95
   safe allowable distance (formula), 298
Conductors, 295-302
   DC resistance, 300
   size of, 295-300
   voltage drop, 296-98
Conduit, 276, 284, 290, 295
Conduit fill, 301-2

396

Conduit hangers, 290
Coning, 152, 201
Connectors, type of, 289
Contactors, 60, 64, 67-68
Contractors, 265
Contracts, 67-68, 254-55
Control panel, 294, *308*
Convective circulation, 70
Conversion efficiency, 17, 121-22, 131, 182
Cor-Ten steel, 223-24
Cost cutting, 228-47
Cost of energy (COE), 137
Costs (of a wind system), 138-40
    installation, 44, 228-29
    interconnection, 58
    legal, 254
    transportation, 225-26
Crane method
    guyed towers, 215, 273, 275-76
    truss towers, 213, 215, 280, 282-85
Crane rental, 280, 282
Crosby clamps, 108, 110, 274-75, 327
Cross girts, *212*, 224, 284
Cube factor, 79, 81, 83

Data loggers, 102-4
DC, hazards of, 26
Dealer, evaluation of, 253-55
Dedicated circuit, 190, 294
Deflection, 167-68, 222
Demand charge, 139
Demand meter, 309
Depreciation deductions, 144-46
Differential heating, 69-70
Disconnect switch, 68, 292-94, 295
Discounts, quantity, 247
Drag devices, 160-62
Drift pin, 282, 284
Drive train, *182*
Dump circuit, 36-38
Dynamometer, 279

Easements, 52
Electric bill, to estimate cost per kWh, 33-35
Electric resistance heating, 29-30
Electric use profile, 33-35
Electrical code, local, 288, 290, 292
Electrical safety hazards, 242-43, 324-25
Electricity, cost of
    deductible business expense, 145
    discounted future value, 143
    estimating cost per kWh, 33-35, 139
    interest compounding, 143, 335
    retail cost, 139, 140
Electricity, low-grade, 28, 29, 37
Electricity, utility-grade, 29, 59
Electricity consumption, 33-36
Emergency conditions, 205, 208-9, 251, 305-6
Energy, 16-18, 75-76

Energy equivalency of common fuels, 147
Energy Pattern Factor (EPF), 83-86
Equations, use of, 19-20
Excavation holes, 265-70, 272

Fall protection, 315-23
Fan tails, 152, *180*
Force-distance relationship, 16-17, 19
Forms (for concrete), 264
Foundation preparation, 261
Foundations, novel, 272-73
Free-standing towers, 211-15, 280-86
Frost heave, 262

Generator-inverter match, 192
Generators, 184-92
    DC, 184-85, 191
    design, 187-88, 191
    direct drive, 185, 187-88
    induction, 32, 59-60, 64, 190-92, 300-301
    power factor of, 61, 185-86
    rating of, 61, 121, 186
    size, 185-86, 192, 386
    swept area of rotor, 78
    transmission type, 183
    use of two, 132, 192
Gin pole method
    guyed towers, 215, 217, 273, 277-81
    hinged towers, 286, *287*
    pole tower, 213
    truss tower, 213, 288
Gin poles, 277-81
    A-frame, 286-87
    attaching to tower, 277, 279
    base-welded, 286-87
    hazards of, 277, 279, 280, 286-87
Giromill, 159, 160, 170, 172
Governors, 199-201
Griggs-Putnam Index, 95-96
Grounding, 303-304
Guy anchors, 265-70, 331
Guy brackets, 108, 274-76, 279
Guy cables, 215, 218, 273-74, 279
    anemometer towers, 107-8
    failure, 215
    length, 273-74
    safety precautions, 110-11, 223, 303-4
    tensioning, 221, 275, *276*, 279
Guy cable attachments, 274-75
Guy levels, 218, 273
Guy radius, 218, 223
Guyed towers, 215-18, 226, 265-80
Guys, temporary, 277, 286

Halladay umbrella mill, 153, 155, 193
Harmonics, dangerous, 222, 230, 279
Harnesses, 317, 319
Heating systems, wind-powered, 26-32

Index

Height and wind speed/power, 86–90, 127–28
Hinged towers, 213–14, 217, 286–87
Hoists, 328–29
Hoists, coffing, 275, *276*, 279, 282
Hot water heating, 26–32, 36–37, 147–48
Hubs, 179–81

Independent power systems, 25–26, 35
Installation,
    anemometer towers, 109–11
    do-it-yourself, 254–55, 258
    of larger machines, 258
    towers, 224, 258–309
    used modern machine, 233–34
    of wind machines on towers, 275–76, 279–84
Insulation, 295–96
Insurance, 67–68, 138–39, 253, 288
Interconnected wind systems, 29, 32–38, 53–59
    dealing with utility, 66–68, 305–7
Inventory control, 260–61
Inverters asynchronous, 189
Inverters, synchonous, 32, 59–60, 66, 189–190, 191–92, *308*
    insulation quality, 296
    voltage drop, 297
Investment, 57–58, 97, 117–18, 136
    return on, 137, 142, 146
Irrigation pumping, wind-assisted, 23–25

Jacobs home light plant, 182, *185*, 187–88, 235–39, 241–43
J-bars, 272, 285
Junction box, 289, 290, 301

Kites, to detect turbulence, 43
Kits, professional, 244–45

Lanyards, 319–20
Levelized value, 57, 143–44
Levers, 18–19
Lifelines, 319–20
Lift/drag ratio, 165–66
Lift vs. drag devices, 160–64
Lightning, 303–4
Lineman's belt, 317–18
Loans, cost of, 142–43, 146

Magnus effect, 162–63
Maintenance, 138, 146, 223–24, 252–53, 309–12
Maintenance contract, 310
Manufacturer evaluation, 48, 248–49, 252–53
Maps, wind power, 93–94, 336–37
Masts, 215, 216, 218, 221, 223, 273, 275, 277

for site surveys, 107–9
Mechanical advantage, 19, 328
Messenger cable, 292
Meters, 59, 325
Mountains and wind speed, 71–73

National Climatic Center (NCC), 91–92
National Electrical Code (NEC), 44, 287–88, 301–2
Neighbors, notification of, 45–46, 48–49
Net energy billing, 59
Noise level, 49–51, 251–52
Nuts, 224, 285

Obstructions, estimating height, 98–99
Odometers, wind-run, 101–2, 104
Operation (of wind system), 305–7
Output, estimating, 119–36. *See also* Annual energy output
Overrunning clutch, 24–25
Overspeed control. *See* Rotor control
Owner's manual, 252, 314

Panemone, 152
Parris-Dunn windcharger, 197–98, 235, 239
Payback, 136–49
Performance monitoring, 101, 104, 307–9
Performance ratings, 132–34, 250–52
Perry, Thomas (designs), 174
Phase angle, 61–62
Piers, 261–66, 272
"Pig tail" outlet, 294–95
Pilot vane, 193–95
Pole towers, 211, 213–15, 225–26, 285–86
Power, 16–18
    apparent vs. true, 60–62
    backup, 24, 56
    sale to utility, 53–59
Power curve, 129, 130, 132, 251
Power curve method, 128–32
Power density, 79–86
Power factor correction, 61–64
Power form, 251–52
Protective covenants, 51–52
Public Utilities Regulatory Policy Act (PURPA), 54–59
Pulleys, 277, *281*, 328

Rated wind speed/power, 130, 132–34, 251–52
Rayleigh distribution, 84–85, 134
Re-bars, 263–64, 272
Relays, 60
Remote power, supplying, 24–26
Rope, 320, 325–27
    hoisting, 277–80
Rotor, 151–72
    bearingless, 201–2
    Darrieus, 25, 123–24, 158–60, 169–71,

     175-76, 202, 220
     downwind, 42, 151-56, *181*, 193
     fixed pitch, 180, 201, 204
     flapping, 181
     H-(Musgrove), 123-24, 158-59, 160, 170, 172
     S-(Savonius), 160, 162, 394
     self-furling, 193, 195-97
     self-starting capability, 169-70
     swept area formula, 77-78, 123-24
     upwind, 42, 151-54, 157, 193, 203
     upwind vs. downwind, 151-57
Rotor axis, offset, 193, 195-96
Rotor controls, 193-209, 251-52, 386
     guidelines, 205, 208-9
     hinged blades, 198
     variable axis, 197-98
     variable blade pitch, 198-99
Rotor diameter
     as measure of output (AEO), 133, 136
Rotor speed, rated, 251-52
Rotor torque, 166-68
Roughness exponent, 87-90

Safety, community concern for, 47-48
Safety belts. See Work belts
Safety procedures, necessity for, 315-16, 319
Sailmill, 230-31
Service drop, 60, 64, 288
Service panel, 32, 44, 294
Service zone, 41
Significant figures, concept of, 21
Silos, 109, 219-20
Site preparation, cost-reducing, 245-47
Site surveys, 94-118
     and airport wind data, 94-95, 112-17
     duration, 111-13
     equipment for, 100-111
     estimating tree height, 98-99
     masts for, 98, 107-9
     need to install anemometer, 97-98
     and wind classification, 94-98
Sites, 40-68, 221
     alternative, 58
     on hills, 43, 86-87
     institutional restrictions, 44-53
     physical restrictions, 40-44
     and wind exposure, 41-43, 79, 90-91
Slings, 275, 285, 321
Small power producers, 54-58
Snap hooks, 320-21
Soil-holding capacity, 261-63
Solidity, 166-69
Space heating, 26, 28-29, 30, 37, 147-48
Speed
     cut-in, 129-30, 251-52
     cut-out, 130, 203, 251-52
     furling, 130, 154-55
Spoilers, 204-5
Stakes, 107-10
Stall, 165-66, 198-99, 201-3
State energy office, 105, 249, 253, 256

Stays, 179-80
Storage batteries, 24-28
Storage, pumped, 39
Storage, thermal, 37
Strand vise, 74-75
Strip chart recorders, 101-4
Surge protection, 303-4
Swept area method, 119-28
     formulas, 77-78, 123-24

Tag lines, 279-80
Tail vanes, 130, 151-53, 193-95
Tax credits, 140-41, 145-46
Temperature inversion, 72-73
Tip-speed ratio, 169
Torque, 19, 166, 168-69
Tower adapter, 224, 280
Tower climbing, 321, 323-24
Tower failure, 41, 47, 222, 285
Tower selection, 210, 215, 219-27
     and cost, 211, 225-27
     ease of installation, 224
     limited by manufacturer, 222
     pleasing appearance, 224-25
     and sites, 222-23
     and transportation costs, 225-26
Tower-turbine match, 118, 210, 221-22
Tower vibrations, 273, 285, 289
Tower work, safety hazards, 316-24
Towers, 210-27, 258-87
     attractive nuisance, 48, 223, 329, 331
     clear zone, 41, 47
     deflection of, 221-22
     design, 48, 221, 316
     height of, 42, 47, 89, 211, 220-21, 225
     leveling of, 279, 285
     novel, 218-19
     pre-assembled, 211-13
     rating of, 221
     strength of, 221-22
     telescopic (nested), 213
     thrust load on, 211, 221-22
Transformer, dedicated, 211
     types of, 64, 66
Transit, surveyor's, 271, 279
Transmissions, 181-84
Trees, 41-42, 95-96
     height estimation methods, 98-99
Troubleshooting, 312-14
Truss towers, 211-13, 280, 282, 284-85
Tube towers, 285-86
Turbine-generator match, 133
Turbines, 224
     aerodynamics of, 164-79
     height above obstruction, 42
     horizontal axis, 151, 153, 288-89
     vertical axis, 122, 151-52, 156-60, 288
Turbulence, 41-43, 86, 108, 220-21

Utilities, buy-back rate, 24, 56-59, 146-47

Value of energy, 139-44
Value of savings, 142
Voltage, 18, 61-65
Voltage drop, 64, 296-301
Voltage spike, 303-4

Warranty, 222, 233, 244, 246, 253-55, 310
Water churn heating, 29, 31
Water pumping, 23-24
Wattmeter, 307-9
Weatherheads, 290, 292
Weights, 193, *196*
Well motor, 24
Well pump, electric, 24, 25
Wincharger, 184, 187, 218, 220, 235-43
Winches, 286, 328-29
Wind(s), 69-75, 112-13
    apparent, 165-66
    frictional effects on, 86
Wind data, analysis, 113-17
    sources of, 90-94, 116, 336-83
Wind energy, 75-76
Wind Energy Resource Atlas, 93-94, 336-83
Wind farm prospecting, 71-72, 133
Wind farms, 38-39, 55, 58, 146, 255
Wind furnace, 28-30, 36-37
Wind machines
    aesthetics of, 250
    characteristics of, 386
    homebuilt, 229-32
    intangible benefits of, 21-22
    test reports on, 249-50, 384
    used, 232-44
Wind machine design, 77, 166, 187-88, 205
Wind machine-tower match, 226
Wind power, 75-90
    average annual, 73-74, 339-83
    calculating from EPF, 83-86
    calculating from site data, 114, 116-17
    equation, 76
    and height, 86-90
    increase with height (formula), 88-90
    ranges from average wind speed, 85-86
    vs. elevation, 76-77
Wind power classes, 93, 336
Wind speed, 41-42, 70, 73-79, 112-15, 165
    average annual, 73-75, 339-83
    to estimate EPF, 83-85
    and height, 86-90
    increases with height, 87, 90
    related to wind power, 76-81
    terrain enhancement of, 71-73
Wind speed class, 81-82, 91, 391-92n
Wind speed distribution, 81-86, 117
Wind speed meters, 100-101, 104
Wind speed profile, 87-88
Wind speed summaries, 82, 91, 112-13, 114, 116-17
Wind systems
    as supplemental power source, 24, 32-33, 57
    DC, 24-27
    product evaluation, 248-52
    safety hazards, 315-16, 324-25
Windchargers, pre-REA, 186, 191-92, 234-44, 328-29
Windmills, 16, 23, 151-53, 160-62, 166-68, 193-95
Wiring, 287-305
Work, 16-17, 75
Work belts, 316-19, *320*

Yaw control, 152-54, 209, 323
Yawing, 121, 169, 224, 323

Zoning restrictions, 44-52